285
Current Topics in Microbiology and Immunology

Editors

R.W. Compans, Atlanta/Georgia
M.D. Cooper, Birmingham/Alabama
T. Honjo, Kyoto · H. Koprowski, Philadelphia/Pennsylvania
F. Melchers, Basel · M.B.A. Oldstone, La Jolla/California
S. Olsnes, Oslo · M. Potter, Bethesda/Maryland
P.K. Vogt, La Jolla/California · H. Wagner, Munich

M. Marsh (Ed.)

Membrane Trafficking in Viral Replication

With 19 Figures

Springer

Dr. Mark Marsh
Cell Biology Unit, MRC-LMCB
University College London
Gower Street
London, WC1E 6BT
United Kingdom

e-mail: m.marsh@ucl.ac.uk

Cover illustration: Lentivirus assembly. The principal image is an electron micrograph of simian immunodeficiency virus (SIV) budding from the surface of an infected T cell line. Budding figures, as well as immature and mature particles can be seen. The inset shows a fluorescence micrograph of a human immunodeficiency virus (HIV) infected macrophage stained to identify the viral gag protein. In these cells, virus assembly occurs intracellularly in late endosomes (see p 219). Micrographs were provided by Dr. Annegret Pelchen-Matthews, Cell Biology Unit, MRC-Laboratory for Molecular Cell Biology, University College London.

Library of Congress Catalog Card Number 72-152360

ISSN 0070-217X
ISBN 3-540-21430-5 Springer Berlin Heidelberg New York

This work is subject to copyright. All rights are reserved, whether the whole or part of the material is concerned, specifically the rights of translation, reprinting, reuse of illustrations, recitation, broadcasting, reproduction on microfilms or in any other way, and storage in data banks. Duplication of this publication or parts thereof is permitted only under the provisions of the German Copyright Law of September 9, 1965, in its current version, and permission for use must always be obtained from Springer-Verlag. Violations are liable for prosecution under the German Copyright Law.

Springer is a part of Springer Science+Business Media
springeronline.com
© Springer-Verlag Berlin Heidelberg 2005
Printed in Germany

The use of general descriptive names, registered names, trademarks, etc. in this publication does not imply, even in the absence of a specific statement, that such names are exempt from the relevant protective laws and regulations and therefore free for general use.
Product liability: The publishers cannot quarantee that accuracy of any information about dosage and application contained in this book. In every individual case the user must check such information by consulting the relevant literature.

Editor: Dr. Rolf Lange, Heidelberg
Desk editor: Anne Clauss, Heidelberg
Production editor: Andreas Gösling, Heidelberg
Cover design: design & production GmbH, Heidelberg
Typesetting: Stürtz AG, Würzburg
Printed on acid-free paper 27/3150/ag   – 5 4 3 2 1 0

Preface

Viruses are major pathogens in humans, and in the organisms with which we share this planet. The massive health and economic burden these agents impose has spurred a huge research effort to understand their most intimate details. One outcome of this effort has been the production, in many but certainly not all cases, of effective vaccines and therapies. Another consequence has been the realization that we can exploit viruses and put them to work on our behalf. Viruses are still seen to have the most potential as vehicles for gene delivery and other therapeutic applications. However, their ability to exploit cellular functions to their own ends makes viruses not only highly effective pathogens but also exquisite experimental tools. Work with viruses underpins much of our current understanding of molecular cell biology and related fields. For membrane traffic in particular, viruses have been crucial in providing insights into key cellular functions and the molecular mechanisms underlying these events.

Viruses can be regarded as enveloped or non-enveloped agents. In enveloped viruses, a membrane derived from a membrane-bound compartment of the host cell protects the genetic material. By contrast to enveloped viruses, non-enveloped viruses protect their nucleic acid with a protein shell assembled from polypeptides encoded in the viral genome. The fact that enveloped viruses contain a membrane has made these agents particularly useful for studies of membrane traffic. Some of the genetically less complex viruses encode just one or two membrane proteins and have the remarkable ability to take over a cell's protein synthetic capacity so that within a few hours of being infected, a cell is dedicated to making only one or two membrane proteins. These properties have made viruses such as the rhabdovirus vesicular stomatitis virus (VSV) and the alphaviruses Semliki Forest virus (SFV) and Sindbis virus extraordinarily useful in early studies of the exocytic pathway. Indeed, the VSV G protein has been used extensively in biochemical and morphological experiments, which have led to key insights into the molecular details of this pathway.

Similarly, analysis of the mechanisms through which these, and other viruses such as the orthomyxovirus influenza virus, generate new virions has led to insights into glycoprotein synthesis and glycosylation, protein folding, endoplasmic reticulum quality control, cell polarity, protein sorting, etc. But these agents have proved to be of use not only in biosynthetic events but in a variety of other processes involving cellular membrane systems. Studying virus entry led to many key insights into clathrin-dependent receptor-mediated endocytosis. Now, nearly 25 years after those initial studies, similar experiments with different viruses are revealing aspects of non-clathrin-mediated endocytosis. Similarly, viruses have provided tractable experimental systems to analyse transport to the nucleus, membrane fusion and other membrane transport events. Currently, the assembly of some enveloped viruses is providing key insights into the function of ESCRT proteins, the formation of multivesicular bodies and the sorting of proteins for degradation in lysosomes.

It is this wealth of applications and the resulting knowledge that has ensued from applying viruses to specific experimental problems that has led to the present volume of *Current Topics in Microbiology and Immunology*. This volume deals with a specific aspect of viral interactions with cellular membrane systems and includes chapters on viral entry, viral membrane fusion, viral membrane protein synthesis and transport, viral replication, viral interaction with cytoskeletal systems and the nucleus, the trafficking of viral membrane proteins and viral perturbation of host cell protein trafficking. These chapters should provide new readers to the field with a broad overview of the cellular membrane trafficking mechanisms and viral interactions with these systems. They should also provide experts in this field with reviews of the current state of each of the fields. Although they incorporate a variety of topics, there are unfortunately areas that are not covered in detail. This perhaps underlines the fact that this dynamic and important field warrants frequent revisits.

It is no coincidence that a number of the contributors to the volume have at one time or another worked with Ari Helenius. The composition of the volume, at least in part, reflects the considerable contribution he has made to these fields of study. I thank all the authors for their timely contributions. I also thank Victoria, Megan, Kathleen and Molly for their support and patience.

Mark Marsh

List of Contents

Viral Entry
S.B. Sieczkarski and G.R. Whittaker 1

The Many Mechanisms of Viral Membrane Fusion Proteins
L.J. Earp, S.E. Delos, H.E. Park, and J.M. White 25

The Role of the Cytoskeleton During Viral Infection
K. Döhner and B. Sodeik 67

Nuclear Import in Viral Infections
U.F. Greber and M. Fornerod 109

Viral RNA Replication in Association with Cellular Membranes
A. Salonen, T. Ahola, and L. Kääriäinen 139

Synthesis and Quality Control of Viral Membrane Proteins
C. Maggioni and I. Braakman 175

Receptor Modulation in Viral replication:
HIV, HSV, HHV-8 and HPV: Same Goal, Different Techniques
to Interfere with MHC-I Antigen Presentation
V. Piguet 199

Trafficking of Viral Membrane Proteins
R. Byland and M. Marsh 219

Subject Index 255

List of Contributors

(Their addresses can be found at the beginning of their respective chapters.)

Ahola, T. 139

Braakman, I. 175

Byland, R. 219

Delos, S.E. 25

Döhner, K. 67

Earp, L.J. 25

Fornerod, M. 109

Greber, U.F. 109

Kääriäinen, L. 139

Maggioni, C. 175

Marsh, M. 219

Park, H.E. 25

Piguet, V. 199

Salonen, A. 139

Sieczkarski, S.B. 1

Sodeik, B. 67

White, J.M. 25

Whittaker, G.R. 1

Viral Entry

S. B. Sieczkarski[1,2] · G. R. Whittaker[1] (✉)

[1] Department of Microbiology and Immunology, Cornell University,
C4127 Veterinary Medical Center, Ithaca, NY 14853, USA
grw7@cornell.edu
[2] Department of Microbiology, School of Medicine and Biomedical Sciences,
University of Buffalo SUNY, Buffalo, NY 14214, USA

1	Introduction	2
2	Virus Receptor Binding	4
2.1	Receptors and Coreceptors	4
2.2	Viruses with Multiple Receptors and Multiple Use of Receptors by Different Viruses	6
2.3	Receptor-Mediated Regulation of Endocytosis	7
2.4	Receptors and Infection of Polarized Epithelia	7
3	Virus Internalization	8
3.1	Direct Entry Through the Plasma Membrane	8
3.2	Internalization of Viruses into Endocytic Compartments	9
3.2.1	Clathrin and Non-Clathrin-, Non-Caveolae-Mediated Endocytosis	9
3.2.2	Caveolae	11
3.2.3	Macropinocytosis	12
4	Postinternalization Trafficking	13
4.1	Rab Proteins and Endocytosis	13
4.2	Kinase-Mediated Regulation of Endocytosis	13
5	Membrane Penetration	14
6	Virus Uncoating	15
7	Virus Entry as a Target for Antiviral Drugs	16
8	Perspectives	18
References		19

Abstract Virus entry is initiated by recognition by receptors present on the surface of host cells. Receptors can be major mediators of virus tropism, and in many cases receptor interactions occur in an apparently programmed series of events utilizing multiple receptors. After receptor interaction, both enveloped and nonenveloped viruses must deliver their genome across either the endosomal or plasma membrane for infection to proceed. Genome delivery occurs either by membrane fusion (in the case of enveloped viruses) or by pore formation or other means of permeabilizing the lipid bilayer (in the case of nonenveloped viruses). For those viruses that enter cells via endosomes, specific receptor interactions (and the signaling events that ensue) may control the particular route of endocytosis and/or the ultimate destination of the incoming virus particles. Our conception of virus entry is increasingly becoming more complex; however, the specificity involved in entry processes, once ascertained, may ultimately lead to the production of effective antiviral agents.

1
Introduction

As obligate intracellular parasites, all viruses must have ways of entering target cells to initiate replication and infection. The first step in virus entry is the recognition of the host cells through cell surface receptors. This initial engagement can both mediate attachment and act as a primer for subsequent conformational alterations in the virus. In many cases, receptor interactions are important for defining the tropism of a virus for a particular organism, tissue, or cell type. One important point to note is that receptors not only serve as attachment points for viruses but can induce conformational alterations in the bound ligand or induce signaling events, which are a prerequisite for subsequent uncoating, trafficking, or fusion events needed for infectious entry. After cell surface attachment, animal viruses can become internalized in two principal ways: by a direct mechanism at the cell surface (plasma membrane) or after their internalization into cellular compartments, for example, endosomes (Fig. 1). In the case of endocytic entry, internalization itself is generally not sufficient for productive infection, as incoming viruses are still part of the extracellular space while in endosomes. Therefore, endocytosed viruses must penetrate the endosomal membrane to be released into the cytoplasm. The endocytic pathway may also be used by viruses that require a specific localization within the cell for successful infection.

This review is intended as a survey of the different routes of entry used by animal viruses and will draw on both studies of the membrane trafficking events required for entry and recent publications. For more detail on entry of other viruses, such as bacteriophages, the reader is referred to the excellent recent review of Bamford and colleagues, which describes common features of virus entry across the three domains of life (Poranen et al. 2002). For more details on the entry mechanisms of individual viruses, many classic reviews are available (Marsh and Helenius 1989; Marsh and Pelchen-Matthews 1994, 2000; Young 2001).

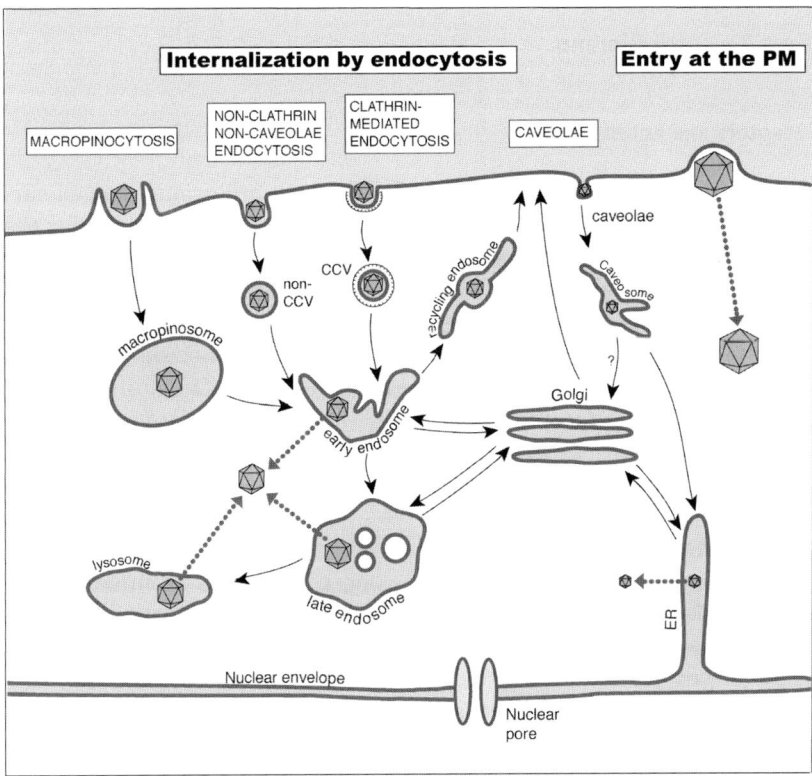

Fig. 1. Routes of virus internalization into host cells. Enveloped viruses may enter host cells by direct fusion at the plasma membrane, depositing the viral nucleocapsid directly in the cytoplasm. Both enveloped and nonenveloped viruses may be internalized into host cells via endocytosis. Four primary endocytic routes have been described, each known to be involved in virus entry. Clathrin-mediated endocytosis can result in viral trafficking either through endosomes toward the lysosome or into the recycling pathway. Non-clathrin-, non-caveolae-mediated endocytosis appears to traffic ligands toward the lysosome-targeted pathway. Viruses entering by caveolae are usually targeted to the ER and possibly the Golgi. Macropinosomes are capable of fusing with other vesicles of the endocytic pathways, such as early endosomes

2
Virus Receptor Binding

2.1
Receptors and Coreceptors

The very first interaction of a virus with its host cell occurs via cell surface receptors. This initiates a chain of events that is responsible for the breach of the cell membrane. There is a wealth of information on cellular receptors for viruses, which we can only summarize here. For more information, the reader is referred to Wimmer's monograph (Wimmer 1994) as well as several recent general reviews (Baranowski et al. 2001; Mettenleiter 2002; Schneider-Schaulies 2000). For a detailed understanding of regulation of receptor expression, see the chapter by Piguet in this volume. In this article we focus on recent findings on virus receptors, especially how they relate to the cell biology of virus entry, membrane trafficking, and infection.

Virus-receptor interactions can be conveniently broken down into defined areas. First, there is the basic concept that viruses interact with cells initially by long-range, possibly nonspecific, electrostatic interactions—based on attraction of the negatively charged cell surface with positive charges on virus particles. A classic example of this is cell surface heparan sulfate, a highly sulfated polysaccharide present in large quantities on the surface of cells and in the extracellular matrix. Numerous viruses, including many herpesviruses, papillomaviruses, paramyxoviruses and dengue virus, use cell surface heparan sulfate to infect a range of target cells (see Liu and Thorp 2002 for a review).

The view of heparan sulfate as a nonspecific receptor may, however, turn out to be an oversimplification—a view exemplified by the herpesviruses. It was originally thought that interaction of cell surface heparan sulfate with herpes simplex virus 1 (HSV-1), via glycoprotein C (gC), was simply the preface to more specific protein-protein interactions with other glycoproteins and coreceptors (see Campadelli-Fiume et al. 2000; Mettenleiter 2002 for reviews). Such specific coreceptors include TNF-R, Prr2, nectin1, and PVR, which bind selectively to different animal herpesviruses, via gD. However, in the case of HSV-1, unique monosaccharide sequences (e.g., 3-O-sulfate) present on the cell surface indicate that heparan sulfate could serve as a specific receptor, also via gD, thus initiating virus entry (Shukla et al. 1999). Once specific cell surface interactions have occurred, it is hypothesized that other herpesvirus glycoprotein-cell surface interactions, possibly including gH and/or gB, then activate the fusion machinery between the virion envelope and the plasma membrane, leading to translocation of the nucleocapsid into the cell (see Campadelli-Fiume et al. 2000; Mettenleiter 2002 for reviews).

In addition to the herpesviruses, the most prominent example of a dual receptor requirement occurs with human immunodeficiency virus type-1 (HIV-1) binding. It has been known for many years that HIV-1 binds to cells of the immune system via cell surface CD4. However, this interaction is not sufficient to mediate fusion. The virus needs to bind so-called chemokine coreceptors, such as CXCR4 and CCR5, for productive infection (Berger et al. 1999; Overbaugh et al. 2001). Whereas CD4 usage is nearly universal, coreceptor usage varies between strains. The interaction of gp120 with the HIV-1 coreceptor is a now-classic example of the induction of a conformational change in the viral glycoprotein priming it for fusion. This coreceptor-induced conformational change is in contrast to the simpler pH-mediated trigger used by viruses, such as influenza, that bind a nonspecific receptor, in this case sialic acid (Skehel and Wiley 2000).

HIV-1 also provides an example demonstrating that virus-receptor binding does not necessarily lead to entry or productive infection of the initially contacted cell, yet can profoundly influence viral pathogenesis. In the case of HIV-1, the virus can bind via a lectinlike interaction to circulating peripheral dendritic cells (DCs), via molecules such as DC-SIGN (Geijtenbeek et al. 2000). Virus binding retains HIV-1 in an active form on the cell surface while the DC undergoes maturation and migration to the regional lymph nodes. The outcome of this interaction is the delivery of the virus to permissive T cells and initiation of infection within the host.

One emerging theme of receptor-coreceptor interactions seems to be that the complex series of conformational changes that follow binding of pH-independent viruses, such as retroviruses, are not shared by pH-dependent viruses, such as influenza virus, for which priming of fusion is a much simpler and more rapid event, without extensive coreceptor involvement. It remains to be seen how this might impact on a virus such as avian leukosis virus (a retrovirus), which seems to be both pH- and coreceptor dependent (Mothes et al. 2000). Considering that fusion is being increasingly targeted as a point for antiviral intervention (see Sect. 7), the speed and complexity of this interaction seem to have profound implications for the ability of antiviral compounds to function.

2.2
Viruses with Multiple Receptors and Multiple Use of Receptors by Different Viruses

Whether or not a virus uses a single protein or other moiety, or whether a combination of receptor and coreceptor is used, the established paradigm is of a simple and unique virus-receptor pairing. This has recently been challenged in several systems, demonstrating that the same virus can use multiple (individual) receptors, and that very different viruses use identical receptors. For instance, as discussed above, it has become apparent that herpesviruses, for example, HSV-1, can share the same receptor yet have a choice of coreceptors. Another important example in which multiple (individual) receptors are used is measles virus (see Oldstone et al. 2002; Yanagi et al. 2002 for reviews). The human CD46 was originally identified as a ubiquitous cellular receptor for the Edmonston and Halle strains of measles virus; however, this finding was challenged as not all cell types seemed to be permissive, using CD46 as a receptor. Subsequently SLAM (CD150) was identified as a principal receptor for measles virus (and morbilliviruses in general), possibly accounting for most cell types involved in measles virus pathology and pathogenesis. The weight of evidence suggests that multiple (different) receptors are important for measles virus, with the possibility that additional receptors remain to be found.

Although in most cases individual viruses have their own distinct receptors, in some cases there can be competition for the same receptor by quite different viruses. Perhaps the best studied example of this is the coxsackie-adenovirus receptor or CAR (Bergelson et al. 1997). CAR is a member of the immunoglobulin superfamily and mediates both attachment and entry of these two viruses. CAR is sufficient for coxsackievirus entry; however, for adenovirus (serogroup C), an additional cooperative event is required after CAR binding—the interaction of the penton base protein with an internalization receptor, the $\alpha_v\beta_3$ and $\alpha_v\beta_5$ integrins (see Nemerow 2000 for a review).

Another example of competition for the same receptor by different viruses is illustrated by the poliovirus receptor, or PVR (CD155), also a member of the immunoglobulin superfamily. As well as acting as a receptor for poliovirus (Racaniello 1996), it is a coreceptor for two herpesviruses, bovine herpesvirus 1 and pseudorabies virus (Mettenleiter 2002). These viruses, along with poliovirus, all have neurotropism but it remains to be seen whether the specific use of PVR is an important factor in pathogenesis. In the case of poliovirus it is clear that species infectivity, but not necessarily tissue tropism, is determined by the receptor (Nomoto et al. 1994).

2.3
Receptor-Mediated Regulation of Endocytosis

One feature of virus-receptor interaction is that the virus may be physically linked to a cellular receptor, which has inherent information for endocytic traffic. This allows the virus to utilize endocytic trafficking signals inherent in the receptor. Whereas in most cases binding of cellular ligands induces signaling events responsible for receptor clustering and internalization into defined endocytic pathways, viruses are inherently multivalent and may be able to mimic ligand-induced receptor internalization in the absence of normal signals. One well-studied example is the transferrin receptor, which undergoes clathrin-mediated internalization and enters the recycling pathway of endocytosis. Viruses such as canine parvovirus bind and enter cells via the transferrin receptor (Parker et al. 2001; Hueffer et al. 2003) and appear to enter a recycling endocytic compartment before genome delivery, possibly via late endosomes, to the cytosol (Suikkanen et al. 2002). Other viruses, such as mouse mammary tumor virus (a retrovirus), that also use transferrin receptor for entry may direct their endocytic uptake in a similar manner (Ross et al. 2002). Such receptor-defined trafficking may not be limited to the transferrin receptor, as other well-characterized, clathrin-dependent receptors, such as LDL receptor-related proteins, are used by rhinoviruses (Hofer et al. 1994). It is presently unclear whether the recent finding that influenza virus may have alternative internalization routes, via clathrin- or non-clathrin-mediated endocytosis (Sieczkarski and Whittaker 2002), is accounted for by the nonspecific receptor (sialic acid) used by the virus.

2.4
Receptors and Infection of Polarized Epithelia

Many viruses infect polarized epithelia, and it has been appreciated for some time that expression of receptor molecules has special significance for viruses that infect epithelial cells (Tucker et al. 1994). One recently appreciated finding is that cell adhesion molecules are used as entry receptors by several viruses, including reovirus (junctional adhesion molecule, JAM), herpes simplex virus (nectin1 and 2), and CAR (see Spear 2002 for a review). As such, cell adhesion molecules localize to junctional complexes and account for the integrity of epithelia; they can have pronounced effects on virus spread within the organism. The use of cell adhesion molecules might be considered unusual in that the receptor is not typically accessible from the apical face of the epithelium, where the virus enters its host. In this case, successful infection might depend on damage to the epithelium. In addition, spread of virus might be compromised without breakdown of the epithelium. Recent studies on adenovirus show that the fiber-knob protein of adenovirus disrupts interactions between the CAR receptors of adjacent cells and allows virus to exit across the airway epithelium (Walters et al. 2002). After infection, human airway epithelia first release adenovirus to the basolateral surface. Virus then travels between epithelial cells by disrupting CAR-CAR interactions, to emerge on the apical surface for viral spread. Thus internalization with adhesion molecules can direct the spread of virus infection within the host and has profound implications for the design of gene therapy vectors.

3
Virus Internalization

3.1
Direct Entry Through the Plasma Membrane

Perhaps the simplest route of entry into the cell is directly through the plasma membrane. Indeed, this route of entry is clearly used by a range of different viruses. Those viruses generally considered to enter through the plasma membrane include paramyxovirus and herpesvirus (Lamb 1993; Spear 1993). Despite the apparent simplicity, there are some disadvantages to this route of entry. First, the cell contains a barrier of cortical actin through which the virus must navigate (see Marsh and Bron 1997 and the chapter by Döhner and Sodeik, this volume, for a discussion). Second, the endosome often carries out the critical function of acidification, needed for triggering of fusion (see the chapter by Earp et al., this volume). By default, therefore, viruses that enter through the plasma membrane must be pH independent. Third, the endosome may also provide a specific chemical environment, such as ion concentration or redox state needed to allow penetration. Finally, endocytic traffic may deliver the genome deep into the interior of the cell, close to the nucleus.

Recently, retroviruses have been actively studied with regard to direct fusion vs. endocytosis. Although classically defined as being pH independent, some retroviruses are now thought to require pH for fusion (see the chapter by Earp et al., this volume) and presumably enter via endocytosis. Of particular note is the finding that the avian leukosis virus, a retrovirus, shows a substantial inhibition of infection in cells expressing dominant-negative dynamin, a cellular molecule essential for endocytosis (Mothes et al. 2000). However, another retrovirus, Moloney murine leukemia virus, showed no effects on expression of dominant-negative dynamin (Lee et al. 1999). It is unclear whether these apparently contradictory data for retroviruses represent differences in the individual viruses or in the assay conditions used, or whether dynamin is required for other membrane traffic events important for virus assembly.

HIV is another retrovirus that has been actively studied in this regard. Whereas the majority of virus does get endocytosed (especially under high-MOI conditions), this pool of virus does not seem to enter the cytosol and is destroyed in lysosomes (Fredericksen et al. 2002). Under certain circumstances this endocytosed virus may be infectious (Fackler and Peterlin 2000); however, the major route of HIV entry is directly through the plasma membrane.

3.2
Internalization of Viruses into Endocytic Compartments

Endocytosis has emerged as the principal route of entry into host cells for the majority of virus families (Russell and Marsh 2001; Sieczkarski and Whittaker 2002). In principle, such viruses have several choices for internalization, based on the available trafficking pathways in the cell (see Fig. 1). These can be broken down into four distinct pathways: (1) clathrin, (2) non-clathrin, non-caveolae pathways, (3) caveolae, and (4) macropinocytosis. However, individual viruses within a family, or the same virus in different cell types, may use different internalization routes.

3.2.1
Clathrin and Non-Clathrin-, Non-Caveolae-Mediated Endocytosis

As befits its major role in endocytosis, clathrin has been shown to play a major role in the internalization of many viruses. Traditionally, several viruses, including influenza virus, VSV, and SFV, were identified in clathrin-coated vesicles at early times of internalization (e.g., 5 min) based on the presence of an electron-dense coat by transmission electron microscopy (Marsh and Helenius 1980; Matlin et al. 1981, 1982).

However, noncoated vesicles were also observed in these experiments, due either to the release of clathrin or the presence of non-clathrin-mediated pathways. Subsequently, a role for clathrin in the case of SFV has been shown by microinjection of anti-clathrin antibodies (Doxsey et al. 1987). The expression of dominant-negative Eps15 (which arrests clathrin-coated pit assembly at the cell surface) has more recently confirmed a role for clathrin in SFV and VSV entry (Sieczkarski and Whittaker 2002; V. Yau and G. Whittaker, unpublished results). Other viruses in which clathrin is required for entry are adenovirus (Meier et al. 2002), parvovirus (Parker and Parrish 2000), Hantaan virus (Jin et al. 2002), and Sindbis virus (Carbone et al. 1997). In the case of influenza virus, the situation is much less clear—in that elimination of clathrin function showed no inhibition of virus infection or replication (Sieczkarski and Whittaker 2002). Thus influenza may have the ability to use either clathrin or non-clathrin endocytic pathways for productive infection.

For many other virus families, the reliance on clathrin is also much less clear-cut. One notable example is the *Picornaviridae*. Potassium depletion has shown a role for clathrin in human rhinovirus infection (Bayer et al. 2001; Madshus et al. 1987). In the case of rhinovirus, it may be that different virus serotypes use different pathways. The role of acidification during rhinovirus entry is also uncertain (Huber et al. 2001). For other family members, such as poliovirus, the data are equally unclear. Poliovirus has been proposed to enter independently of clathrin (DeTulleo and Kirchausen 1998), and a pH-independent route of entry seems likely (Perez and Carrasco 1993). However, it is still unclear whether endosomes are required for entry, or whether penetration can occur directly through the plasma membrane (for discussion, see Hogle 2002). Poliovirus illustrates two important points that must be remembered when thinking about virus entry. First, the use of an endocytic route is not necessarily tied to pH dependence, and second, a pH-independent virus may choose to enter through the cell surface or through endosomes, depending on specific circumstances.

Enteroviruses, on the other hand, appear to use non-clathrin pathways that are dependent on cell surface lipid rafts (Stuart et al. 2002). In this case, the use of a novel pathway seems to be directly related to the receptor used by the virus (decay accelerating factor, or DAF). One newly-added group in the *Picornaviridae* are the echoviruses. Human parechovirus-2 was shown to enter cells via clathrin-dependent endocytosis but to traffic to the ER (Joki-Korpela et al. 2001). Thus these echoviruses may show features similar to simian virus 40 (SV40). Echovirus-1 shows no colocalization with markers of the clathrin pathway but appears to use caveolae for entry (Marjomaki et al. 2002); see Sect. 3.2.2.

Another virus family that makes use of both clathrin and non-clathrin pathways for entry is the *Polyomaviridae*. SV40 is well established to use caveolae as a route of entry (Kartenbeck et al. 1989; Pelkmans et al. 2001; see Sect. 3.2.2). However, one recent study reports incoming murine polyomavirus having no colocalization with caveolin and being independent of both clathrin and dynamin (Gilbert and Benjamin 2000). In glial cells, however, the human polyomavirus JC virus enters by pH-sensitive, clathrin-dependent endocytosis, in contrast to the pH-independent, caveolar pathway used by SV40 in the same cell (Ashok and Atwood 2003; Pho et al. 2000). Clearly, more investigation is required to clarify the role of clathrin and non-clathrin pathways in the entry of the different members of the *Polyomavirinae*.

Although generally considered to use direct fusion at the plasma membrane, herpesviruses can also use endocytosis. A role for clathrin has been implicated in the uptake of the Epstein-Barr virus (EBV) into B cells of the immune system, but not in epithelial cells (Miller and Hutt-Fletcher 1992). Other evidence for herpesviruses utilizing the endocytic pathway is currently emerging, with HSV-1 also believed to utilize pH-dependent endocytosis to enter certain cell types (Nicola et al. 2003).

3.2.2
Caveolae

Caveolae have emerged as a somewhat specialized route of virus entry, notably for SV40. The uptake of SV40 is somewhat unusual, in that incoming virions accumulate in the smooth ER (Kartenbeck et al. 1989). Entry via caveolae was recently shown by virus colocalization with endogenous caveolin, a major component of caveolae (Norkin 1999), and more specifically with the use of GFP-tagged caveolin-1 combined with video microscopy—which showed the delivery of SV40 from caveolae directly to the ER, utilizing dynamin and local actin recruitment but bypassing the traditional endosome/lysosome system (Pelkmans et al. 2001, 2002). However, it is also possible that virus moves transiently through the Golgi en route to the ER (Norkin et al. 2002; Richards et al. 2002). Other notable features of SV40 entry via caveolae are that the kinetics are slow compared to the clathrin-coated vesicle pathway, that the vesicles do not become acidified, and that internalization is an active process driven by virus binding.

Colocalization with endogenous caveolin has also implicated caveolae in the entry of mouse polyoma virus (Richterova et al. 2001); however, incoming murine polyomavirus was shown not to colocalize with caveolin (Gilbert and Benjamin 2000). Despite extensive investigation, it appears that the caveolar route of virus entry may be a relatively specialized event, limited to SV40 as well as other viruses that are routed to the ER/Golgi during entry, such as echoviruses (Marjomaki et al. 2002), although caveolae may not be obligatory in the latter case (Joki-Korpela et al. 2001; see Sect. 3.2.1).

3.2.3
Macropinocytosis

Macropinocytosis is generally considered to be a nonspecific mechanism for internalization, in that it is not reliant on ligand binding to a specific receptor. Instead, formation of endocytic vesicles occurs as a response to cell stimulation, particularly at sites of membrane ruffling. The finding that macropinosomes have the ability to become acidified and can intersect with endocytic vesicles makes them possible routes of entry for a wide variety of viruses; however, detailed mechanisms have not emerged in most cases. Experiments have recently been used to demonstrate macropinocytic uptake of HIV-1 into macrophages (Marechal et al. 2001), although most viruses internalized by this route were probably noninfectious. Although the reasons remain unclear, virus-induced endocytic activity is required for adenovirus type 2 penetration even though the virions themselves are internalized in clathrin-coated vesicles (Meier et al. 2002).

Macropinocytosis is known to be a major route of entry into antigen-presenting cells like dendritic cells, and the ultimate importance of macropinocytosis in viral infection may therefore turn out to be in the presentation of the invading virus to the immune system, rather than in its primary infection of epithelial surfaces.

4
Postinternalization Trafficking

4.1
Rab Proteins and Endocytosis

One class of molecules that regulate membrane traffic events involved in endocytosis is the Rab family of small GTPases (Somsel Rodman and Wandinger-Ness 2000). The involvement of Rab GTPases in virus entry is at a preliminary stage of investigation. Adenovirus uptake was increased by overexpression of wild-type Rab5 (which controls entry into the early endosome) and decreased by dominant-negative Rab5 (Rauma et al. 1999). We recently carried out a more comprehensive study of virus entry, using influenza virus, SFV, and VSV. All of these viruses were sensitive to Rab5 inhibition, but only influenza was affected by Rab7 inhibition (which controls formation of late endosomes and lysosomes) (Sieczkarski and Whittaker 2003). These data suggest that influenza virus has a requirement for functional late endosomes during entry.

The above viruses seem to rapidly enter and escape from their respective endocytic vesicles (typically within 20–30 min). Although the role of Rab proteins was not addressed, parvovirus is suggested to have a different type of endocytic interaction. Canine parvovirus enters a slow recycling pathway, normally typified by the presence of Rab11, and eventually breaks through to the late endosome for productive infection (Parker and Parrish 2000; Suikkanen et al. 2002).

Rab proteins are also convenient markers for colocalization studies, as they tend to be organelle specific in their localization. For example, Miyazawa et al. (Miyazawa et al. 2001) used Rab7 to show the localization of adenovirus subgroup B (Ad7) to late endosomes. In contrast, these authors showed that the related (and more commonly studied) adenovirus subgroup C does not enter the late endosome but seems to penetrate from the early endosome. Thus very similar viruses may show selective endocytic traffic, which could have profound implications for their pathogenesis.

4.2
Kinase-Mediated Regulation of Endocytosis

As befits their role as major signaling pathways in cells, virus endocytic events are regulated by phosphorylation. Attention has focused on two kinase families, the PI 3-kinases (PI3K) and protein kinase C (PKC). For adenovirus, the initial interaction of the virus with cell surface integrins activates PI3K, which in turn affects virus endocytosis. Inhibition of PI3K reduces adenovirus internalization to approximately 30% of control samples (Li et al. 1998). It has not yet been shown at which point in entry adenovirus internalization is arrested by PI3K inhibition.

For many years, PKC has also been implicated in virus entry processes. The entry of several enveloped viruses, including rhabdoviruses, alphaviruses, poxviruses, and herpesviruses, has been proposed to require PKC based on the action of protein kinase inhibitors such as H7 and staurosporine (Constantinescu et al. 1991). More recently, it was shown that the successful entry of adenovirus type 2 requires PKC. In the presence of calphostin C, an inhibitor of the classic and novel PKC isoforms, adenovirus is prevented from escaping endosomes and accumulates in cytoplasmic vesicles near the cell periphery. Bisindolylmaleimide I, a broad-spectrum, highly specific PKC inhibitor, as well calphostin C, prevents influenza virus entry and subsequent infection (Root et al. 2000; S. Sieczkarski and G. Whittaker, unpublished results). Other viruses seem to require PKC for infection, but not specifically for virus entry (S. Sieczkarski and G. Whittaker, unpublished results). A specific role for PKC in influenza virus entry is reinforced by recent data from our laboratory in cells overexpressing a kinase-dead form of the PKCβII isotype, but not PKCα. In these cells, influenza virus entry was arrested at the level of the late endosome without any apparent defect in endosome acidification (Sieczkarski et al. 2003). It remains to be seen which specific isoforms of PKC are involved in the entry of different viruses, and at which point in endocytic trafficking they act.

5
Membrane Penetration

To replicate, viruses must deliver their genomes across a cellular membrane system to the cytoplasm. For enveloped viruses, this occurs by fusion of the virion envelope with a cellular membrane. As virus fusion is covered extensively in the chapter by Earp et al., this volume, it will not be discussed here. Nonenveloped viruses deliver their genomes by a less obvious means and breach the hydrophobic barrier of the membrane by pore formation or other mechanism(s). In some cases this process is somewhat understood, but for many nonenveloped viruses it is still a big mystery.

Picornaviridae are the best-studied nonenveloped virus for penetration. In the case of poliovirus, many molecular events are now recognized—based on extensive structural studies of the molecules involved (reviewed in Hogle 2002). After receptor interaction, it is generally thought that conformational changes in the virus result in the formation of a functional intermediate (the A particle), where the hydrophobic N-terminus of VP1, and possibly a myristate group on VP4, possibly combined with changes in Ca^{2+} concentrations, allow membrane binding and pore formation. It has been proposed that there are common elements for entry of both enveloped (such as the extensively characterized influenza HA) and nonenveloped viruses (such as poliovirus), based on similar mechanisms for receptor binding leading to release of the virus from a metastable state and exposure of hydrophobic sequences (Hogle 2002).

One distinguishing feature of membrane penetration for picornaviruses appears to be the presence of a small, defined pore, through which the genome translocates. This localized event may be crucial, especially in the case of plasma membrane penetration, where a more generalized perforation or rupture would damage the host cell. Although low pH per se may not be a key factor in penetration of picornaviruses, combinations of concanamycin (a vacuolar H^+-ATPase inhibitor) with valinomycin (an ionophore that promotes K^+ efflux from cells) can prevent poliovirus entry into cells (Iruzun and Carrasco 2001). Therefore, it appears that the virus requires an intact K^+ gradient across the membrane in order to uncoat and enter cells. In a similar manner, K^+ concentrations have been proposed to play a role in the exposure of a hydrophobic conformer of the µ1 protein, which mediates membrane disruption or perforation during entry of reoviruses (Chandran et al. 2002).

Another intriguing finding is that parvoviruses encode a sequence with similarity to cellular phospholipases within their VP1 capsid protein unique region (Girod et al. 2002; Zadori et al. 2001). It is proposed that conformation changes occurring within the acidic environment expose the phospholipase activity, which can mediate localized disruption of the endosomal membrane and allow capsid entry into the cytosol. This seems to be a specialized event for these viruses, and a similar mechanism has not yet been described for other viruses.

6
Virus Uncoating

After membrane penetration, viruses must deliver their genome contents to the site of replication. As this event is often tied to the processes of cytoplasmic and nuclear transport, it will not be covered extensively here, and the reader is referred to other chapters in this volume and to previous reviews (Greber et al. 1994; Smyth and Martin 2002). However, one virus family that does not make direct use of either the cytoplasmic or nuclear transport machineries are reoviruses. These viruses are somewhat unusual in that they require low pH during virus entry, not for membrane penetration but for nucleocapsid uncoating (Sturzenbecker et al. 1987). Reoviruses have a very stable double-shelled capsid, which is disrupted by proteases during virus entry. The requirement for low pH is generally not thought to be a direct effect, but rather low pH is needed for activation of proteases, such as lysosomal cathepsin L and B (Ebert et al. 2002), that cleave the outer capsid shell, allowing genome release. In some cases, however, reoviruses can bypass the lysosomal requirement and utilize exogenous proteases (Golden et al. 2002). It is believed that in vivo such proteases present in the lumen of the gut can profoundly influence virus pathogenesis, which leads to the concept of two modes of virus entry for reovirus—cell surface or lysosomal (Borsa et al. 1979).

7
Virus Entry as a Target for Antiviral Drugs

As a target for therapeutic intervention, virus entry offers the advantage that the number of incoming particles is often small, maximizing any inhibitory action of an antiviral drug. Indeed, the newest generation of anti-HIV drugs target entry into host cells (O'Hara and Olson 2002). HIV entry inhibitors fall into two main groups: those targeting fusion (discussed in the chapter by Earp et al., this volume) and those designed as coreceptor inhibitors. In the latter group, two small molecules, SCH-C and AMD 3100 (specific for CCR5 and CXCR4, respectively), are currently in clinical trials, along with a humanized monoclonal antibody to CCR5.

Other antiviral agents targeting virus entry include amantadine and the WIN compounds. Amantadine was one of the first antiviral drugs to be used in patients, and it targets the influenza A virus M2 ion channel (Hay et al. 1985; Pinto et al. 1992). Amantadine and its derivative rimantadine are still in use today but suffer from problems due to the emergence of resistant viruses. The WIN compounds target the "canyon" present as part of the receptor-binding pocket of picornaviruses and prevent virus uncoating (Lewis et al. 1998). Although powerful antivirals

Table 1. Known receptors, coreceptors, and utilized entry pathways for selected viruses

Virus	Receptor	Coreceptor	Entry pathway(s)
Herpesviruses	Heparin sulfate	None, TNF-R, Prr2, nectin 1 and 2, PVR	Fusion at PM; clathrin-mediated endocytosis?
HIV	CD4	CXCR4, CCR5	Fusion at PM; macropinocytosis
Measles virus	CD46, SLAM (CD150)	?	Fusion at PM
Coxsackievirus	CAR	None	Clathrin-mediated endocytosis
Adenovirus	CAR	$\alpha_v\beta_3$-, $\alpha_v\beta_5$-integrin	Clathrin-mediated endocytosis
Poliovirus; bovine herpesvirus 1, pseudorabies virus	PVR (CD155)	?	Endocytosis or uncoating at PM; Fusion at PM
Parvovirus	Transferrin receptor	?	Clathrin-mediated endocytosis
Rhinoviruses	LDL receptor	?	Clathrin- and non-clathrin-mediated endocytosis
Influenza virus	Sialic acid	?	Clathrin- and non-clathrin-mediated endocytosis
Reovirus	JAM	?	Clathrin-mediated endocytosis
VSV	Glycolipid	?	Clathrin-mediated endocytosis
SFV	MHC class I	?	Clathrin-mediated endocytosis
Sindbis virus	Laminin receptor	?	Clathrin-mediated endocytosis
Enteroviruses	DAF	?	Non-clathrin-mediated endocytosis
Echoviruses	$\alpha_2\beta_1$, $\alpha_v\beta_3$, CD55	None, β_2 microglobulin	Clathrin-mediated endocytosis; caveolae
SV40	Sialic acid	?	Caveolae

in experimental systems, they have not been successful therapeutically, because of the many possible serotypes of picornaviruses and variable binding affinities of the drug with the hydrophobic pocket of the canyon. However, an advance on the WIN compounds has occurred with the development of a new generation of capsid-function inhibitors, such as Pleconaril (Rotbart 2002).

8
Perspectives

The classic notion of a virus binding to a single receptor to enter cells through a single defined internalization mechanism is quickly being overtaken by a more complex picture of virus entry. New findings, such as specific coreceptor usage and virus attachment to multiple (different) receptors (summarized in Table 1), have called into question our notions of how other viruses function. For instance, viruses known to bind to a nonspecific receptor may turn out to also have a more specific coreceptor. The use of multiple receptors may allow understanding of the pathogenic differences among virus strains. Beyond the initial entry steps, viruses also appear to follow more complex intracellular trafficking routes than imagined previously. Through kinase cascades, viruses may be able to alter cellular signaling pathways to control their endocytic movement. The recognition of new endocytic entry pathways, such as non-clathrin-, non-caveolae-mediated endocytosis and macropinocytosis, has coincided with the discovery that many virus families appear to use multiple routes of entry. The specifics of these entry decisions, be they based on cell type, receptor availability, or extracellular conditions, have yet to be determined. The complexity involved in virus entry may make discovering treatments targeting this stage of the infectious cycle more challenging, but the specificity involved in the processes, once ascertained, may ultimately lead to the production of effective antiviral agents.

Acknowledgements We thank Ruth Collins and Karsten Hüffer for critical reading of this manuscript. Work in the authors' laboratory is supported by the American Lung Association and the National Institutes of Health.

References

Ashok A, Atwood WJ (2003) Contrasting roles of endosomal pH and the cytoskeleton in infection of human glial cells by JC virus and Simian Virus 40. J Virol 77:1347–1356

Baranowski E, Ruiz-Jarabo CM, Domingo E (2001) Evolution of cell recognition by viruses. Science 292:1102–1105

Bayer N, Schober D, Huttinger M, Blaas D, Fuchs R (2001) Inhibition of clathrin-dependent endocytosis has multiple effects on human rhinovirus serotype 2 cell entry. J Biol Chem 276:3952–3962

Bergelson JM, Cunningham JA, Droguett G, Kurt-Jones EA, Krithivas A, Hong JS, Horwitz MS, Crowell RL, Finberg RW (1997) Isolation of a common receptor for coxsackie B viruses and adenoviruses 2 and 5. Science 275:1320–1323

Berger EA, Murphy PM, Farber JM (1999) Chemokine receptors as HIV-1 coreceptors: roles in viral entry, tropism and disease. Annu Rev Immunol 17:657–700

Borsa J, Morash BD, Sargent MD, Copps TP, Lievaart PA, Szekely JG (1979) Two modes of entry of reovirus particles into L cells. J Gen Virol 45:161–170

Campadelli-Fiume G, Cocchi F, Menotti L, Lopez M (2000) The novel receptors that mediate the entry of herpes simplex viruses and animal alphaherpesviruses into cells. Rev Med Virol 10:305–319

Carbone R, Fre S, Iannolo G, Belleudi F, Mancini P, Pelicci PG, Torrisi MR, Di Fiore PP (1997) Eps15 and Eps15R are essential components of the endocytic pathway. Cancer Res 57:5498–5504

Chandran K, Farsetta DL, Nibert ML (2002) Strategy for nonenveloped virus entry: a hydrophobic conformer of the reovirus membrane penetration protein micro 1 mediates membrane disruption. J Virol 76:9920–9933

Constantinescu SN, Cernescu CD, Popescu LM (1991) Effects of protein kinase C inhibitors on viral entry and infectivity. FEBS Lett 292:31–33

DeTulleo L, Kirchausen T (1998) The clathrin endocytic pathway in viral infection. EMBO J 17:4585–4593

Doxsey SJ, Brodsky FM, Blank GS, Helenius A (1987) Inhibition of endocytosis by anti-clathrin antibodies. Cell 50:453–463

Ebert DH, Deussing J, Peters C, Dermody TS (2002) Cathepsin L and cathepsin B mediate reovirus disassembly in murine fibroblast cells. J Biol Chem 277:24609–24617

Fackler OT, Peterlin BM (2000) Endocytic entry of HIV-1. Curr Biol 10:1005–1008

Fredericksen BL, Wei BL, Yao J, Luo T, Garcia JV (2002) Inhibition of endosomal/lysosomal degradation increases the infectivity of human immunodeficiency virus. J Virol 76:11440–11446

Geijtenbeek TB, Kwon DS, Torensma R, van Vliet SJ, van Duijnhoven GC, Middel J, Cornelissen IL, Nottet HS, KewalRamani VN, Littman DR, Figdor CG, van Kooyk Y (2000) DC-SIGN, a dendritic cell-specific HIV-1-binding protein that enhances trans-infection of T cells. Cell 100:587–597

Gilbert JM, Benjamin TL (2000) Early steps of polyomavirus entry into cells. J Virol 74:8582–8588

Girod A, Wobus CE, Zadori Z, Ried M, Leike K, Tijssen P, Kleinschmidt JA, Hallek M (2002) The VP1 capsid protein of adeno-associated virus type 2 is carrying a phospholipase A2 domain required for virus infectivity. J Gen Virol 83:973–978

Golden JW, Linke J, Schmechel S, Thoemke K, Schiff LA (2002) Addition of exogenous protease facilitates reovirus infection in many restrictive cells. J Virol 76:7430–7443

Greber UF, Singh I, Helenius A (1994) Mechanisms of virus uncoating. Trends Microbiol 2:52–56

Hay AJ, Wolstenholme AJ, Skehel JJ, Smith MH (1985) The molecular basis of the specific anti-influenza action of amantadine. EMBO J 4:3021–3024

Hofer F, Gruenberger M, Kowalski H, Machat H, Huettinger M, Kuechler E, Blass D (1994) Members of the low density lipoprotein receptor family mediate cell entry of a minor-group common cold virus. Proc Natl Acad Sci U S A 91:1839–1842

Hogle JM (2002) Poliovirus cell entry: Common structural themes in viral cell entry pathways. Annu Rev Microbiol 56:677–702

Huber M, Brabec M, Bayer N, Blaas D, Fuchs R (2001) Elevated endosomal pH in HeLa cells overexpressing mutant dynamin can affect infection by pH-sensitive viruses. Traffic 2:727–736

Hueffer K, Parker JS, Weichert WS, Geisel RE, J-Y. S, Parrish CR (2003) The natural host range shift and subsequent evolution of canine parvovirus resulted from virus-specific binding to the canine transferrin receptor. J Virol in press:

Iruzun A, Carrasco L (2001) Entry of poliovirus into cells is blocked by valinomycin and concanamycin A. Biochemistry 40:3589–3600

Jin M, Park J, Lee S, Park B, Shin J, Song KJ, Ahn TI, Hwang SY, Ahn BY, Ahn K (2002) Hantaan virus enters cells by clathrin-dependent receptor-mediated endocytosis. Virology 294:60–69

Joki-Korpela P, Marjomaki V, Krogerus C, Heino J, Hyypia T (2001) Entry of human parechovirus 1. J Virol 75:1958–1967

Kartenbeck J, Stukenbrok H, Helenius A (1989) Endocytosis of simian virus 40 into the endoplasmic reticulum. J Cell Biol 109:2721–2729

Lamb RA (1993) Paramyxovirus fusion: a hypothesis for changes. Virology 197:1–11

Lee S, Zhao Y, Anderson WF (1999) Receptor-mediated Moloney murine leukemia virus entry can occur independently of the clathrin-coated-pit-mediated endocytic pathway. J Virol 73:5994–6005

Lewis JK, Bothner B, Smith TJ, Siuzdak G (1998) Antiviral agent blocks breathing of the common cold virus. Proc Natl Acad Sci U S A 95:6774–6778

Li E, Stupak D, Klemke R, Cheresh DA, Nemerow GR (1998) Adenovirus endocytosis via α_v integrins requires phosphoinositide-3-OH kinase. J Virol 72:2055–2061

Liu J, Thorp SC (2002) Cell surface heparan sulfate and its roles in assisting viral infections. Med Res Rev 22:1–25

Madshus IH, Sandvig K, Olsnes S, van Deurs B (1987) Effect of reduced endocytosis induced by hypotonic shock and potassium depletion on the infection of Hep 2 cells by picornaviruses. J Cell Physiol 131:14–22

Marechal V, Prevost MC, Petit C, Perret E, Heard JM, Schwartz O (2001) Human immunodeficiency virus Type 1 entry into macrophages mediated by macropinocytosis. J Virol 75:11166–11177

Marjomaki V, Pietiainen V, Matilainen H, Upla P, Ivaska J, Nissinen L, Reunanen H, Huttunen P, Hyypia T, Heino J (2002) Internalization of echovirus 1 in caveolae. J Virol 76:1856–1865

Marsh M, Bron R (1997) SFV infection in CHO cells: cell-type specific restrictions to productive virus entry at the cell surface. J Cell Sci 110:95–103

Marsh M, Helenius A (1980) Adsorptive endocytosis of Semliki Forest virus. J Mol Biol 142:439–454

Marsh M, Helenius A (1989) Virus entry into animal cells. Adv Virus Res 36:107–151

Marsh M, Pelchen-Matthews A (1994), The endocytic pathway and virus entry, in Cellular Receptors for Animal Viruses, Ed. Wimmer E, 215–240, Cold Spring Harbor Press, Cold Spring Harbor, NY

Marsh M, Pelchen-Matthews A (2000) Endocytosis in viral replication. Traffic 1:525–532

Matlin KS, Reggio H, Helenius A, Simons K (1981) Infectious entry pathway of influenza virus in a canine kidney cell line. J Cell Biol 91:601–613

Matlin KS, Reggio H, Helenius A, Simons K (1982) Pathway of vesicular stomatitis virus leading to infection. J Mol Biol 156:609–631

Meier O, Boucke K, Hammer SV, Keller S, Stidwill RP, Hemmi S, Greber UF (2002) Adenovirus triggers macropinocytosis and endosomal leakage together with its clathrin-mediated uptake. J Cell Biol 158:1119–1131

Mettenleiter TC (2002) Brief overview on cellular virus receptors. Virus Res 82:3–8

Miller N, Hutt-Fletcher LM (1992) Epstein-Barr virus enters B cells and epithelial cells by different routes. J Virol 66:3409–3414

Miyazawa N, Crystal RG, Leopold PL (2001) Adenovirus serotype 7 retention in a late endosomal compartment prior to cytosol escape is modulated by fiber protein. J Virol 75:1387–1400

Mothes W, Boerger AL, Narayan S, Cunningham JM, Young JAT (2000) Retroviral entry mediated by receptor priming and low pH triggering of and envelope glycoprotein. Cell 103:679–689

Nemerow GR (2000) Cell receptors involved in adenovirus entry. Virology 274:1–4

Nicola AV, McEvoy AM, Straus SE (2003) Roles for endocytosis and low pH in herpes simplex virus entry into HeLa and Chinese hamster ovary cells. J Virol 77:5324–5332

Nomoto A, Koike S, Aoki J (1994) Tissue tropism and species specificity of poliovirus infection. Trends Microbiol 2:47–51

Norkin LC (1999) Simian virus 40 infection via MHC class I molecules and caveolae. Immunol Rev 168:13–22

Norkin LC, Anderson HA, Wolfrom SA, Oppenheim A (2002) Caveolar endocytosis of simian virus 40 is followed by brefeldin A-sensitive transport to the endoplasmic reticulum, where the virus disassembles. J Virol 76:5156–5166

O'Hara B, Olson W (2002) HIV entry inhibitors in clinical development. Curr Opin Pharmacol 2:523

Oldstone MB, Homann D, Lewicki H, Stevenson D (2002) One, two, or three step: measles virus receptor dance. Virology 299:162–163

Overbaugh J, Miller AD, Eiden MV (2001) Receptors and entry cofactors for retroviruses include single and multiple transmembrane-spanning proteins as well as newly described glycophosphatidylinositol-anchored and secreted proteins. Microbiol Mol Biol Rev 65:371-389

Parker JS, Murphy WJ, Wang D, O'Brien SJ, Parrish CR (2001) Canine and feline parvoviruses can use human or feline transferrin receptors to bind, enter, and infect cells. J Virol 75:3896-3902

Parker JS, Parrish CR (2000) Cellular uptake and infection by canine parvovirus involves rapid dynamin-regulated clathrin-mediated endocytosis, followed by slower intracellular trafficking. J Virol 74:1919-1930

Pelkmans L, J. K, Helenius A (2001) Caveolar endocytosis of simian virus 40 reveals a new two-step vesicular transport pathway to the ER. Nat Cell Biol 3:473-483

Pelkmans L, Puntener D, Helenius A (2002) Local actin polymerization and dynamin recruitment in SV40-induced internalization of caveolae. Science 296:535-539

Perez L, Carrasco L (1993) Entry of poliovirus into cells does not require a low-pH step. J Virol 67:4543-4548

Pho MT, Ashok A, Atwood WJ (2000) JC virus enters human glial cells by clathrin-dependent receptor-mediated endocytosis. J Virol 74:2288-2292

Pinto LH, Holsinger LJ, Lamb RA (1992) Influenza virus M2 protein has ion channel activity. Cell 69:517-528

Poranen MM, Daugelavicius R, Bamford DH (2002) Common principles in viral entry. Annu Rev Microbiol 56:521-538

Racaniello VR (1996) Early events in poliovirus infection: virus-receptor interactions. Proc Natl Acad Sci U S A 93:11378-11381

Rauma T, Tuukkanen J, Bergelson JM, Denning G, Hautala T (1999) rab5 GTPase regulates adenovirus endocytosis. J Virol 73:9664-9668

Richards AA, Stang E, Pepperkok R, Parton RG (2002) Inhibitors of COP-mediated transport and cholera toxin action inhibit simian virus 40 infection. Mol Biol Cell 13:1750-1764

Richterova Z, Liebl D, Horak M, Palkova Z, Stokrova J, Hozak P, Korb J, Forstova J (2001) Caveolae are involved in the trafficking of mouse polyomavirus virions and artificial VP1 pseudocapsids toward cell nuclei. J Virol 75:10880-10891

Root CR, Wills EG, McNair LL, Whittaker GR (2000) Entry of influenza viruses into cells is inhibited by a highly specific protein kinase C inhibitor. J Gen Virol 81:2697-2705

Ross SR, Schofield JJ, Farr CJ, Bucan M (2002) Mouse transferrin receptor 1 is the cell entry receptor for mouse mammary tumor virus. Proc Natl Acad Sci USA 99:12386-12390

Rotbart HA (2002) Treatment of picornavirus infections. Antiviral Res 53:83-98

Russell DG, Marsh M (2001), Endocytosis in pathogen entry and replication, in Endocytosis, Ed. Marsh M, 247-280, Oxford University Press, Oxford

Schneider-Schaulies J (2000) Cellular receptors for viruses: links to tropism and pathogenesis. J Gen Virol 81:1413-1429

Shukla D, Liu J, Blaiklock P, Shworak NW, Bai X, Esko JD, Cohen GH, Eisenberg RJ, Rosenburg RD, Spear PG (1999) A novel role for 3-O-sulfated heparan sulfate in herpes simplex virus 1 entry. Cell 99:13-22

Sieczkarski SB, Brown HA, Whittaker GR (2003) The role of protein kinase C βII in influenza virus entry via late endosomes. J Virol 77:460-469

Sieczkarski SB, Whittaker GR (2002) Dissecting virus entry via endocytosis. J Gen Virol 83:1535-1545

Sieczkarski SB, Whittaker GR (2002) Influenza virus can enter and infect cells in the absence of clathrin-mediated endocytosis. J Virol 76:10455-10464

Sieczkarski SB, Whittaker GR (2003) Differential requirements of Rab5 and Rab7 for endocytosis of influenza and other enveloped viruses. Traffic 4:333-343

Skehel JJ, Wiley DC (2000) Receptor binding and membrane fusion in virus entry: the influenza hemagglutinin. Annu Rev Biochem 69:531-569

Smyth MS, Martin JH (2002) Picornavirus uncoating. Mol Pathol 55:214-219

Somsel Rodman J, Wandinger-Ness A (2000) Rab GTPases coordinate endocytosis. J Cell Sci 113:183-192

Spear P (2002) Viral interactions with receptors in cell junctions and effects on junctional stability. Dev Cell 3:462-464

Spear PG (1993) Entry of alphaherpesviruses into cells. Semin Virol 4:167-180

Stuart AD, Eustace HE, McKee TA, Brown TD (2002) A novel cell entry pathway for a DAF-using human enterovirus is dependent on lipid rafts. J Virol 76:9307-9322

Sturzenbecker LJ, Nibert M, Furlong D, Fields BN (1987) Intracellular digestion of reovirus particles requires a low pH and is an essential step in the viral infectious cycle. J Virol 61:2351-2361

Suikkanen S, Saajarvi K, Hirsimaki J, Valilehto O, Reunanen H, Vihinen-Ranta M, Vuento M (2002) Role of recycling endosomes and lysosomes in dynein-dependent entry of canine parvovirus. J Virol 76:4401-4411

Tucker SP, Wimmer E, Compans RW (1994), Expression of viral receptors and the vectorial release of viruses in polarized cells, in Cellular Receptors for Animal Viruses, Ed. Wimmer E, Cold Spring Harbor Laboratory Press, Cold Spring Harbor, NY

Walters RW, Freimuth P, Moninger TO, Ganske I, Zabner J, Welsh MJ (2002) Adenovirus fiber disrupts CAR-mediated intercellular adhesion allowing virus escape. Cell 110:789-799

Wimmer E (1994), Cellular Receptors for Animal Viruses, Cold Spring Harbor Laboratory Press, Cold Spring Harbor, NY

Yanagi Y, Ono N, Tatsuo H, Hashimoto K, Minagawa H (2002) Measles virus receptor SLAM (CD150). Virology 299:155-161

Young JAT (2001), Virus Entry and Uncoating, in Fields Virology, Ed. Knipe DM Howley PM, Lippincott Williams and Wilkins, Philadelphia, PA

Zadori Z, Szelei J, Lacoste MC, Li Y, Gariepy S, Raymond P, Allaire M, Nabi IR, Tijssen P (2001) A viral phospholipase A2 is required for parvovirus infectivity. Dev Cell 1:291-302

The Many Mechanisms of Viral Membrane Fusion Proteins

L. J. Earp[1] · S. E. Delos[2] · H. E. Park[1] · J. M. White[2] (✉)

[1] Department of Microbiology, University of Virginia, Charlottesville, VA, USA
[2] Department of Cell Biology, School of Medicine, UVA Health System,
P.O. Box 800732, Charlottesville, VA 22908-0732, USA
jw7g@virginia.edu

1	Introduction	26
2	Activation of Viral Fusion Proteins	27
2.1	Low pH Activation	28
2.2	Receptor Activation at Neutral pH	29
2.3	"Two-Step" Activation	29
3	Classification of Fusion Proteins Based on Structural Criteria	30
4	Examples of Fusion Activation Mechanisms	32
4.1	Influenza HA (Class I Fusion Protein, Low pH)	33
4.2	HIV Env (Class I Fusion Protein, Neutral pH)	34
4.3	Paramyxovirus F Proteins (Class I Fusion Protein, Neutral pH, Attachment Protein Assisted)	36
4.4	TBE E and SFV E1 (Class II Fusion Proteins, Low pH)	38
5	Membrane Dynamics During Fusion	40
6	Membrane-Interacting Regions of Viral Fusion Proteins	41
6.1	The Fusion Peptide	42
6.1.1	Structure of N-terminal Fusion Peptides	43
6.1.2	Structure of Internal Fusion Peptides	44
6.1.3	Roles of Fusion Peptides	45
6.2	The Transmembrane Domain	45
6.3	The Juxtamembrane Region of the Ectodomain	46
6.4	The Cytoplasmic Tail	47
7	Rafts in Viral Membrane Fusion	48
8	Inhibitors of Viral Fusion	49
8.1	Inhibition of Helix Bundle Formation	50
8.2	Inhibition of Other Steps in Fusion	51
9	Perspectives	51
References		53

Abstract Every enveloped virus fuses its membrane with a host cell membrane, thereby releasing its genome into the cytoplasm and initiating the viral replication cycle. In each case, one or a small set of viral surface transmembrane glycoproteins mediates fusion. Viral fusion proteins vary in their mode of activation and in structural class. These features combine to yield many different fusion mechanisms. Despite their differences, common principles for how fusion proteins function are emerging: In response to an activating trigger, the metastable fusion protein converts to an extended, in some cases rodlike structure, which inserts into the target membrane via its fusion peptide. A subsequent conformational change causes the fusion protein to fold back upon itself, thereby bringing its fusion peptide and its transmembrane domain—and their attached target and viral membranes—into intimate contact. Fusion ensues as the initial lipid stalk progresses through local hemifusion, and then opening and enlargement of a fusion pore. Here we review recent advances in our understanding of how fusion proteins are activated, how fusion proteins change conformation during fusion, and what is happening to the lipids during fusion. We also briefly discuss the therapeutic potential of fusion inhibitors in treating viral infections.

Keywords Membrane fusion protein · Class I fusion protein · Class II fusion protein · Influenza HA · HIV Env · Low-pH activation · Receptor activation · Conformational changes · Membrane dynamics · Anti-fusion antivirals

1
Introduction

Fusion of enveloped viruses with host cells remains an important topic of research for two major reasons. First, it has recently become clear that fusion is a good target for therapeutic intervention (Kilby et al. 1998). Second, viral fusion reactions continue to serve as models for cellular fusion events. Although several viral fusion proteins, such as influenza hemagglutinin (HA) and the human immunodeficiency virus (HIV) envelope glycoprotein (Env), have emerged as paradigms, it is important to realize that there are many distinguishing features among viral fusion proteins (Table 1). Viral fusion proteins can be activated for fusion by different mechanisms. They have also been classified according to structural criteria. For some viruses, the viral receptor does not actively participate in fusion, whereas for others, one or more receptors are essential players. The location of the fusion peptide, critical for fusion, can vary. Finally, whereas some viruses require a single viral glycoprotein to mediate fusion, others require multiple viral glycoproteins. There are many excellent recent reviews on viral fusion and the glycoproteins that mediate this process (Durell et al. 1997; Eckert and Kim 2001; Heinz and

Table 1. Viral membrane fusion proteins

Family	Viral proteins needed	pH of fusion	Class	Fusion peptide
Orthomyxovirus	HA	Low	I	N-terminal
Alphavirus	E1	Low	II	Internal
Flavivirus	E	Low	II	Internal
Rhabdovirus	G	Low	?	Internal
Bunyavirus	G1/G2	Low	?	?
Arenavirus	GP	Low	?	?
Filovirus	GP	Low[a]	I	Internal
Retrovirus	Env	Neutral[b]	I	N-terminal, internal
Paramyxovirus	F, HN	Neutral	I	N-terminal
Herpesvirus	gB, gD, gH, gL	Neutral	?	?
Coronavirus	S	Neutral	I	Internal
Poxvirus	N.D.	Neutral	?	?
Hepadnavirus	S	Neutral?[d]	?	?[e]
Iridovirus	N.D.	N.D.	?	?

[a] Inferred from infectivity assays.
[b] Most retroviruses fuse at neutral pH. MMTV appears to require low pH [Ross et al. (2002) PNAS 99:12386-90] to fuse. Avian retroviruses require receptor priming at neutral pH followed by exposure to low pH [Mothes et al. (2000) Cell 103:679-89; see text for a discussion of this model].
[c] Coronaviruses possess heptad repeats [Chambers et al. (1990) J Gen Virol 71:3075-80] characteristic of class I viral fusion proteins. Recent work indicates that they are, indeed, class I fusion proteins [Bosch et al. (2003) J Virol 77:8801-11].
[d] With infectivity assays, hepadnavirus uptake was shown to be pH-independent [Hagelstein et al. (1997) Virology 229:292-4]. However, recent studies have shown that duck hepatitis B virus may require low pH [Grgacic et al. (2000) J Virol 74:5116-22].
[e] The S protein contains a stretch of amino acids predicted to be a fusion peptide but has not been further characterized.

Allison 2001; Skehel and Wiley 2000; Weissenhorn et al. 1999). The goal of this review is to give the reader an appreciation for the diversity of viral fusion mechanisms.

2
Activation of Viral Fusion Proteins

All fusion proteins exist on virion surfaces in a metastable state in which the fusion peptide, a critical hydrophobic sequence, is hidden or shielded within the glycoprotein oligomer (Carr et al. 1997; Hernandez et al.

1996; Rey et al. 1995; Skehel and Wiley 2000; Wilson et al. 1981). After activation, the fusion peptides are rendered accessible for interaction with a target membrane. A major distinction among viral fusion proteins is the "trigger" for activation. There are two well-recognized mechanisms: (1) exposure to low pH and (2) specific interactions with target cell receptors at neutral pH. A third mechanism involving receptor priming at neutral pH followed by further activation at low pH was recently proposed (Mothes et al. 2000).

2.1
Low pH Activation

Orthomyxoviruses, togaviruses, flaviviruses, rhabdoviruses, bunyaviruses, arenaviruses, and, apparently, filoviruses require low pH to fuse with target membranes (Table 1) (Doms et al. 1985; Gaudin et al. 1999b; Stegmann et al. 1987; White and Helenius 1980). These viruses are endocytosed after binding to the target cell surface. The low-pH environment of the endosome activates the viral fusion protein to convert from a metastable state to one that is capable of driving fusion. Although the presence of a receptor may modulate the rate or extent of fusion (Ohuchi et al. 2002; Stegmann et al. 1996; White et al. 1982), receptors are not essential for low-pH-dependent fusion. Low-pH-dependent fusion generally occurs within seconds to minutes at 37°C but can also occur, albeit more slowly, at T<22°C.

Four main techniques have been used to assess whether a virus requires low pH to fuse. The first technique is testing the effects of agents, such as bafilomycin, that inhibit endosomal acidification. In some studies of this type, fusion has been measured directly by assessing the transfer of fluorescent probes from the virus to the target cell (Earp et al. 2003; Irurzun et al. 1997; Zarkik et al. 1997). In others, fusion has been inferred by monitoring postfusion events, such as the synthesis of viral DNA (Mothes et al. 2000).

A second test is to assess whether fusion of bound virions can be induced by briefly warming virus-cell complexes in low-pH medium (Mothes et al. 2000; White et al. 1980). A third test is to assess whether pretreatment of virions at low pH (in the absence of target membranes) inactivates the virus for fusion. Some (Bron et al. 1993; Corver et al. 2000; Di Simone and Buchmeier 1995; Korte et al. 1999; Nir et al. 1990; Stegmann et al. 1987), but not all (Puri et al. 1988), viruses that fuse at low pH can be inactivated by this method. Viral fusion proteins that are inactivated by low pH undergo irreversible conformational changes. In the case of X:31 HA, this results in insertion of the fusion peptide into the viral membrane (Korte et al. 1999; Weber et al. 1994).

The fourth test is to assess whether cells expressing the viral fusion protein can fuse. Cell-cell fusion can be observed by light or fluorescence microscopy (Frey et al. 1995; Melikyan et al. 1997b; Mothes et al. 2000), or it can be scored with gene reporter assays that monitor interactions of components from the fusing cells (Delos and White 2000; Earp et al. 2003; Feng et al. 1996; Nussbaum et al. 1994). Although cell-cell fusion assays are relatively simple to perform, the results do not always correlate with virus-cell fusion or infection (Earp et al. 2003; Lavillette et al. 1998; Schmid et al. 2000).

2.2
Receptor Activation at Neutral pH

Many enveloped viruses do not require low pH to fuse with target cells. This has generally been established in controlled experiments using the approaches described in Sect. 2.1. Viruses that can fuse at neutral pH include paramyxoviruses, herpesviruses, coronaviruses, poxviruses, and most retroviruses (Table 1) (Hernandez et al. 1997; McClure et al. 1990; Stein et al. 1987; Taguchi and Matsuyama 2002). The fusion proteins of these viruses are activated via specific interactions with one or more receptors in the target cell membrane (Hernandez et al. 1996; Hunter 1997; Stein et al. 1987). Viruses that can fuse at neutral pH are thought to do so at the plasma membrane. However, they may also be able to fuse with neutral-pH intracellular compartments (e.g., caveosomes) that can be accessed through newly recognized endocytic pathways (Pelkmans and Helenius 2003; Shin and Abraham 2001) (see also the chapter by Sieczkarski and Whittaker, this volume). It is important to note, however, that viruses that can fuse at neutral pH may also possess the ability to fuse at low pH (Earp et al. 2003; Fackler and Peterlin 2000). To date, neutral-pH fusion has been found to display a sharp temperature threshold, with little or no fusion occurring at $T<20°C$.

2.3
"Two-Step" Activation

Recently, a third model was proposed for the activation of alpharetroviruses. In this model, activation of the alpharetroviral Env begins with receptor binding at neutral pH (at $T>22°C$) but is only complete after exposure to low pH (Mothes et al. 2000). The role of low pH in this "two-step" model is derived from two key observations: (1) The continuous presence of endosomal acidification inhibitors prevents production of alpharetroviral reverse transcripts, and (2) cells expressing Env and cells expressing the viral receptor only form large syncytia after exposure to low pH (Mothes et al. 2000). Our recent work indicates that alpharetrovirus fusion can proceed to the lipid mixing stage at neutral pH (Earp et al. 2003), and that receptor binding and low pH sequentially induce distinct conformational changes in the alpharetrovial Env (Matsuyama et al. 2004). Current work is now focused on determining the precise role of low pH in the fusion cascade.

3
Classification of Fusion Proteins Based on Structural Criteria

All viral fusion proteins contain a relatively large ectodomain, generally a single transmembrane domain, and all contain a cytoplasmic tail. So far, two major groups (class I and class II) have been defined based on structural criteria (Heinz and Allison 2001; Lescar et al. 2001) (Tables 1 and 2).

Class I fusion proteins are synthesized as precursors that are cleaved into two subunits by host cell proteases. In some cases (e.g., influenza HA), the two subunits remain associated through a disulfide bond; in others (e.g., HIV Env), the two subunits remain associated through non-covalent interactions. The proteolytic processing event that generates the two subunits is critical, as it creates the metastable state of the fusion protein (Chen et al. 1998). Class I fusion proteins exist as relatively long trimeric spikes in both their metastable and activated states. In their metastable states, they project perpendicularly to the viral membrane. The activated forms of the fusion subunits of known class I fusion proteins are highly α-helical (Skehel and Wiley 2000), and the final lowest-energy (which we will refer to as "postfusion") forms (Fig. 1) contain "six-helix bundles" (Bullough et al. 1994; Carr and Kim 1993). All six-helix bundles contain a relatively long (65–115 Å) central N-terminal trimeric coiled-coil. Some (e.g., HIV Env, SIV Env, and paramyxovirus F) form six-helix bundles that extend to their membrane proximal ends [i.e., three C-terminal helices (Fig. 1A, green) pack in the grooves of the central coiled-coil (Fig. 1A, blue)]. Others display a mixture of helical and nonhelical segments that pack into the grooves of the central coiled-coil. For example, the HA2 subunit of influenza HA contains a relatively small six-helix bundle (Fig. 1, green/blue) at its membrane distal end,

Table 2. Class I vs. class II viral membrane fusion proteins

Property	Class I	Class II
Type of integral membrane protein	Type I	Type I[a]
Synthesized as	Inactive precursor	Inactive precursor[b]
Exist on virion in	Metastable state	Metastable state
Orientation in virion (to membrane)	Perpendicular	Parallel
Converted to metastable state by	Proteolytic processing within fusion protein precursor	Proteolytic processing of an associated protein
No. of subunits in fusion protein	2	1
Major secondary structure of fusion subunit	α-Helix[c]	β-Sheet
Activated to fusogenic form by	Low pH or cell receptor(s)[d]	Low pH
Oligomeric state of metastable protein	Trimer	Dimer
Oligomeric state of fusion active protein	Trimer	Trimer
Location of fusion peptide	N-terminal or internal	Internal loop
Structure of final fusogenic form	Trimer of hairpins (coiled-coil)	Trimer of hairpins (non-coiled-coil)

[a] The TBE E glycoprotein has two membrane anchoring segments near its C-terminal end [Heinz and Allison (2001) Curr Opin Microbiol 4:450–5].

[b] Known class II fusion proteins are activated by proteolytic cleavage of an accessory protein.

[c] The postfusion forms of all known class I fusion proteins are α-helical. The fusion subunit of metastable influenza HA is also highly α-helical, and this appears to be the case for a paramyxovirus F protein [Chen et al. (2001a) Structure 9:255–66]. Comparable information is not available for the metastable forms of other class I fusion proteins.

[d] In the case of paramyxoviruses, the receptor binding protein relays the information of receptor binding to the fusion subunit [Lamb 1993; Colman and Lawrence 2003]

followed by an extended chain (Fig. 1, yellow) that packs in the groove and extends to the N-terminal (membrane proximal) end of its central coiled-coil (Fig. 1B). Because of these variations, the postfusion forms of class I fusion proteins are often referred to as "trimers of hairpins" (Eckert and Kim 2001).

The general structure of class II fusion proteins is quite different from that of class I fusion proteins. A well-characterized example is the envelope glycoprotein (E) of tick-borne encephalitis (TBE) virus. During biosynthesis, TBE E and a second viral membrane glycoprotein, the precursor to the membrane protein (prM), form heterodimers. As virions ma-

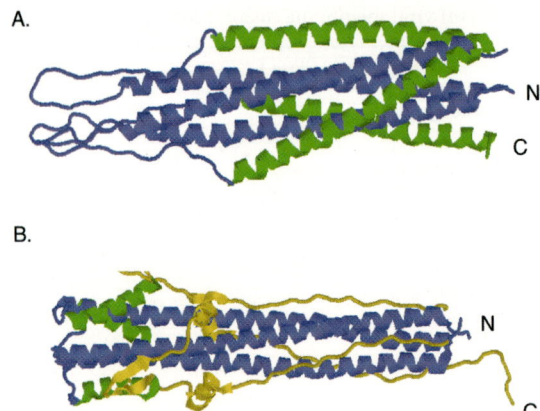

Fig. 1. Structures of the postfusion forms of SIV Env (**A**) and influenza HA (**B**). A NMR structure of the postfusion form of SIV Env gp41 subunit (PDB accession number 2EZO). **B** Crystal structure of the postfusion form of influenza HA2 subunit (PDB accession number 1QU1). Coiled-coil regions are *blue*. C-terminal helices are *green*. For influenza HA2, the C-terminal extended region is *yellow*. *N* and *C* indicate the points where the fusion peptide and the transmembrane domain, respectively, attach

ture, a host cell protease cleaves prM, resulting in reorganization of proteins on the viral surface (Allison et al. 1995). After prM cleavage, the E proteins exist as metastable homodimers. The ectodomains of the dimer are oriented antiparallel to one another. In further contrast to the trimeric class I fusion protein spikes, the ectodomains of the E homodimer lie parallel to the viral membrane and close to the surface. The TBE E protein is composed mostly of β-strand structure (Heinz and Allison 2001; Rey et al. 1995). The architecture of the Semliki Forest virus (SFV) spike, another well-characterized class II fusion protein, is similar to that of TBE E, but in this case, the metastable oligomer is a heterodimer of two membrane-anchored proteins, E1 and E2, with an associated small protein (E3).

4
Examples of Fusion Activation Mechanisms

In Sects. 4.1–4.4, we discuss a few examples of viral fusion proteins that employ different fusion mechanisms in more detail. These will include examples of class I and class II fusion proteins, activated by low pH or by receptor interactions at neutral pH.

4.1
Influenza HA (Class I Fusion Protein, Low pH)

High-resolution structures are available for both the complete native (metastable) (Wilson et al. 1981) and activated (Bullough et al. 1994; Chen et al. 1999) forms of the influenza HA. On the viral surface, HA exists as a trimer of heterodimers (Fig. 2A). Each heterodimer consists of HA1, which contains the receptor binding domain (Fig. 2, gray), and HA2, which contains the fusion peptide (Fig. 2, red) and the transmembrane domain (located at the C-terminus). In the native (neutral pH) structure, the fusion peptide is buried within the HA oligomer. Three long helices, one from each monomer, come together to form the triple-stranded coiled-coil of the metastable trimer (Fig. 2A and B, blue and green).

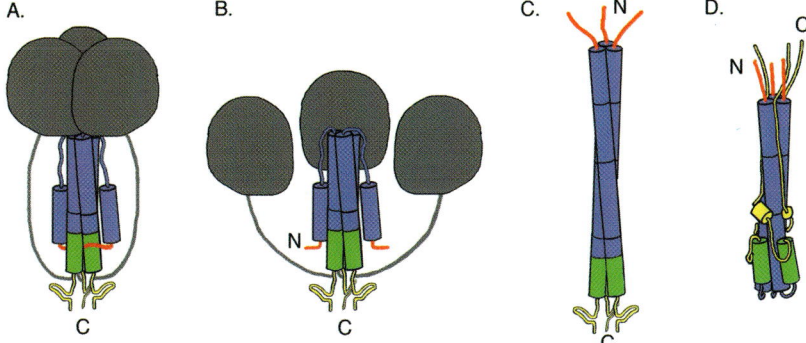

Fig. 2A–D. Low-pH-induced conformational changes within influenza HA. HA1 is depicted in *gray*. The fusion peptide is *red* (HA2 residues 1–24). The coiled-coil is *blue*, with the C-terminal helix colored *green*. The C-terminal extended region is *yellow*. The transmembrane domain (not shown) attaches to the C-terminal end, indicated by "*C*", of HA2. A model for conformational changes: **A** In the native, metastable, structure of HA, the fusion peptides are buried within the trimer interface. HA1 acts as a clamp to hold HA2 in a metastable state. HA2 is largely shielded by HA1. To illuminate the HA2 core, we have cartooned the portion of HA1 that covers HA2 as a simple (*gray*) line. **B** On exposure to low pH, the HA1 headgroups separate, allowing expulsion of the fusion peptide. **C** A loop-to-helix transition causes the fusion peptide to be repositioned to one end of HA2, where it can bind to the target membrane. **D** A helix-to-loop transition causes the C-terminal helix and the C-terminal extended region to reverse direction and bind to the grooves of the coiled-coil in an antiparallel orientation

On exposure to low pH, HA undergoes dramatic conformational changes. The globular head domains separate, releasing the clamp that holds HA2 in its metastable state (Fig. 2B). As a result, the fusion peptide is exposed (Fig. 2C, red) at the top of an extended triple-stranded coiled-coil, in a position where it can interact with the target membrane. A helix-to-loop transition causes a short helix (Fig. 2D, green) and the C-terminal extended region (yellow) to flip up and run antiparallel to the central coiled-coil (Bullough et al. 1994). As a result, the fusion peptide and transmembrane domain are brought into close proximity at the same end of the molecule (Fig. 2D).

Many regions of HA are important for fusion. The fusion peptide is critical for hydrophobic attachment of the virus to the target membrane (Sect. 6.1). Mutations that prevent (1) globular head domain separation (Godley et al. 1992; Kemble et al. 1992), (2) the "B-loop"-to helix transition (Gruenke et al. 2002; Qiao et al. 1998), or (3) the C-terminal extended region from packing into the grooves of the final coiled-coil (Borrego-Diaz et al. 2003; Park et al. 2003) ablate the ability of HA to reach the lipid mixing stage of fusion. In our model (Gruenke et al. 2002), conversion of HA to a prehairpin intermediate (Fig. 2C) allows HA to bind to the target membrane. Further conversion to the hairpin structure (Fig. 2D) then drives the formation and opening of a fusion pore.

4.2
HIV Env (Class I Fusion Protein, Neutral pH)

Like influenza HA, HIV Env is synthesized as a single-chain precursor and cleaved during biosynthesis to yield gp120 and gp41. Native (metastable) HIV Env is a trimer of the heterodimers of gp120 (the receptor binding subunit) and gp41 (the fusion subunit). Env is activated for fusion (at neutral pH) after sequential binding to CD4 and a coreceptor (a chemokine receptor). Binding of Env to CD4 causes conformational changes in Env that permit binding to the coreceptor. After coreceptor binding, additional conformational changes occur in Env that lead to fusion (Eckert and Kim 2001).

Crystal structures exist for the core of the gp120 subunit (Kwong et al. 1998) as well as for the postfusion (Fig. 3, Step 6) form of gp41 (Chan et al. 1997; Weissenhorn et al. 1997). However, there is not yet a crystal structure of the native (metastable) Env trimer. Therefore, a detailed picture of HIV Env activation via receptor interaction is not available. We presume that the first steps of Env activation are separation of the

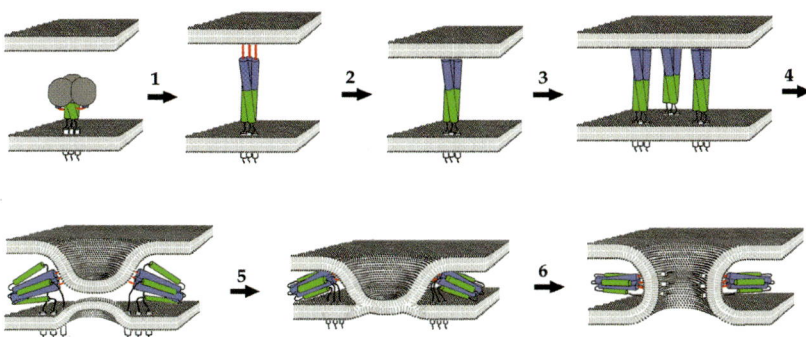

Fig. 3. Model of HIV fusion. Env exists as a trimer in the surface of the native viral membrane, with fusion peptides (*red*) presumably buried within the trimer interface. SU domains (pictured as *gray* globular domains at the top of the trimer) provide the receptor-binding function. For clarity, SU domains are omitted after *Step 1*. Target cell receptors are not pictured in this model. On exposure to receptor and coreceptor at T $\geq 22°C$ and neutral pH, Env undergoes conformational changes that result in exposure of the fusion peptides (*Step 1*), which then insert into the target membrane (*Step 2*). Multiple Envs may cluster (*Step 3*) to form a fusion site. Additional conformational changes (*Steps 4* and *5*) lead to the formation of a six-helix bundle, resulting in hemifusion (*Step 5*) (defined as mixing of the outer leaflets of the viral and cellular membranes). Eventually a fusion pore forms (*Step 6*) and enlarges (not shown)

globular head domains, expulsion of the fusion peptide, and extension of gp41 into a prehairpin intermediate (Fig. 3, Step 1). Several lines of evidence indicate the existence of the prehairpin intermediate. For example, peptide analogs of the C-terminal helix (Fig. 3, green) strongly inhibit HIV fusion and infection (Chan and Kim 1998; Kilby et al. 1998). Also, a synthetic peptide corresponding to the C-terminal helix coimmunoprecipitates with HIV Env after engagement of receptors (Furuta et al. 1998; He et al. 2003). The C-terminal helix then packs, in an antiparallel fashion, into the groove of the N-terminal coiled-coil (Fig. 3, Step 5). Because the C-terminal helices of gp41 extend along the entire length of the N-terminal coiled-coil, this packing would bring the fusion peptide and transmembrane domain very close together. The transition to the six-helix bundle drives membrane merger (Melikyan et al. 2000a). Moreover, complete six-helix bundles are needed to form "robust" fusion pores (Markosyan et al. 2003).

As mentioned above, HIV studies, primarily using epitope accessibility assays, have indicated that engagement of HIV receptors induces conformational changes in gp120 and gp41 (Eckert and Kim 2001; Xiang et al. 2002). A remaining issue for all receptor-activated viral fusion proteins is how information is transmitted (after receptor binding) through the receptor binding subunit to the fusion subunit. Such transmission is essential to allow rearrangements in the fusion subunit (e.g., six-helix bundle formation) that drive fusion. For HIV, part of the mechanism may involve reduction of one or more disulfide bonds in gp120 (Abrahamyan et al. 2003; Barbouche et al. 2003; Fenouillet et al. 2001; Gallina et al. 2002). In murine retroviral Envs, a proline-rich hinge region appears to relay receptor binding information from the N-terminal to the C-terminal region of the receptor binding subunits (SU) (Barnett and Cunningham 2001; Lavillette et al. 2001). Because the proline-rich region of SU is linked to TM by a disulfide bond (Pinter et al. 1997), this may provide a relay system to trigger conformational changes in the fusion subunit. Clearly, the molecular pathways by which receptor-activated fusion proteins change from their metastable to their activated forms need to be defined.

In Fig. 3, we show a working model for HIV Env-mediated fusion. It is derived in part from studies with influenza HA, and it is similar to other HIV fusion models (Eckert and Kim 2001). Our hypothesis is that all class I fusion proteins will employ similar mechanisms. We note, however, that even in the case of influenza HA, alternate models are still entertained (see Fig. 2 in Jahn et al. 2003). Furthermore, others have suggested that different class I fusion proteins may use fundamentally different mechanisms (Chen et al. 2001a).

The features that we predict will be common to the fusion mechanisms of all class I fusion proteins (Fig. 3) include: (1) conversion from a metastable state to an activated state, (2) exposure and repositioning of the fusion peptide for binding to the target bilayer, (3) recruitment of several activated fusion proteins to a fusion site (Blumenthal et al. 1996; Danieli et al. 1996; Markovic et al. 2001; Markovic et al. 1998), and (4) subsequent conformational changes that result in close apposition of the fusion peptide and the transmembrane domain.

4.3
Paramyxovirus F Proteins (Class I Fusion Protein, Neutral pH, Attachment Protein Assisted)

The viral fusion proteins that have thus far been discussed in detail contain a receptor binding domain (e.g., the gp120 subunit of HIV Env) within the fusion protein spike. In other cases, the receptor binding domain resides in a separate viral spike. Paramyxoviruses have an attachment protein spike and a separate fusion (F) protein spike. Most, but not all, paramyxoviruses require both the attachment protein and the F protein for fusion (Bagai and Lamb 1995; Paterson et al. 2000). In most cases, the attachment protein must come from the same paramyxovirus as the fusion protein (Bossart et al. 2002). In the few cases in which the F protein is sufficient, fusion is enhanced if the attachment protein is also expressed (Bagai and Lamb 1995). The need for the attachment protein can be overcome by mutations in the F protein (Paterson et al. 2000; Seth et al. 2003) or by conducting fusion reactions at T>37°C (Paterson et al. 2000; Wharton et al. 2000). Paramyxovirus fusion proteins thus represent special cases of receptor-activated fusion proteins, in which receptor activation is communicated from one viral spike glycoprotein to another.

F proteins are proteolytically cleaved during biosynthesis to generate two disulfide-bonded subunits, F_1 and F_2 (Begona Ruiz-Arguello et al. 2002; Gonzalez-Reyes et al. 2001; Lamb 1993), found as metastable trimers of dimers (Baker et al. 1999) on virions. It has been suggested that binding of the attachment protein to a host cell receptor causes conformational changes in this protein, which in turn cause activating conformational changes in the metastable F protein (Lamb 1993; Russell et al. 2001; Takimoto et al. 2002). The exact mechanism by which attachment proteins activate F proteins is not known, but several groups have provided evidence for cross talk between attachment and F proteins (Bossart et al. 2002; Deng et al. 1999; McGinnes et al. 2002; Stone-Hulslander and Morrison 1997; Takimoto et al. 2002; Yao et al. 1997).

The post-fusion form of the F protein from the paramyxovirus SV5 contains a six-helix bundle (Baker et al. 1999). Similar to HIV Env (He et al. 2003; Kilby et al. 1998; Munoz-Barroso et al. 1998) and other retroviral fusion proteins (Earp et al. 2003; Netter 2002), peptide analogs of the N- and C-terminal helices of paramyxovirus six-helix bundles are potent inhibitors of fusion and infection (Bossart et al. 2002; Joshi et al. 1998; Lambert et al. 1996; Young et al. 1999). As is also the case for HIV Env (Markosyan et al. 2003; Melikyan et al. 2000a), a recent study showed that conversion of the SV5 F protein to a six-helix bundle drives membrane fusion (Russell et al. 2001).

Issues yet to be addressed for paramyxoviruses are the structure of the complete native (metastable) F trimer and how it is converted to its activated form. The first glimpses at the metastable and postfusion states of the F trimer came from EM observations of the respiratory syncytial virus (RSV) F protein. Preparations of purified recombinant F protein contained both cone-shaped rods and "lollipop"-shaped structures. On storage, there appeared to be a shift from the cone-shaped to the "lollipop"-shaped structures (Calder et al. 2000). Examination of F complexed with specific monoclonal antibodies suggested that the "lollipop" structures contained six-helix bundles composed of N- and C-terminal heptad repeats (Calder et al. 2000).

A high-resolution structure of an F protein ectodomain was recently presented (Chen et al. 2001a). The protein used for the analysis contained a mixture of precursor F_0 and proteolytically cleaved F. It also apparently lacked the second heptad repeat, which forms the C-helix in the postfusion form. This trimeric F protein structure is fundamentally different from that of influenza HA; the N-terminal end of its coiled-coil is positioned near the viral membrane end of the molecule (i.e., opposite the orientation of the coiled-coil in the metastable HA trimer). If this F protein structure represents the native metastable F trimer, then it suggests a mechanism of fusion activation for F fundamentally different from that for HA (Chen et al. 2001a). Additional work is needed to test this idea.

4.4
TBE E and SFV E1 (Class II Fusion Proteins, Low pH)

All known class II fusion proteins are activated by low pH. However, the mechanism by which class II fusion proteins are initially activated is quite different than the mechanism by which class I fusion proteins are initially activated. For example, the ectodomain of the TBE glycoprotein forms an antiparallel dimer that lies parallel and close to the viral membrane (Fig. 4B). At low pH, the TBE E homodimer converts to an E homotrimer (Allison et al. 1995; Heinz and Allison 2001; Stiasny et al. 2001). This transformation is thought to occur in two steps: dissociation of the E homodimer, followed by reassociation of E trimers (Stiasny et al. 1996). Membrane binding occurs after dimer dissociation and promotes the formation of E homotrimers (Stiasny et al. 2002). Homotrimer formation may involve interactions between α-helices in the stem region of the E protein (Allison et al. 1999).

The SFV fusion protein also converts from a dimer to a trimer during fusion activation. On native virions, E1 exists as a tight heterodimeric complex with a second membrane protein, E2. On exposure to low pH, E1 dissociates from E2, changes conformation, and forms highly stable E1 homotrimers (Ahn et al. 1999; Kielian 1995; Wahlberg et al. 1992; Wahlberg and Garoff 1992). During this process, E1 binds hydrophobically through its fusion peptide to target membranes and mediates fu-

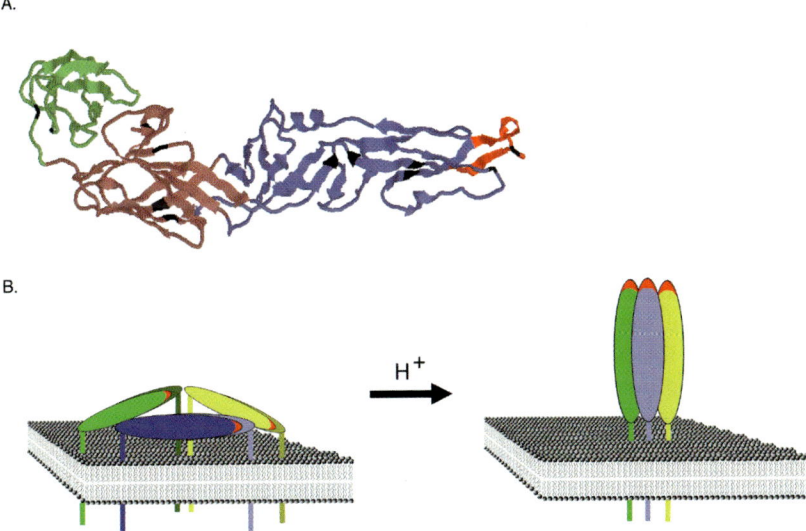

Fig. 4. Structure and cartoon of conformational changes of the TBE E protein. A Crystal structure of TBE E (PDB accession number 1SVB). The fusion peptide is *red*. Domains I, II, and III are *pink*, *blue*, and *green*, respectively. Disulfide bonds are *black*. B Cartoon depicting possible rearrangements during the dimer to trimer transition upon exposure to low pH (Allison et al. 1995). Each of the three dimers (*blue*, *green*, *yellow*; *left*) supplies one monomer (*light shaded subunits*) to the homotrimer (*right*). The organization of the dimers is as found in TBE recombinant subviral particles (Ferlenghi et al. 2001); it may represent an intermediate arrangement (from that on native virions) found during fusion activation (Kuhm et al. 2002). Note that other possibilities for the dimer to trimer transition exist (for example involving relative movements of domains about hinge regions). For very recent developments regarding the fusion mechanism of class II fusion proteins, see Bressanelli et al. (2004), Gibbons et al. (2004) and Modis et al. (2004)

sion. Similar to TBE E, it appears that binding to the target bilayer fosters formation of the activated E1 homotrimer (Kielian 1995; Kielian et al. 2000); the fusion peptide and the transmembrane domain of E1 appear to be important for E1 homotrimer formation (Kielian et al. 1996, 2000; Sjoberg and Garoff 2003). Thus both class I and class II viral fusion proteins appear to function as trimers during fusion. It has been proposed that activated TBE E (Helenius 1995) and SFV E1 stand up as trimeric spikes and present their fusion peptides to the target membrane (Fig. 4B, right). This would be analogous to Fig. 3, Step 1. If this occurs, then the spike would have to refold to bring the viral and cellular membranes together (e.g., analogous to Steps 4 and 5 in Fig. 3). Very recent evidence indicates that this is, indeed, the case (Bressanelli et al. 2004; Gibbons et al. 2004; Modis et al. 2004).

5
Membrane Dynamics During Fusion

Thus far we have focused on the conditions that elicit viral fusion reactions and the conformational changes in viral fusion proteins necessary for fusion. However, it is the viral and cellular bilayer membranes that merge during fusion. Lipid bilayers are stable structures that do not fuse spontaneously. Fusion proteins have evolved to catalyze the necessary lipid rearrangements. We now review a lipid rearrangement model and focus on the roles of different regions of viral fusion proteins in choreographing the structural changes that the membranes undergo throughout the fusion cascade (Fig. 3).

The favored model for the lipid transition state during membrane fusion is the stalk model. In this model, two opposing membranes bend toward each other, creating "dimples" (when viewed from the *trans* surface) or "nipples" (when viewed from the *cis* surface) (Fig. 3, Step 4). Nipples continue to bend until they meet. The two *cis* leaflets then merge, creating a lipid stalk (see Fig. 2 in Kozlovsky and Kozlov 2002) that proceeds to a state of local hemifusion (Fig. 3, Step 5). In a second step, transient fusion pores form, which give rise to stable pores (Fig. 3, Step 6).

The first direct visualization of a lipid stalk intermediate was achieved by electron diffraction studies of the effect of sequential dehydration on lipid bilayers composed of a lipid that has negative spontaneous curvature (Yang and Huang 2002). The stalk intermediate was stable at intermediate relative humidities. The results suggested that both the formation of a lipid stalk and its transition to a conformation that can be equated with pore formation require external forces.

Cellular membranes do not have spontaneous negative curvature and are highly hydrated. Membrane curvature can be promoted by introducing defects into the contacting bilayers. Thus roles for the fusion protein include pulling the fusing bilayers toward one another (dimpling), dehydrating the membranes, and creating membrane defects that lower the energy barrier for stalk and pore formation. Two intermediates in HIV fusion have been trapped: one in which the two membranes are joined by activated Envs, but are not yet fused (Melikyan et al. 2000a), and one in which small, "labile" pores have formed that can either expand into stable, "robust" pores or return to the prefusion state (Markosyan et al. 2003). These observations suggest a role for the fusion protein in formation and stabilization of both the fusion stalk and the fusion pore.

The mechanism by which a small pore enlarges is not known. However, several possibilities have been proposed. One is that the initial fusion pore is formed by a small number of activated fusion proteins. Additional activated fusion proteins then move into the fusion site to buttress and stabilize the pore, thereby allowing it to expand (Kozlov and Chernomordik 2002). Another possibility is that multiple small fusion pores coalesce to form larger ones. This was supported by EM visualization of HA-mediated fusion, in which multiple dimples/nipples were arranged circularly and lipid fragments were seen at the center of a fusion ring (Kanaseki et al. 1997).

6
Membrane-Interacting Regions of Viral Fusion Proteins

As discussed above, roles for the fusion protein in the fusion cascade (Fig. 3) include pulling the fusing bilayers toward one another (dimpling) and creating membrane defects that lower the energy barriers for stalk formation and fusion pore opening/enlargement. The fusion peptide and the transmembrane domain must remain stably associated with the target and viral membranes, respectively, for fusion to occur. Once the fusion peptide is stably associated with the target bilayer (Fig. 3, Step 2), we envision that rearrangements in the fusion protein ectodomain that bring the fusion peptide and transmembrane domains close together (Fig. 3, Step 4) result in dimpling of membranes toward one another. In addition to serving as critical membrane anchors, the fusion peptide and the transmembrane domain likely create membrane defects that facilitate the next stages of fusion. Here, we review information about the structure and function of the fusion peptide and the transmembrane domain during fusion. We also review evidence that juxtamembrane sequences, on both sides of the transmembrane domain, participate in fusion.

6.1
The Fusion Peptide

Fusion peptides are relatively apolar sequences that interact with membranes and are central to viral fusion reactions (Martin and Ruysschaert 2000; Martin et al. 1999; Skehel et al. 2001; White 1990). They have been

A.

```
N-terminal
   Class I      Influenza HA2:   GLFGAIAGFIENGWEG
                Sendai F1:       FFGAVIGTIALGVATA
                Resp. Syn. F1:   FLGFLLGVGSAIASGV
                HIV gp41:        AAIGALFLFGLGAAGSTMGAA

Internal
   Class I      Ebola GP:        GAAIGLAWIPYFGPAA
                ASLV gp37:       IFASILAPGVAAAQAL

   Class II     SFV E1:          DYQCKVYTGVYPFMWGGAYCFCD
                TBE E:           DRGWGNHCGLFGKGSIVA

unclassified    VSV G:           QGTWLNPGFPPQSCGYATV
```

B.

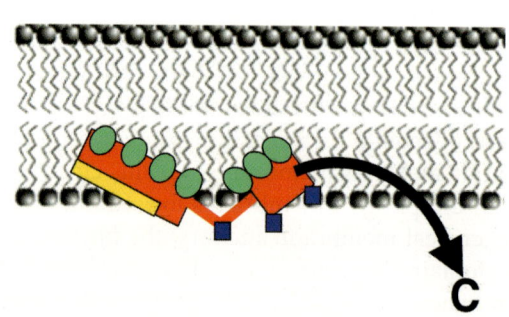

Fig. 5A, B. Characteristics of viral fusion peptides. A Selected viral fusion peptide sequences. N-terminal (Skehel et al. 2001) and internal (Delos et al. 2000) fusion peptide sequences are aligned according to their first noncharged residue. B Model of HA fusion peptide structure in target membrane at pH 5 (adapted from Tamm et al. 2002). The fusion peptide (*red*) resides in the target membrane in a kinked structure composed of two α-helices, each penetrating the outer leaflet. The glycine ridge is depicted by a *yellow box*, the hydrophobic interior face by *cyan ovals*, and the surface charged residues by *blue squares*. "C" denotes the direction of the HA2 ectodomain

classified as N-terminal or internal depending on their location within the fusion subunit (Table 1). Although fusion peptides are highly conserved within each virus family, there is little sequence similarity between fusion peptides of different families (Fig. 5A). Generally, however, fusion peptides contain a high percentage of glycines and/or alanines, as well as several critical bulky hydrophobic residues (Martin and Ruysschaert 2000; Martin et al. 1999; Skehel et al. 2001; Tamm and Han 2000; Tamm et al. 2002).

6.1.1
Structure of N-terminal Fusion Peptides

A significant body of work has emerged on the structure and function of synthetic fusion peptides (Martin and Ruysschaert 2000; Martin et al. 1999; Skehel et al. 2001; Tamm and Han 2000; Tamm et al. 2002). Synthetic fusion peptides are disordered in solution but ordered (α-helix and/or β-sheet) when they associate with membranes. The N-terminal fusion peptides that have been studied insert into membranes at oblique angles and do not penetrate the inner leaflet of the membrane. In general, mutations that abrogate fusion reduce the ability of synthetic fusion peptides to insert at oblique angles and to disrupt membranes (Martin et al. 1999). Contradictory conclusions on the precise structure of synthetic fusion peptides in membranes likely stem from the general low solubility of the peptides in aqueous solution and the different experimental methods employed (Tamm et al. 2002).

To circumvent solubility problems, a polar sequence was added to the C-terminal end of the influenza HA fusion peptide, rendering it soluble in both aqueous and hydrophobic environments (Han et al. 2001). At pH 5, the HA fusion peptide consists of an N-terminal helix, a kink, and a short C-terminal helix (Fig. 5B). Both the N- and C-terminal helices penetrate the outer leaflet of the target bilayer. The kink remains at the phospholipid surface; the interior (lipid-facing surface) of the kink is lined with hydrophobic residues. The conserved glycines form a ridge along the outer face of the N-terminal helix. Three charged residues are also found on the outer face (Fig. 5B). An HA in which the conserved glycine at the beginning of the fusion peptide (Gly1) has been changed to valine cannot mediate fusion. If Gly1 is changed to serine, HA mediates only hemifusion or only forms small nonexpanding fusion pores (Qiao et al. 1999; Skehel et al. 2001). Interestingly, these mutant fusion peptides have membrane-associated structures and orientations significantly different from those of the wild-type fusion peptide (Li et al.

2003). Simulations suggested similar membrane penetrating orientations for the HIV fusion peptide and two fusion-defective mutants (Kamath and Wong 2002).

6.1.2
Structure of Internal Fusion Peptides

In addition to a significant number of apolar residues, many internal fusion peptides contain a conserved proline at or near their centers (Fig. 5A). Mutagenesis of this proline in the avian sarcoma/leukosis virus (ASLV) EnvA fusion peptide suggested that it stabilizes a β-turn (Delos et al. 2000). This, coupled with the observation that mutating two cysteines that flank the fusion peptide abolishes fusion activity (Delos and White 2000), suggested that the internal EnvA fusion peptide exists as a looped structure stabilized by a disulfide bond. The ability of the Ebola virus fusion protein, which also contains an internal fusion peptide, to support infection was similarly inhibited when its central proline and flanking cysteines were mutated (Ito et al. 1999; Jeffers et al. 2002). A similar mutation of a proline within the predicted turn segment of the candidate fusion peptide of VSV G also significantly decreased fusion and abolished infectivity (Fredericksen and Whitt 1995). The idea of loop structures for internal fusion peptides is further supported by the known looped structure of the TBE E and SFV EI fusion peptides (Rey et al. 1995; Allison et al. 2001; Lescar et al. 2001). In some cases, two or more noncontiguous sequence loops may function as a collective fusion peptide (Gaudin et al. 1999a; Li et al. 1993).

Like N-terminal fusion peptides, internal fusion peptides contain a significant number of glycines and hydrophobic residues (Fig. 5A). Changing either of two glycines within the SFV E1 fusion peptide to alanines altered the pH threshold for fusion, and changing one of the glycines to aspartic acid abolished fusion (Duffus et al. 1995; Kielian et al. 1996). Alteration of hydrophobic residues at the beginning, middle, or end of the (internal) ASLV EnvA fusion peptide to charged residues impaired the ability of EnvA to mediate fusion (Hernandez and White 1998). Similarly, a tryptophan and a glycine are critical for Ebola GP-mediated infection (Ito et al. 1999). Also, a bulky hydrophobic residue is needed at the tip of the TBE E fusion peptide loop (Rey et al. 1995) (Fig. 4A, red). Collectively, these results suggest that internal fusion peptides function as loops that require a mixture of hydrophobic and flexible residues, similar to those found in N-terminal fusion peptides.

6.1.3
Roles of Fusion Peptides

Fusion peptides appear to act at several steps along the fusion pathway. As demonstrated by mutants in which apolar fusion peptide residues were changed to charged residues (Freed et al. 1992; Gething et al. 1986; Hernandez and White 1998; Schoch and Blumenthal 1993), fusion peptides clearly play an important role in anchoring the fusion protein to the target membrane (Fig. 3, Step 2). The energy provided by inserting the fusion peptides of a single HA trimer into a membrane would be sufficient to initiate stalk formation (Gunther-Ausborn et al. 2000). The fusion peptide may also assist in creating the stalk by displacing water from the lipid-water interface, thus decreasing the repulsive force between the two fusing membranes (Tamm and Han 2000). Fusion peptides may also function in fusion pore opening. In support of this possibility is the observation that an HA mutant in which Gly1 was changed to serine mediates extensive lipid, but not content mixing (Qiao et al. 1999). Furthermore, defects in syncytium formation and infectivity were observed for HIV Env harboring the mutation V2E in its fusion peptide (Freed et al. 1992). Biophysical studies comparing a synthetic fusion peptide harboring this mutation with the wild-type peptide suggested a requirement for fusion peptide aggregation in the creation of the HIV fusion pore (Kliger et al. 1997; Pereira et al. 1995).

6.2
The Transmembrane Domain

Studies with chimeric fusion proteins have suggested that the transmembrane domains of some viral fusion proteins do not require a specific sequence to support fusion (Armstrong et al. 2000 and references therein). In contrast, studies with glycosylphosphatidylinositol (GPI)-anchored fusion proteins have demonstrated that there is a strict requirement for a proteinaceous membrane anchor for fusion proteins to efficiently mediate the transition from hemifusion to full fusion (Kemble et al. 1994; Melikyan et al. 1997a; Tong and Compans 2000). There also appears to be a minimum length for the fusion protein transmembrane domain to be able to support this transition (Armstrong et al. 2000; West et al. 2001). Therefore, it has been suggested that fusion protein transmembrane domains must span both leaflets of the viral bilayer to mediate fusion pore opening (Armstrong et al. 2000).

The transmembrane domains of some fusion proteins appear to have specific amino acid requirements for fusion function. For example, a conserved positively charged residue in the middle of the transmembrane domains of certain retroviral Envs appears to be important for the ability to mediate fusion and infection (Einfeld and Hunter 1994; Owens et al. 1994; Pietschmann et al. 2000; West et al. 2001). Two glycine residues in the transmembrane domain of VSV-G appear to be important for the transition from hemifusion to full fusion (Cleverley and Lenard 1998). Studies using a synthetic peptide corresponding to the mutant VSV-G transmembrane domain (Dennison et al. 2002) suggested that the VSV-G transmembrane domain lowers the energy barrier for fusion and stabilizes the transient fusion pore, thereby promoting its conversion to a stable fusion pore. Two glycines may allow the VSV-G transmembrane domain to adopt alternative conformations under different conditions, and such flexibility may be important for function. The transmembrane domain of HA from the Japan (Melikyan et al. 2000b), but not the X:31 (Armstrong et al. 2000), strain of influenza appears to require a glycine near the middle. An ability to adopt alternative conformations was also invoked to explain the requirement for a proline near the middle of the transmembrane domain of the murine leukemia virus (MLV) Env glycoprotein (Taylor and Sanders 1999).

The observations that mutations in fusion peptides or transmembrane domains (Armstrong et al. 2000; Baker et al. 1999; Tamm et al. 2002) can impair the ability to mediate full fusion (Fig. 3, Step 6) have suggested that both of these apolar domains function in the transition from hemifusion to full fusion. Initially, the fusion peptide appears to insert only into the outer leaflet of the target membrane (Tamm and Han 2000). It has been proposed that the transmembrane domain and the fusion peptide, which are close to each other after membrane merger, may interact to stabilize the fusion pore (Tamm et al. 2002; Zhou et al. 1997). If this is the case, the fusion peptide might span both leaflets of the fused membrane in its final conformation (Tamm et al. 2002).

6.3
The Juxtamembrane Region of the Ectodomain

Several lines of evidence suggest that ectodomain sequences that lie just before the transmembrane domains of certain viral fusion proteins may be important for fusion. These sequences tend to have a high proportion of tryptophans or other aromatic residues and are predicted to partition into the interfacial regions of membranes (Suarez et al. 2000). Indeed, synthetic peptides containing juxtamembrane ectodomain sequences from HIV Env and Ebola GP partition into the interfacial region of target membranes (Saez-Cirion et al. 2003; Saez-Cirion et al. 2002; Schibli et al. 2001; Suarez et al. 2000). Mutation of three tryptophans within this region of HIV gp41 abrogated infection (Salzwedel et al. 1999), apparently by inhibiting fusion pore enlargement (Munoz-Barroso et al. 1999). Extending the HIV gp41 C-terminal heptad repeat peptide to include the tryptophan-rich juxtamembrane ectodomain sequence appeared to increase the potency of the peptide as an inhibitor of fusion (Kliger et al. 2001). It was suggested that the extended heptad repeat peptide was more potent because it could bind to two sites on HIV Env (the N-terminal coiled-coil and a second, as yet unidentified site) (Kliger et al. 2001). A likely effect of these peptides on late stages of fusion is to prevent formation of a required structure in Env that provides additional membrane destabilization. In this manner, partitioning of juxtamembrane sequences into the interfacial region of membranes may promote the transition from a stalk intermediate to a fusion pore (see Sect. 5).

6.4
The Cytoplasmic Tail

A specific cytoplasmic tail sequence does not appear to be essential, but it can modulate late stages of fusion. The cytoplasmic tail has been shown to influence the transition from hemifusion to full fusion (Sakai et al. 2002) or fusion pore enlargement (Dutch and Lamb 2001; Kozerski et al. 2000) in some viruses. The mechanism by which cytoplasmic tails may influence these later stages of fusion is not known. Some studies using synthetic peptides have suggested a direct interaction between the cytoplasmic tail and the viral membrane (Chen et al. 2001b; Fujii et al. 1992; Gawrisch et al. 1993; Haffar et al. 1991; Kliger and Shai 1997). Others have shown that the cytoplasmic tail can influence the structure of the ectodomain of the fusion protein (Aguilar et al. 2003; Edwards et al. 2002).

The ability of the cytoplasmic tail to affect ectodomain structure is most clearly manifested for those viral fusion proteins that harbor fusion-suppressing sequences. These sequences have been found in the fusion proteins of MLV (Ragheb and Anderson 1994), other type C retroviruses (Bobkova et al. 2002), some lentiviruses (Kim et al. 2003; Luciw et al. 1998), and a paramyxovirus F protein (Tong et al. 2002). The cytoplasmic tail of MLV Env is cleaved during virus budding (Schultz and Rein 1985). MLV Envs harboring uncleaved cytoplasmic tails do not induce fusion (Yang and Compans 1996). Although viruses lacking fusion-suppressing sequences display increased cell-cell fusion, they are more susceptible to neutralizing antibodies (Januszeski et al. 1997; Li et al. 2001; Rein et al. 1994; Yang and Compans 1996) and are impaired in their ability to sustain multiple rounds of infection (Cathomen et al. 1998; Freed and Martin 1996; Piller et al. 2000).

Acylation of cytoplasmic tails can also affect fusion, apparently at a late stage. For example, a mutant HA from the Japan strain of influenza in which three (normally palmitoylated) cysteine residues were mutated appeared to fuse normally when monitored by dye redistribution assays (Melikyan et al. 1997b). However, electrophysiological measurements revealed that fusion pores formed by the mutant HA did not flicker like those formed by wt-HA (Melikyan et al. 1997b). Similar mutations in HA from the A/USSR/77 (H1N1) and A/FPV/Rostock/34 (H7N1) influenza subtypes were shown, respectively, to inhibit syncytia formation (Fischer et al. 1998) and the transition from hemifusion to full fusion (Sakai et al. 2002). Palmitoylation of HIV Env was also shown to be important for Env incorporation into virions and for infectivity (Rousso et al. 2000). Thus acylation of cytoplasmic tails appears to have multiple effects on viral fusion reactions, the details of which are not completely understood.

7
Rafts in Viral Membrane Fusion

Lipid rafts are plasma membrane microdomains that are enriched in cholesterol and glycosphingolipids with saturated acyl chains. They are organizational platforms for a variety of cellular functions including sorting of membrane proteins and signaling (Brown and London 2000). Although there is growing evidence that certain viruses employ rafts, or raftlike membrane microdomains, during virus assembly (Suomalainen 2002), the question of whether these structures are required at the site of fusion in the target cell is less clear. Here, we consider the role of rafts in the fusion of two enveloped viruses, SFV and HIV. It is important to consider whether cholesterol and/or sphingolipids are required for fusion because they are found in lipid rafts, or if they serve some other purpose. For example, cholesterol may interact directly with the fusion protein, thereby facilitating its insertion into the target membrane. Alternatively, a need for cholesterol and sphingolipids could reflect an ability of raft structures to concentrate viral receptors. A third possibility is that the cholesterol imparts the membrane fluidity (or other biophysical properties) needed to lower the energy barrier for fusion.

SFV requires cholesterol and sphingolipids in the target membrane for fusion. These moieties enable the SFV spike protein to undergo conformational changes and bind to the target membrane (Ahn et al. 2002; Kielian et al. 2000). In a recent study, it was shown that after hydrophobic association with target bilayers, the SFV glycoprotein ectodomain associates with membrane structures with properties similar to rafts. However, careful studies using liposomes prepared with specific cholesterol and sphingolipid analogs demonstrated that the cholesterol and sphingolipid requirements in the target membrane did *not* correlate with their ability to form lipid rafts (Ahn et al. 2002). A related conclusion was drawn based on the fusion activities of both SFV and Sindbis virus with liposomes (Waarts et al. 2002). For both viruses, the requirement for cholesterol and sphingolipids in the target membrane appears to be for insertion of the fusion peptide (Vashishtha et al. 1998).

In the case of HIV, several studies have suggested a need for raftlike membrane microdomains for virus entry (Kozak et al. 2002; Popik et al. 2002). Depleting plasma membrane cholesterol from target cells resulted in reduced levels of virus infectivity or cell-cell fusion. Other studies have concluded that rafts are not necessary for HIV entry (Percherancier et al. 2003; Viard et al. 2002). In one study, depleting cholesterol from cells that express low levels of virus receptors inhibited HIV Env-mediated cell-cell fusion, but depleting cholesterol from cells that express high levels of virus receptors did not (Viard et al. 2002). Therefore, it was concluded that rafts per se are not needed for fusion. Rather, the presence of raftlike structures in the plasma membrane may concentrate virus receptors. Previous work has shown that a critical density of HIV receptors is required for fusion and infection (Reeves et al. 2002). Clearly more work is needed to clarify the role of rafts in virus-cell fusion and entry.

8
Inhibitors of Viral Fusion

It has recently become apparent, largely because of the success of T-20 in the inhibition of HIV infection in patients (Jiang et al. 2002; Kilby et al. 1998), that fusion is a good target for antiviral intervention. This was originally conceptualized because fusion is an essential early step in the virus infectious cycle, it happens in an exoplasmic space, and strategies can be designed to inhibit fusion without interfering with host cell proteins. Some fusion inhibitors function by inhibiting six-helix bundle formation. Others function by preventing earlier conformational changes in viral fusion proteins.

8.1
Inhibition of Helix Bundle Formation

The peptide T-20 (also known as Fuzeon) corresponds to the C-terminal helix of HIV Env (Fig. 3, green). T-20 works by preventing six-helix bundle formation. T-20 is a potent inhibitor of infections in tissue culture. Peptides corresponding to equivalent regions of other retroviruses as well as several paramyxoviruses function similarly (Earp et al. 2003; Russell et al. 2001). Notably, all of the viruses that have been shown to be highly susceptible to "C-helix" peptide inhibitors function at neutral pH, at least up to the lipid interacting stage of virus-cell fusion (Earp et al. 2003). Peptides corresponding to the N-terminal helices of HIV Env and the SV5 F protein also inhibit fusion, although with lower potency (Lu et al. 1995; Russell et al. 2001). The mechanism of inhibition by N-terminal peptides is still under consideration (He et al. 2003). In the case of the SV5 F, the N-peptide appears to target an earlier intermediate than the C-peptide (Russell et al. 2001). Other strategies are being considered to stabilize the prehairpin intermediate (Fig. 3, Step 1) and thereby prevent six-helix bundle formation. One strategy is the development of antibodies that recognize the prehairpin intermediate (Golding et al. 2002). Another, exemplified in three studies, is the development of small molecules that prevent six-helix bundle formation (Debnath et al. 1999; Eckert et al. 1999; Ferrer et al. 1999). All three studies targeted a hydrophobic pocket in the groove of the central coiled-coil of HIV gp41 that is important for interaction with the C-terminal helix in the post-fusion form. In the first approach, two organic compounds were identified from a screening effort conducted in conjunction with molecular docking, a method to identify small molecules that fit in a target site (Debnath et al. 1999). The second approach replaced three residues of the C-terminal helix that bind to the hydrophobic pocket with organic moieties, generated by combinatorial chemistry (Ferrer et al. 1999). The third approach used a mirror image phage display library to identify small, D-amino acid containing peptides that bind to the pocket (Eckert et al. 1999). Although none of the small molecules identified to date is as potent as T-20, the precedent has been set for attaining this goal.

8.2
Inhibition of Other Steps in Fusion

An effort to block the fusion activity of influenza was based on the idea that maintaining HA in its native metastable state should prevent fusion and infection. The first trial targeted a site in X:31 HA that includes part of the fusion peptide. With the use of an antibody-based assay to monitor fusion peptide exposure, a compound, *tert*-butylhydroquinone (TBHQ), that prevents the first stages of the HA conformational change and inhibits infectivity was discovered (Bodian et al. 1993). A follow-up study, targeting a site near the B-loop in HA, yielded additional inhibitors. Whereas some functioned like TBHQ, a second class was identified that appeared to push HA to an inactive state (Hoffman et al. 1997). A random screen against an H1 influenza virus identified an inhibitor that appears to function similarly to TBHQ. In the latter case, the binding site for the inhibitor was mapped to the vicinity of the fusion peptide (Cianci et al. 1999). Other small molecules that inhibit conformational changes in HA have been identified (Staschke et al. 1998). To date, none of the HA inhibitors has blocked all HA subtypes and none has an IC_{50} value in the submicromolar range. It is not yet clear whether the latter limitation represents a fundamental difficulty in inhibiting viral fusion proteins that function at low pH.

In addition to the small molecule approaches described above, antibodies that prevent fusion-inducing changes in viral glycoproteins have been described. The first example was an antibody that prevents low-pH-induced fusion of West Nile virus with model liposomes (Gollins and Porterfield 1986). Recently, a Fab fragment that binds to two HA1 monomers was shown to prevent an early conformational change in influenza HA (Barbey-Martin et al. 2002), separation of the globular head domains (Fig. 2B). As described above, antibodies have been developed that likely block six-helix bundle formation in the case of HIV Env (Golding et al. 2002).

9
Perspectives

The goal of this review was to give the reader an appreciation for the diversity of viral fusion mechanisms while recognizing their common underlying principles. We also summarized what is known about the lipid dynamics and lipid structures involved in fusion, and we also briefly overviewed recent developments in targeting viral fusion as an antiviral strategy. There is clearly much more we need to know about viral fusion proteins, viral fusion reactions, and the design of antifusion agents.

We end this review by enumerating some pressing issues and questions that remain about viral fusion. A major goal is to determine high-resolution structures for the complete ectodomains of the metastable trimers of class I viral fusion proteins in addition to the influenza HA. Structures of a complete paramyxovirus F and a complete retroviral Env ectodomain will be highly informative because we currently lack a detailed molecular description of how a receptor activates any viral fusion protein at neutral pH. A second goal will be to further delineate the mechanisms of class II viral fusion proteins. Do the transitions to their recently described low pH forms (Bressanelli et al. 2004; Gibbons et al. 2004; Modis et al. 2004) mediate hemifusion or fusion pore opening? What about the mechanisms of the as yet unclassified viral fusion proteins? These include viruses such as rhabdoviruses (e.g., VSV) that need only one protein to promote fusion, as well as more complicated viruses such as herpesviruses and poxviruses that require multiple viral glycoproteins.

The ensuing years should also bring a more complete understanding of how viral fusion proteins interact with target membrane bilayers. Class II fusion proteins insert their internal fusion peptides into target membranes as loops (Bressanelli et al. 2004; Gibbons et al. 2004; Modis et al. 2004). It has been predicted that the internal fusion proteins of the class I fusion proteins from Ebola and avian retroviruses form disulfide-bonded loop structures (Weisenhorn et al. 1998), and mutagenesis work has supported this prediction (Delos et al. 2000; Delos and White 2000; Jeffers et al. 2002). It remains to be seen, from high resolution structural studies, whether all internal fusion peptides, be they from class I, class II, or other classes of fusion proteins, interact with target bilayers as (disulfide bond) stabilized loops. Finally, we expect that there will be major developments in furthering the concept of targeting fusion as a weapon against pathogenic enveloped viruses. Particular emphasis will likely be on the development of small molecule inhibitors through the use of combinatorial chemistry in conjunction with high-throughput screens. It will be interesting to learn whether small molecule fusion inhibitors can be identified that block the entry of viruses that fuse in endosomes in response to low pH. This is a challenge for low-pH-activated class I fusion proteins such as influenza HA as well as for all known class II fusion proteins. Stay tuned. There are likely to be exciting developments in our understanding of viral fusion mechanisms as well as in the development of antifusion antivirals in the years ahead.

Acknowledgements Work in the authors' laboratory was supported by the NIH (AI22470). We thank Dr. Margaret Kielian for helpful comments on the manuscript.

References

Abrahamyan LG, Markosyan RM, Moore JP, Cohen FS, and Melikyan GB (2003) Human immunodeficiency virus type 1 Env with an intersubunit disulfide bond engages coreceptors but requires bond reduction after engagement to induce fusion. J Virol 77:5829-36

Aguilar HC, Anderson WF, and Cannon PM (2003) Cytoplasmic tail of moloney murine leukemia virus envelope protein influences the conformation of the extracellular domain: implications for the mechanism of action of the R peptide. J Virol 77:1281-1291

Ahn A, Gibbons DL, and Kielian M (2002) The fusion peptide of Semliki Forest virus associates with sterol-rich membrane domains. J Virol 76:3267-75

Ahn A, Klimjack MR, Chatterjee PK, and Kielian M (1999) An epitope of the Semliki Forest virus fusion protein exposed during virus-membrane fusion. J Virol 73:10029-39

Allison SL, Schalich J, Stiasny K, Mandl CW, Kunz C, and Heinz FX (1995) Oligomeric rearrangement of tick-borne encephalitis virus envelope proteins induced by an acidic pH. J Virol 69:695-700

Allison SL, Stiasny K, Stadler K, Mandl CW, and Heinz FX (1999) Mapping of functional elements in the stem-anchor region of tick-borne encephalitis virus envelope protein E. J Virol 73:5605-12

Allison SL, Schalich J, Stiasny K, Mandl CW, and Heinz FX (2001) Mutational evidence for an internal fusion peptide in flavivirus envelope protein E. J Virol 75:4268-75

Armstrong RT, Kushnir AS, and White JM (2000) The transmembrane domain of influenza hemagglutinin exhibits a stringent length requirement to support the hemifusion to fusion transition. J Cell Biol 151:425-38

Bagai S and Lamb RA (1995) Quantitative measurement of paramyxovirus fusion: differences in requirements of glycoproteins between simian virus 5 and human parainfluenza virus 3 or Newcastle disease virus. J Virol 69:6712-9

Baker KA, Dutch RE, Lamb RA, and Jardetzky TS (1999) Structural basis for paramyxovirus-mediated membrane fusion. Mol Cell 3:309-19

Barbey-Martin C, Gigant B, Bizebard T, Calder LJ, Wharton SA, Skehel JJ, and Knossow M (2002) An antibody that prevents the hemagglutinin low pH fusogenic transition. Virology 294:70-4

Barbouche R, Miquelis R, Jones IM, and Fenouillet E (2003) Protein-disulfide isomerase-mediated reduction of two disulfide bonds of HIV envelope glycoprotein 120 occurs post-CXCR4 binding and is required for fusion. J Biol Chem 278:3131-3136

Barnett AL, and Cunningham JM (2001) Receptor binding transforms the surface subunit of the mammalian C-type retrovirus envelope protein from an inhibitor to an activator of fusion. J Virol 75:9096-105

Begona Ruiz-Arguello M, Gonzalez-Reyes L, Calder LJ, Palomo C, Martin D, Saiz MJ, Garcia-Barreno B, Skehel JJ, and Melero JA (2002) Effect of proteolytic processing at two distinct sites on shape and aggregation of an anchorless fusion protein of human respiratory syncytial virus and fate of the intervening segment. Virology 298:317-26

Blumenthal R, Sarkar DP, Durell S, Howard DE, and Morris SJ (1996) Dilation of the influenza hemagglutinin fusion pore revealed by the kinetics of individual cell-cell fusion events. J Cell Biol 135:63-71

Bobkova M, Stitz J, Engelstadter M, Cichutek K, and Buchholz CJ (2002) Identification of R-peptides in envelope proteins of C-type retroviruses. J Gen Virol 83:2241-6

Bodian DL, Yamasaki RB, Buswell RL, Stearns JF, White JM, and Kuntz ID (1993) Inhibition of the fusion-inducing conformational change of influenza hemagglutinin by benzoquinones and hydroquinones. Biochemistry 32:2967-78

Borrego-Diaz E, Peeples ME, Markosyan RM, Melikyan GB, and Cohen FS (2003) Completion of trimeric hairpin formation of influenza virus hemagglutinin promotes fusion pore opening and enlargement. Virology 316:234-44

Bosch BJ, Zee R van der, Haan CA de, and Rottier PJ (2003) The coronavirus spike protein is a class I virus fusion protein: structural and functional characterization of the fusion core complex. J Virol 77:8801-11

Bossart KN, Wang LF, Flora MN, Chua KB, Lam SK, Eaton BT, and Broder CC (2002) Membrane fusion tropism and heterotypic functional activities of the nipah virus and hendra virus envelope glycoproteins. J Virol 76:11186-11198

Bressanelli S, Stiasny K, Allison SL, Stura EA, Duquerroy S, Lescar J, Heinz FX, and Ray FA (2004) Structure of a flavivirus envelope glycoprotein in its low-pH-induced membrane fusion conformation. EMBO J 23:728-38

Bron R, Wahlberg JM, Garoff H, and Wilschut J (1993) Membrane fusion of Semliki Forest virus in a model system: correlation between fusion kinetics and structural changes in the envelope glycoprotein. EMBO J 12:693-701

Brown DA and London E (2000) Structure and function of sphingolipid- and cholesterol-rich membrane rafts. J Biol Chem 275:17221-4

Bullough PA, Hughson FM, Skehel JJ, and Wiley DC (1994) Structure of influenza haemagglutinin at the pH of membrane fusion. Nature 371:37-43

Calder LJ, Gonzalez-Reyes L, Garcia-Barreno B, Wharton SA, Skehel JJ, Wiley DC, and Melero JA (2000) Electron microscopy of the human respiratory syncytial virus fusion protein and complexes that it forms with monoclonal antibodies. Virology 271:122-31

Carr CM, Chaudhry C, and Kim PS (1997) Influenza hemagglutinin is spring-loaded by a metastable native conformation. Proc Natl Acad Sci USA 94:14306-13

Carr CM and Kim PS (1993) A spring-loaded mechanism for the conformational change of influenza hemagglutinin. Cell 73:823-32

Cathomen T, Naim HY, and Cattaneo R (1998) Measles viruses with altered envelope protein cytoplasmic tails gain cell fusion competence. J Virol 72:1224-34

Chan DC, Fass D, Berger JM, and Kim PS (1997) Core structure of gp41 from the HIV envelope glycoprotein. Cell 89:263-73

Chan DC and Kim PS (1998) HIV entry and its inhibition. Cell 93:681-4

Chen J, Lee KH, Steinhauer DA, Stevens DJ, Skehel JJ, and Wiley DC (1998) Structure of the hemagglutinin precursor cleavage site, a determinant of influenza pathogenicity and the origin of the labile conformation. Cell 95:409-17

Chen J, Skehel JJ, and Wiley DC (1999) N- and C-terminal residues combine in the fusion-pH influenza hemagglutinin HA(2) subunit to form an N cap that terminates the triple- stranded coiled coil. Proc Natl Acad Sci USA 96:8967-72

Chen L, Gorman JJ, McKimm-Breschkin J, Lawrence LJ, Tulloch PA, Smith BJ, Colman PM, and Lawrence MC (2001a) The structure of the fusion glycoprotein of Newcastle disease virus suggests a novel paradigm for the molecular mechanism of membrane fusion. Structure 9:255-66

Chen SS, Lee SF, and Wang CT (2001b) Cellular membrane-binding ability of the C-terminal cytoplasmic domain of human immunodeficiency virus type 1 envelope transmembrane protein gp41. J Virol 75:9925-38

Cianci C, Yu KL, Dischino DD, Harte W, Deshpande M, Luo G, Colonno RJ, Meanwell NA, and Krystal M (1999) pH-dependent changes in photoaffinity labeling patterns of the H1 influenza virus hemagglutinin by using an inhibitor of viral fusion. J Virol 73:1785-94

Cleverley DZ and Lenard J (1998) The transmembrane domain in viral fusion: essential role for a conserved glycine residue in vesicular stomatitis virus G protein. Proc Natl Acad Sci USA 95:3425-30

Colman PM and Lawrence MC (2003) The structural biology of type I viral membrane fusion. Nat Rev Mol Cell Biol 4:309-19

Corver J, Ortiz A, Allison SL, Schalich J, Heinz FX, and Wilschut J (2000) Membrane fusion activity of tick-borne encephalitis virus and recombinant subviral particles in a liposomal model system. Virology 269:37-46

Damico RL, Crane J, and Bates P (1998) Receptor-triggered membrane association of a model retroviral glycoprotein. Proc Natl Acad Sci USA 95:2580-5

Danieli T, Pelletier SL, Henis YI, and White JM (1996) Membrane fusion mediated by the influenza virus hemagglutinin requires the concerted action of at least three hemagglutinin trimers. J Cell Biol 133:559-69

Debnath AK, Radigan L, and Jiang S (1999) Structure-based identification of small molecule antiviral compounds targeted to the gp41 core structure of the human immunodeficiency virus type 1. J Med Chem 42:3203-9

Delos SE, Gilbert JM, and White JM (2000) The central proline of an internal viral fusion peptide serves two important roles. J Virol 74:1686-93

Delos SE and White JM (2000) Critical role for the cysteines flanking the internal fusion peptide of avian sarcoma/leukosis virus envelope glycoprotein. J Virol 74:9738-41

Deng R, Wang Z, Mahon PJ, Marinello M, Mirza A, and Iorio RM (1999) Mutations in the Newcastle disease virus hemagglutinin-neuraminidase protein that interfere with its ability to interact with the homologous F protein in the promotion of fusion. Virology 253:43-54

Dennison SM, Greenfield N, Lenard J, and Lentz BR (2002) VSV transmembrane domain (TMD) peptide promotes PEG-mediated fusion of liposomes in a conformationally sensitive fashion. Biochemistry 41:14925–14934

Di Simone C and Buchmeier MJ (1995) Kinetics and pH dependence of acid-induced structural changes in the lymphocytic choriomeningitis virus glycoprotein complex. Virology 209:3–9

Doms RW, Helenius A, and White J (1985) Membrane fusion activity of the influenza virus hemagglutinin. The low pH-induced conformational change. J Biol Chem 260:2973–81

Duffus WA, Levy-Mintz P, Klimjack MR, and Kielian M (1995) Mutations in the putative fusion peptide of Semliki Forest virus affect spike protein oligomerization and virus assembly. J Virol 69:2471–9

Durell SR, Martin I, Ruysschaert JM, Shai Y, and Blumenthal R (1997) What studies of fusion peptides tell us about viral envelope glycoprotein-mediated membrane fusion. Mol Membr Biol 14:97–112

Dutch RE and Lamb RA (2001) Deletion of the cytoplasmic tail of the fusion protein of the paramyxovirus simian virus 5 affects fusion pore enlargement. J Virol 75:5363–9

Earp LJ, Delos SE, Netter RC, Bates P, and White JM (2003) The avian retrovirus avian sarcoma/leukosis virus subtype A reaches the lipid mixing stage of fusion at neutral pH. J Virol 77:3058–3066

Eckert DM and Kim PS (2001) Mechanisms of viral membrane fusion and its inhibition. Annu Rev Biochem 70:777–810

Eckert DM, Malashkevich VN, Hong LH, Carr PA, and Kim PS (1999) Inhibiting HIV-1 entry: discovery of D-peptide inhibitors that target the gp41 coiled-coil pocket. Cell 99:103–15

Edwards TG, Wyss S, Reeves JD, Zolla-Pazner S, Hoxie JA, Doms RW, and Baribaud F (2002) Truncation of the cytoplasmic domain induces exposure of conserved regions in the ectodomain of human immunodeficiency virus type 1 envelope protein. J Virol 76:2683–91

Einfeld DA and Hunter E (1994) Expression of the TM protein of Rous sarcoma virus in the absence of SU shows that this domain is capable of oligomerization and intracellular transport. J Virol 68:2513–20

Fackler OT and Peterlin BM (2000) Endocytic entry of HIV-1. Curr Biol 10:1005–8

Feng Y, Broder CC, Kennedy PE, and Berger EA (1996) HIV-1 entry cofactor: functional cDNA cloning of a seven-transmembrane, G protein-coupled receptor. Science 272:872–7

Fenouillet E, Barbouche R, Courageot J, and Miquelis R (2001) The catalytic activity of protein disulfide isomerase is involved in human immunodeficiency virus envelope-mediated membrane fusion after CD4 cell binding. J Infect Dis 183:744–52

Ferlenghi I, Clarke M, Ruttan T, Allison SL, Schalich J, Heinz FX, Harrison SC, Rey F, and Fuller SD (2001) Molecular organization of a recombinant subviral particle from tick-borne encephalitis virus. Mol Cell 7:593–602

Ferrer M, Kapoor TM, Strassmaier T, Weissenhorn W, Skehel JJ, Oprian D, Schreiber SL, Wiley DC, and Harrison SC (1999) Selection of gp41-mediated HIV-1 cell entry inhibitors from biased combinatorial libraries of non-natural binding elements. Nat Struct Biol 6:953–60

Fischer C, Schroth-Diez B, Herrmann A, Garten W, and Klenk HD (1998) Acylation of the influenza hemagglutinin modulates fusion activity. Virology 248:284–94

Fredericksen BL and Whitt MA (1995) Vesicular stomatitis virus glycoprotein mutations that affect membrane fusion activity and abolish virus infectivity. J Virol 69:1435–43

Freed EO, Delwart EL, Buchschacher GLJ, and Panganiban AT (1992) A mutation in the human immunodeficiency virus type 1 transmembrane glycoprotein gp41 dominantly interferes with fusion and infectivity. Proc Natl Acad Sci USA 89: 70–4

Freed EO and Martin MA (1996) Domains of the human immunodeficiency virus type 1 matrix and gp41 cytoplasmic tail required for envelope incorporation into virions. J Virol 70:341–51

Frey S, Marsh M, Gunther S, Pelchen-Matthews A, Stephens P, Ortlepp S, and Stegmann T (1995) Temperature dependence of cell-cell fusion induced by the envelope glycoprotein of human immunodeficiency virus type 1. J Virol 69:1462–72

Fujii G, Horvath S, Woodward S, Eiserling F, and Eisenberg D (1992) A molecular model for membrane fusion based on solution studies of an amphiphilic peptide from HIV gp41. Protein Sci 1:1454–64

Furuta RA, Wild CT, Weng Y, and Weiss CD (1998) Capture of an early fusion-active conformation of HIV-1 gp41. Nat Struct Biol 5:276–9

Gallina A, Hanley TM, Mandel R, Trahey M, Broder CC, Viglianti GA, and Ryser HJ (2002) Inhibitors of protein-disulfide isomerase prevent cleavage of disulfide bonds in receptor-bound glycoprotein 120 and prevent HIV-1 entry. J Biol Chem 277:50579–88

Gaudin Y, de Kinkelin P, and Benmansour A (1999a) Mutations in the glycoprotein of viral haemorrhagic septicaemia virus that affect virulence for fish and the pH threshold for membrane fusion. J Gen Virol 80:1221–9

Gaudin Y, Tuffereau C, Durrer P, Brunner J, Flamand A, and Ruigrok R (1999b) Rabies virus-induced membrane fusion. Mol Membr Biol 16:21–31

Gawrisch K, Han KH, Yang JS, Bergelson LD, and Ferretti JA (1993) Interaction of peptide fragment 828–848 of the envelope glycoprotein of human immunodeficiency virus type I with lipid bilayers. Biochemistry 32:3112–8

Gething MJ, Doms RW, York D, and White J (1986) Studies on the mechanism of membrane fusion: site-specific mutagenesis of the hemagglutinin of influenza virus. J Cell Biol 102:11–23

Gibbons DL, Vaney MC, Roussel A, Vigouroux A, Reilly B, Lepault J, Kielian M, and Rey FA (2004) Conformational change and protein-protein interactions of the fusion protein of Semliki Forest virus. Nature 427:320–25

Gilbert JM, Mason D, and White JM (1990) Fusion of Rous sarcoma virus with host cells does not require exposure to low pH. J Virol 64:5106–13

Godley L, Pfeifer J, Steinhauer D, Ely B, Shaw G, Kaufmann R, Suchanek E, Pabo C, Skehel JJ, and Wiley DC (1992) Introduction of intersubunit disulfide bonds in the membrane-distal region of the influenza hemagglutinin abolishes membrane fusion activity. Cell 68:635–45

Golding H, Zaitseva M, de Rosny E, King LR, Manischewitz J, Sidorov I, Gorny MK, Zolla-Pazner S, Dimitrov DS, and Weiss C (2002) Dissection of human immunodeficiency virus type 1 entry with neutralizing antibodies to gp41 fusion intermediates. J Virol 76:6780–90

Gollins SW and Porterfield JS (1986) A new mechanism for the neutralization of enveloped viruses by antiviral antibody. Nature 321:244–6

Gonzalez-Reyes L, Ruiz-Arguello MB, Garcia-Barreno B, Calder L, Lopez JA, Albar JP, Skehel JJ, Wiley DC, and Melero JA (2001) Cleavage of the human respiratory syncytial virus fusion protein at two distinct sites is required for activation of membrane fusion. Proc Natl Acad Sci USA 98:9859–64

Gruenke JA, Armstrong RT, Newcomb WW, Brown JC, and White JM (2002) New insights into the spring-loaded conformational change of influenza hemagglutinin. J Virol 76:4456–66

Gunther-Ausborn S, Schoen P, Bartoldus I, Wilschut J, and Stegmann T (2000) Role of hemagglutinin surface density in the initial stages of influenza virus fusion: lack of evidence for cooperativity. J Virol 74:2714–20

Haffar OK, Dowbenko DJ, and Berman PW (1991) The cytoplasmic tail of HIV-1 gp160 contains regions that associate with cellular membranes. Virology 180:439–41

Han X, Bushweller JH, Cafiso DS, and Tamm LK (2001) Membrane structure and fusion-triggering conformational change of the fusion domain from influenza hemagglutinin. Nat Struct Biol 8:715–20

He Y, Vassell R, Zaitseva M, Nguyen N, Yang Z, Weng Y, and Weiss C (2003) Peptides trap the human immunodeficiency virus type 1 envelope glycoprotein fusion intermediate at two sites. J Virol 77:1666–1671

Heinz FX and Allison SL (2001) The machinery for flavivirus fusion with host cell membranes. Curr Opin Microbiol 4:450–5

Helenius A (1995) Alphavirus and flavivirus glycoproteins: structures and functions Cell 81:651–3

Hernandez LD, Hoffman LR, Wolfsberg TG, and White JM (1996) Virus-cell and cell-cell fusion. Annu Rev Cell Dev Biol 12:627–61

Hernandez LD, Peters RJ, Delos SE, Young JA, Agard DA, and White JM (1997) Activation of a retroviral membrane fusion protein: soluble receptor-induced liposome binding of the ALSV envelope glycoprotein. J Cell Biol 139:1455–64

Hernandez LD and White JM (1998) Mutational analysis of the candidate internal fusion peptide of the avian leukosis and sarcoma virus subgroup A envelope glycoprotein. J Virol 72:3259–67

Hoffman LR, Kuntz ID, and White JM (1997) Structure-based identification of an inducer of the low-pH conformational change in the influenza virus hemagglutinin: irreversible inhibition of infectivity. J Virol 71:8808–20

Hunter E (1997) Viral entry and receptors. In J. M. Coffin (ed.), Retroviruses. Cold Spring Harbor Laboratory Press, Plainview, NY, p. 71–121

Irurzun A, Nieva JL, and Carrasco L (1997) Entry of Semliki forest virus into cells: effects of concanamycin A and nigericin on viral membrane fusion and infection. Virology 227:488–92

Ito H, Watanabe S, Sanchez A, Whitt MA, and Kawaoka Y (1999) Mutational analysis of the putative fusion domain of Ebola virus glycoprotein. J Virol 73:8907–12

Jahn R, Lang T, and Sudhof TC (2003) Membrane fusion. Cell 112:519–533
Januszeski MM, Cannon PM, Chen D, Rozenberg Y, and Anderson WF (1997) Functional analysis of the cytoplasmic tail of Moloney murine leukemia virus envelope protein. J Virol 71:3613–9
Jeffers SA, Sanders DA, and Sanchez A (2002) Covalent modifications of the ebola virus glycoprotein. J Virol 76:12463–72
Jiang S, Zhao Q, and Debnath AK (2002) Peptide and non-peptide HIV fusion inhibitors. Curr Pharm Des 8:563–80
Joshi SB, Dutch RE, and Lamb RA (1998) A core trimer of the paramyxovirus fusion protein: parallels to influenza virus hemagglutinin and HIV-1 gp41. Virology 248:20–34
Kamath S and Wong TC (2002) Membrane structure of the human immunodeficiency virus gp41 fusion domain by molecular dynamics simulation. Biophys J 83:135–43
Kanaseki T, Kawasaki K, Murata M, Ikeuchi Y, and Ohnishi S (1997) Structural features of membrane fusion between influenza virus and liposome as revealed by quick-freezing electron microscopy. J Cell Biol 137:1041–56
Kemble GW, Bodian DL, Rose J, Wilson IA, and White JM (1992) Intermonomer disulfide bonds impair the fusion activity of influenza virus hemagglutinin. J Virol 66:4940–50
Kemble GW, Danieli T, and White JM (1994) Lipid-anchored influenza hemagglutinin promotes hemifusion, not complete fusion. Cell 76:383–91
Kielian M (1995) Membrane fusion and the alphavirus life cycle. Adv Virus Res 45:113–51
Kielian M, Klimjack MR, Ghosh S, and Duffus WA (1996) Mechanisms of mutations inhibiting fusion and infection by Semliki Forest virus. J Cell Biol 134:863–72
Kielian M, Chatterjee PK, Gibbons DL, and Lu YE (2000) Specific roles for lipids in virus fusion and exit. Examples from the alphaviruses. Subcell Biochem 34:409–55
Kilby JM, Hopkins S, Venetta TM, DiMassimo B, Cloud GA, Lee JY, Alldredge L, Hunter E, Lambert D, Bolognesi D, Matthews T, Johnson MR, Nowak MA, Shaw GM, and Saag MS (1998) Potent suppression of HIV-1 replication in humans by T-20, a peptide inhibitor of gp41-mediated virus entry. Nat Med 4:1302–7
Kim FJ, Manel N, Boublik Y, Battini JL, and Sitbon M (2003) Human T-cell leukemia virus type 1 envelope-mediated syncytium formation can be activated in resistant mammalian cell lines by a carboxy-terminal truncation of the envelope cytoplasmic domain. J Virol 77:963–969
Kliger Y, Aharoni A, Rapaport D, Jones P, Blumenthal R, and Shai Y (1997) Fusion peptides derived from the HIV type 1 glycoprotein 41 associate within phospholipid membranes and inhibit cell-cell fusion. Structure-function study. J Biol Chem 272:13496–505
Kliger Y, Gallo SA, Peisajovich SG, Munoz-Barroso I, Avkin S, Blumenthal R, and Shai Y (2001) Mode of action of an antiviral peptide from HIV-1. Inhibition at a post-lipid mixing stage. J Biol Chem 276:1391–7
Kliger Y and Shai Y (1997) A leucine zipper-like sequence from the cytoplasmic tail of the HIV-1 envelope glycoprotein binds and perturbs lipid bilayers. Biochemistry 36:5157–69

Korte T, Ludwig K, Booy FP, Blumenthal R, and Herrmann A (1999) Conformational intermediates and fusion activity of influenza virus hemagglutinin. J Virol 73: 4567-74

Kozak SL, Heard JM, and Kabat D (2002) Segregation of CD4 and CXCR4 into distinct lipid microdomains in T lymphocytes suggests a mechanism for membrane destabilization by human immunodeficiency virus. J Virol 76:1802-15

Kozerski C, Ponimaskin E, Schroth-Diez B, Schmidt MF, and Herrmann A (2000) Modification of the cytoplasmic domain of influenza virus hemagglutinin affects enlargement of the fusion pore. J Virol 74:7529-37

Kozlov MM and Chernomordik LV (2002) The protein coat in membrane fusion: lessons from fission. Traffic 3:256-67

Kozlovsky Y and Kozlov MM (2002) Stalk model of membrane fusion: solution of energy crisis. Biophys J 82:882-95

Kuhn RJ, Zhang W, Rossmann MG, Pletnev SV, Corver J, Lenches E, Jones CT, Mukhopadhyay S, Chipman PR, Strauss EG, Baker TS, and Strauss JH (2002) Structure of dengue virus: implications for flavivirus organization, maturation, and fusion. Cell 108:717-25

Kwong PD, Wyatt R, Robinson J, Sweet RW, Sodroski J, and Hendrickson W (1998) Structure of an HIV gp120 envelope glycoprotein in complex with the CD4 receptor and a neutralizing human antibody. Nature 393:648-59

Lamb RA (1993) Paramyxovirus fusion: a hypothesis for changes. Virology 197:1-11

Lambert DM, Barney S, Lambert AL, Guthrie K, Medinas R, Davis DE, Bucy T, Erickson J, Merutka G, and Petteway SRJ (1996) Peptides from conserved regions of paramyxovirus fusion (F) proteins are potent inhibitors of viral fusion. Proc Natl Acad Sci USA 93:2186-91

Lavillette D, Boson B, Russell SJ, and Cosset FL (2001) Activation of membrane fusion by murine leukemia viruses is controlled in cis or in trans by interactions between the receptor-binding domain and a conserved disulfide loop of the carboxy terminus of the surface glycoprotein. J Virol 75:3685-95

Lavillette D, Maurice M, Roche C, Russell SJ, Sitbon M, and Cosset FL (1998) A proline-rich motif downstream of the receptor binding domain modulates conformation and fusogenicity of murine retroviral envelopes. J Virol 72:9955-65

Lescar J, Roussel A, Wien MW, Navaza J, Fuller SD, Wengler G, and Rey FA (2001) The fusion glycoprotein shell of Semliki Forest virus: an icosahedral assembly primed for fusogenic activation at endosomal pH. Cell 105:137-48

Li M, Yang C, and Compans RW (2001) Mutations in the cytoplasmic tail of murine leukemia virus envelope protein suppress fusion inhibition by R peptide. J Virol 75:2337-44

Li Y, Drone C, Sat E, and Ghosh HP (1993) Mutation analysis of the vesicular stomatitis virus glycoprotein G for membrane fusion domains. J Virol 67:4070-7

Li Y, Han X, and Tamm LK (2003) Thermodynamics of fusion peptide-membrane interactions. Biochemistry 42:7245-51

Lu M, Blacklow SC, and Kim PS (1995) A trimeric structural domain of the HIV-1 transmembrane glycoprotein. Nat Struct Biol 2:1075-82

Luciw PA, Shaw KE, Shacklett BL, and Marthas ML (1998) Importance of the intracytoplasmic domain of the simian immunodeficiency virus (SIV) envelope glycoprotein for pathogenesis. Virology 252:9-16

Markosyan RM, Cohen FS, and Melikyan GB (2003) HIV-1 envelope proteins complete their folding into six-helix bundles immediately after fusion pore formation. Mol Biol Cell 14:926-38

Markovic I, Leikina E, Zhukovsky M, Zimmerberg J, and Chernomordik LV (2001) Synchronized activation and refolding of influenza hemagglutinin in multimeric fusion machines. J Cell Biol 155:833-44

Markovic I, Pulyaeva H, Sokoloff A, and Chernomordik LV (1998) Membrane fusion mediated by baculovirus gp64 involves assembly of stable gp64 trimers into multiprotein aggregates. J Cell Biol 143:1155-66

Martin I and Ruysschaert JM (2000) Common properties of fusion peptides from diverse systems. Biosci Rep 20:483-500

Martin II, Ruysschaert J, and Epand RM (1999) Role of the N-terminal peptides of viral envelope proteins in membrane fusion. Adv Drug Deliv Rev 38:233-255

Matsuyama S, Delos SE, and White JM (2004) Sequential roles of receptor binding and low pH in forming prehairpin and hairpin conformations of an avian retroviral envelope glycoprotein. J Virol (in press)

McClure MO, Sommerfelt MA, Marsh M, and Weiss RA (1990) The pH independence of mammalian retrovirus infection. J Gen Virol 71:767-73

McGinnes LW, Gravel K, and Morrison TG (2002) Newcastle disease virus HN protein alters the conformation of the F protein at cell surfaces. J Virol 76:12622-33

Melikyan GB, Brener SA, Ok DC, and Cohen FS (1997a) Inner but not outer membrane leaflets control the transition from glycosylphosphatidylinositol-anchored influenza hemagglutinin-induced hemifusion to full fusion. J Cell Biol 136:995-1005

Melikyan GB, Jin H, Lamb RA, and Cohen FS (1997b) The role of the cytoplasmic tail region of influenza virus hemagglutinin in formation and growth of fusion pores. Virology 235:118-28

Melikyan GB, Markosyan RM, Hemmati H, Delmedico MK, Lambert DM, and Cohen FS (2000a) Evidence that the transition of HIV-1 gp41 into a six-helix bundle, not the bundle configuration, induces membrane fusion. J Cell Biol 151:413-23

Melikyan GB, Markosyan RM, Roth MG, and Cohen FS (2000b) A point mutation in the transmembrane domain of the hemagglutinin of influenza virus stabilizes a hemifusion intermediate that can transit to fusion. Mol Biol Cell 11:3765-75

Modis Y, Ogata S, Clements D, and Harrison SC (2004) Structure of the dengue virus envelope protein after membrane fusion. Nature 427:313-9

Mothes W, Boerger AL, Narayan S, Cunningham JM, and Young JA (2000) Retroviral entry mediated by receptor priming and low pH triggering of an envelope glycoprotein. Cell 103:679-89

Munoz-Barroso I, Durell S, Sakaguchi K, Appella E, and Blumenthal R (1998) Dilation of the human immunodeficiency virus-1 envelope glycoprotein fusion pore revealed by the inhibitory action of a synthetic peptide from gp41. J Cell Biol 140:315-23

Munoz-Barroso I, Salzwedel K, Hunter E, and Blumenthal R (1999) Role of the membrane-proximal domain in the initial stages of human immunodeficiency virus type 1 envelope glycoprotein-mediated membrane fusion. J Virol 73:6089-92

Netter RC (2002) Ph.D Dissertation. University of Pennsylvania.

Nir S, Duzgunes N, de Lima MC, and Hoekstra D (1990) Fusion of enveloped viruses with cells and liposomes. Activity and inactivation. Cell Biophys 17:181–201

Nussbaum O, Broder CC, and Berger EA (1994) Fusogenic mechanisms of enveloped-virus glycoproteins analyzed by a novel recombinant vaccinia virus-based assay quantitating cell fusion-dependent reporter gene activation. J Virol 68:5411–22

Ohuchi M, Ohuchi R, Sakai T, and Matsumoto A (2002) Tight binding of influenza virus hemagglutinin to its receptor interferes with fusion pore dilation. J Virol 76:1405–13

Owens RJ, Burke C, and Rose JK (1994) Mutations in the membrane-spanning domain of the human immunodeficiency virus envelope glycoprotein that affect fusion activity. J Virol 68:570–4

Park HE, Gruenke JA, and White JM (2003) Leash in the groove mechanism of membrane fusion. Nat Struct Biol 10:1048–53

Paterson RG, Russell CJ, and Lamb RA (2000) Fusion protein of the paramyxovirus SV5: destabilizing and stabilizing mutants of fusion activation. Virology 270:17–30

Pelkmans L and Helenius A (2003) Insider information: what viruses tell us about endocytosis. Curr Opin Cell Biol 15:414–22

Percherancier Y, Lagane B, Planchenault T, Staropoli I, Altmeyer R, Virelizier JL, Arenzana-Seisdedos F, Hoessli DC, and Bachelerie F (2003) HIV-1 entry into T-cells is not dependent on CD4 and CCR5 localization to sphingolipid-enriched, detergent-resistant, raft membrane domains. J Biol Chem 278:3153–3161

Pereira FB, Goni FM, and Nieva JL (1995) Liposome destabilization induced by the HIV-1 fusion peptide effect of a single amino acid substitution. FEBS Lett 362:243–6

Pietschmann T, Zentgraf H, Rethwilm A, and Lindemann D (2000) An evolutionarily conserved positively charged amino acid in the putative membrane-spanning domain of the foamy virus envelope protein controls fusion activity. J Virol 74:4474–82

Piller SC, Dubay JW, Derdeyn CA, and Hunter E (2000) Mutational analysis of conserved domains within the cytoplasmic tail of gp41 from human immunodeficiency virus type 1: effects on glycoprotein incorporation and infectivity. J Virol 74:11717–23

Pinter A, Kopelman R, Li Z, Kayman SC, and Sanders DA (1997) Localization of the labile disulfide bond between SU and TM of the murine leukemia virus envelope protein complex to a highly conserved CWLC motif in SU that resembles the active-site sequence of thiol-disulfide exchange enzymes. J Virol 71:8073–7

Popik W, Alce TM, and Au WC (2002) Human immunodeficiency virus type 1 uses lipid raft-colocalized CD4 and chemokine receptors for productive entry into CD4(+) T cells. J Virol 76:4709–22

Puri A, Winick J, Lowy RJ, Covell D, Eidelman O, Walter A, and Blumenthal R (1988) Activation of vesicular stomatitis virus fusion with cells by pretreatment at low pH. J Biol Chem 263:4749–53

Qiao H, Armstrong RT, Melikyan GB, Cohen FS, and White JM (1999) A specific point mutant at position 1 of the influenza hemagglutinin fusion peptide displays a hemifusion phenotype. Mol Biol Cell 10:2759–69

Qiao H, Pelletier SL, Hoffman L, Hacker J, Armstrong RT, and White JM (1998) Specific single or double proline substitutions in the "spring-loaded" coiled-coil region of the influenza hemagglutinin impair or abolish membrane fusion activity. J Cell Biol 141:1335-47

Ragheb JA and Anderson WF (1994) pH-independent murine leukemia virus ecotropic envelope-mediated cell fusion: implications for the role of the R peptide and p12E TM in viral entry. J Virol 68:3220-31

Reeves JD, Gallo SA, Ahmad N, Miamidian JL, Harvey PE, Sharron M, Pohlman S, Sfakianos JN, Derdeyn CA, Blumenthal R, Hunter E, and Doms RW (2002) Sensitivity of HIV-1 to entry inhibitors correlates with envelope/coreceptor affinity, receptor density, and fusion kinetics. Proc Natl Acad Sci USA 99:16249-54

Rein A, Mirro J, Haynes JG, Ernst SM, and Nagashima K (1994) Function of the cytoplasmic domain of a retroviral transmembrane protein: p15E-p2E cleavage activates the membrane fusion capability of the murine leukemia virus Env protein. J Virol 68:1773-81

Rey FA, Heinz FX, Mandl C, Kunz C, and Harrison SC (1995) The envelope glycoprotein from tick-borne encephalitis virus at 2 Å resolution. Nature 375:291-8

Rousso I, Mixon MB, Chen BK, and Kim PS (2000) Palmitoylation of the HIV-1 envelope glycoprotein is critical for viral infectivity. Proc Natl Acad Sci USA 97:13523-5

Russell CJ, Jardetzky TS, and Lamb RA (2001) Membrane fusion machines of paramyxoviruses: capture of intermediates of fusion. EMBO J 20:4024-34

Saez-Cirion A, Gomara MJ, Agirre A, and Nieva JL (2003) Pre-transmembrane sequence of Ebola glycoprotein. Interfacial hydrophobicity distribution and interaction with membranes. FEBS Lett 533:47-53

Saez-Cirion A, Nir S, Lorizate M, Agirre A, Cruz A, Perez-Gil J, and Nieva JL (2002) Sphingomyelin and cholesterol promote HIV-1 gp41 pretransmembrane sequence surface aggregation and membrane restructuring. J Biol Chem 277:21776-85

Sakai T, Ohuchi R, and Ohuchi M (2002) Fatty acids on the A/USSR/77 influenza virus hemagglutinin facilitate the transition from hemifusion to fusion pore formation. J Virol 76:4603-11

Salzwedel K, West JT, and Hunter E (1999) A conserved tryptophan-rich motif in the membrane-proximal region of the human immunodeficiency virus type 1 gp41 ectodomain is important for Env-mediated fusion and virus infectivity. J Virol 73:2469-80

Schibli DJ, Montelaro RC, and Vogel HJ (2001) The membrane-proximal tryptophan-rich region of the HIV glycoprotein, gp41, forms a well-defined helix in dodecylphosphocholine micelles. Biochemistry 40:9570-8

Schmid E, Zurbriggen A, Gassen U, Rima B, ter Meulen V, and Schneider-Schaulies J (2000) Antibodies to CD9, a tetraspan transmembrane protein, inhibit canine distemper virus-induced cell-cell fusion but not virus-cell fusion. J Virol 74:7554-61

Schoch C and Blumenthal R (1993) Role of the fusion peptide sequence in initial stages of influenza hemagglutinin-induced cell fusion. J Biol Chem 268:9267-74

Schultz A and Rein A (1985) Maturation of murine leukemia virus Env proteins in the absence of other viral proteins. Virology 145:335-9

Seth S, Vincent A, and Compans RW (2003) Mutations in the cytoplasmic domain of a paramyxovirus fusion glycoprotein rescue syncytium formation and eliminate the hemagglutinin-neuraminidase protein requirement for membrane fusion. J Virol 77:167-78

Shin JS and Abraham SN (2001) Caveolae as portals of entry for microbes. Microbes Infect 3:755-61

Sjoberg M and Garoff H (2003) Interactions between the transmembrane segments of the alphavirus E1 and E2 proteins play a role in virus budding and fusion. J Virol 77:3441-3450

Skehel JJ, Cross K, Steinhauer D, and Wiley DC (2001) Influenza fusion peptides. Biochem Soc Trans 29:623-6

Skehel JJ and Wiley DC (2000) Receptor binding and membrane fusion in virus entry: the influenza hemagglutinin. Annu Rev Biochem 69:531-69

Staschke KA, Hatch SD, Tang JC, Hornback WJ, Munroe JE, Colacino JM, and Muesing MA (1998) Inhibition of influenza virus hemagglutinin-mediated membrane fusion by a compound related to podocarpic acid. Virology 248:264-274

Stegmann T, Bartoldus I, and Zumbrunn J (1995) Influenza hemagglutinin-mediated membrane fusion: influence of receptor binding on the lag phase preceding fusion. Biochemistry 34:1825-32

Stegmann T, Booy FP, and Wilschut J (1987) Effects of low pH on influenza virus. Activation and inactivation of the membrane fusion capacity of the hemagglutinin. J Biol Chem 262:17744-9

Stein BS, Gowda SD, Lifson JD, Penhallow RC, Bensch KG, and Engleman EG (1987) pH-independent HIV entry into CD4-positive T cells via virus envelope fusion to the plasma membrane. Cell 49:659-68

Stiasny K, Allison SL, Mandl CW, and Heinz FX (2001) Role of metastability and acidic pH in membrane fusion by tick-borne encephalitis virus. J Virol 75:7392-7398

Stiasny K, Allison SL, Marchler-Bauer A, Kunz C, and Heinz FX (1996) Structural requirements for low-pH-induced rearrangements in the envelope glycoprotein of tick-borne encephalitis virus. J Virol 70:8142-8147

Stiasny K, Allison SL, Schalich J, and Heinz FX (2002) Membrane interactions of the tick-borne encephalitis virus fusion protein E at low pH. J Virol 76:3784-90

Stone-Hulslander J and Morrison TG (1997) Detection of an interaction between the HN and F proteins in Newcastle disease virus-infected cells. J Virol 71:6287-95

Suarez T, Gallaher WR, Agirre A, Goni FM, and Nieva JL (2000) Membrane interface-interacting sequences within the ectodomain of the human immunodeficiency virus type 1 envelope glycoprotein: putative role during viral fusion. J Virol 74:8038-47

Suomalainen M (2002) Lipid rafts and assembly of enveloped viruses. Traffic 3:705-9

Taguchi F and Matsuyama S (2002) Soluble receptor potentiates receptor-independent infection by murine coronavirus. J Virol 76:950-8

Takimoto T, Taylor GL, Connaris HC, Crennell SJ, and Portner A (2002) Role of the hemagglutinin-neuraminidase protein in the mechanism of paramyxovirus-cell membrane fusion. J Virol 76:13028-33

Tamm LK and Han X (2000) Viral fusion peptides: a tool set to disrupt and connect biological membranes. Biosci Rep 20:501-18

Tamm LK, Han X, Li Y, and Lai AL (2002) Structure and function of membrane fusion peptides. Biopolymers 66:249-60

Taylor GM and Sanders DA (1999) The role of the membrane-spanning domain sequence in glycoprotein-mediated membrane fusion. Mol Biol Cell 10:2803-15

Tong S and Compans RW (2000) Oligomerization, secretion, and biological function of an anchor-free parainfluenza virus type 2 (PI2) fusion protein. Virology 270:368-76

Tong S, Li M, Vincent A, Compans RW, Fritsch E, Beier R, Klenk C, Ohuchi M, and Klenk HD (2002) Regulation of fusion activity by the cytoplasmic domain of a paramyxovirus F protein. Virology 301:322-333

Vashishtha M, Phalen T, Marquardt MT, Ryu JS, Ng AC, and Kielian M (1998) A single point mutation controls the cholesterol dependence of Semliki Forest virus entry and exit. J Cell Biol 140:91-9

Viard M, Parolini I, Sargiacomo M, Fecchi K, Ramoni C, Ablan S, Ruscetti FW, Wang JM, and Blumenthal R (2002) Role of cholesterol in human immunodeficiency virus type 1 envelope protein-mediated fusion with host cells. J Virol 76:11584-95

Waarts BL, Bittman R, and Wilschut J (2002) Sphingolipid and cholesterol dependence of alphavirus membrane fusion. Lack of correlation with lipid raft formation in target liposomes. J Biol Chem 277:38141-38147

Wahlberg JM, Bron R, Wilschut J, and Garoff H (1992) Membrane fusion of Semliki Forest virus involves homotrimers of the fusion protein. J Virol 66:7309-18

Wahlberg JM and Garoff H (1992) Membrane fusion process of Semliki Forest virus. I: Low pH-induced rearrangement in spike protein quaternary structure precedes virus penetration into cells. J Cell Biol 116:339-48

Weber T, Paesold G, Galli C, Mischler R, Semenza G, and Brunner J (1994) Evidence for H^+-induced insertion of influenza hemagglutinin HA2 N-terminal segment into viral membrane. J Biol Chem 269:18353-8

Weissenhorn W, Carfi A, Lee KH, Skehel JJ, and Wiley DC (1998) Crystal structure of the Ebola virus membrane fusion subunit, GP2, from the envelope glycoprotein ectodomain. Mol Cell 2:605-16

Weissenhorn W, Dessen A, Calder LJ, Harrison SC, Skehel JJ, and Wiley DC (1999) Structural basis for membrane fusion by enveloped viruses. Mol Membr Biol 16:3-9

Weissenhorn W, Dessen A, Harrison SC, Skehel JJ, and Wiley DC (1997) Atomic structure of the ectodomain from HIV-1 gp41. Nature 387:426-30

West JT, Johnston PB, Dubay SR, and Hunter E (2001) Mutations within the putative membrane-spanning domain of the simian immunodeficiency virus transmembrane glycoprotein define the minimal requirements for fusion, incorporation, and infectivity. J Virol 75:9601-12

Wharton SA, Skehel JJ, and Wiley DC (2000) Temperature dependence of fusion by sendai virus. Virology 271:71-78

White J and Helenius A (1980) pH-dependent fusion between the Semliki Forest virus membrane and liposomes. Proc Natl Acad Sci USA 77:3273-7

White J, Kartenbeck J, and Helenius A (1980) Fusion of Semliki forest virus with the plasma membrane can be induced by low pH. J Cell Biol 87:264-72

White J, Kartenbeck J, and Helenius A (1982) Membrane fusion activity of influenza virus. EMBO J 1:217–22

White JM (1990) Viral and cellular membrane fusion proteins. Annu Rev Physiol 52:675–97

Wilson IA, Skehel JJ, and Wiley DC (1981) Structure of the haemagglutinin membrane glycoprotein of influenza virus at 3 Å resolution. Nature 289:366–373

Xiang SH, Kwong PD, Gupta R, Rizzuto CD, Casper DJ, Wyatt R, Wang L, Hendrickson WA, Doyle ML, and Sodroski J (2002) Mutagenic stabilization and/or disruption of a CD4-bound state reveals distinct conformations of the human immunodeficiency virus type 1 gp120 envelope glycoprotein. J Virol 76:9888–9899

Yang C and Compans RW (1996) Analysis of the cell fusion activities of chimeric simian immunodeficiency virus-murine leukemia virus envelope proteins: inhibitory effects of the R peptide. J Virol 70:248–54

Yang L and Huang HW (2002) Observation of a membrane fusion intermediate structure. Science 297:1877–9

Yao Q, Hu X, and Compans RW (1997) Association of the parainfluenza virus fusion and hemagglutinin-neuraminidase glycoproteins on cell surfaces. J Virol 71:650–6

Young JK, Li D, Abramowitz MC, and Morrison TG (1999) Interaction of peptides with sequences from the Newcastle disease virus fusion protein heptad repeat regions. J Virol 73:5945–56

Zarkik S, Defrise-Quertain F, Portetelle D, Burny A, and Ruysschaert JM (1997) Fusion of bovine leukemia virus with target cells monitored by R18 fluorescence and PCR assays. J Virol 71:738–40

Zhou J, Dutch RE, and Lamb RA (1997) Proper spacing between heptad repeat B and the transmembrane domain boundary of the paramyxovirus SV5 F protein is critical for biological activity. Virology 239:327–39

The Role of the Cytoskeleton During Viral Infection

K. Döhner · B. Sodeik (✉)

Department of Virology, OE 5320, Hannover Medical School,
Carl-Neuberg-Str. 1, 30625 Hannover, Germany
Sodeik.Beate@MH-HANNOVER.DE

1	Introduction	68
2	The Cytoskeleton	70
2.1	Intermediate Filaments	71
2.2	Actin Filaments	71
2.3	Microtubules	73
3	Molecular Motors	74
3.1	Myosins	74
3.2	Kinesins	75
3.3	Cytoplasmic Dynein	76
3.3.1	Binding Partners of Dynein Light Chains	77
4	Experimental Approaches	78
4.1	The Tool Kit	78
4.2	Video Microscopy	79
4.3	Virological Assays, Cytoskeletal Cooperation, and Viral Modifications	80
5	Neurotropic Herpesvirus and Cytoplasmic Transport	82
5.1	Microtubule Transport During Entry	82
5.2	Microtubule Transport During Egress	84
5.3	Biochemical Analysis of Viral Capsid Transport	86
5.4	Interaction of Herpesvirus with Cytoskeletal Proteins	87
6	Retrovirus Entry and Budding	88
7	Adenovirus Entry and Microtubules	90
8	Parvovirus Entry and Microtubules	91
9	Poxviruses—Multiple Cargos for Microtubules and Actin Tails	92
9.1	Cytosolic Vaccinia Virus Cores	92
9.2	Particle Transport During Virus Egress	93
10	Actin Remodelling During Baculovirus Infection	94
11	Perspectives	95
	References	96

Abstract Upon infection, virions or subviral nucleoprotein complexes are transported from the cell surface to the site of viral transcription and replication. During viral egress, particles containing viral proteins and nucleic acids again move from the site of their synthesis to that of virus assembly and further to the plasma membrane. Because free diffusion of molecules larger than 500 kDa is restricted in the cytoplasm, viruses as well as cellular organelles employ active, energy-consuming enzymes for directed transport. This is particularly evident in the case of neurotropic viruses that travel long distances in the axon during retrograde or anterograde transport. Viruses use two strategies for intracellular transport: Viral components either hijack the cytoplasmic membrane traffic or they interact directly with the cytoskeletal transport machinery. In this review we describe how viruses—particularly members of the *Herpesviridae, Adenoviridae, Parvoviridae, Poxviridae,* and *Baculoviridae*—make use of the microtubule and the actin cytoskeleton. Analysing the underlying principles of viral cytosolic transport will be helpful in the design of viral vectors to be used in research as well as human gene therapy, and in the identification of new antiviral target molecules.

1
Introduction

As obligate intracellular parasites, viruses use and manipulate the cell's machinery for membrane trafficking, transcription, splicing, nuclear pore transport and protein synthesis. In fact, these cellular processes were elucidated to a large extent by studying viral systems. Here, we discuss the molecular interactions of viruses with the host cytoskeleton and the mechanisms of viral cytoplasmic transport.

Any viral life cycle can be divided into the following phases: (1) adsorption to the cell, (2) penetration of the plasma or endosome membrane, (3) genome uncoating and release from a viral nucleoprotein complex or capsid, (4) early viral protein synthesis, (5) genome replication, (6) late viral protein synthesis, (7) virus assembly/maturation and finally (8) viral release or egress. We define phases (1)–(3), up to the stage at which the viral genome has been unpacked for transcription and replication, collectively as "viral cell entry".

Virions, subviral particles or nucleoprotein complexes are transported during cell entry from the cell surface to the site of viral transcription and replication, as well as from the site of synthesis to that of virus assembly and back to the plasma membrane for virus egress (for recent reviews see Cudmore et al. 1997; Sodeik 2000; Ploubidou and Way 2001; Smith and Enquist 2002). Because free diffusion of molecules larger than 500 kDa is restricted in the cytoplasm compared to dilute solutions, viruses as well as cellular organelles and chromosomes depend on active

The Role of the Cytoskeleton During Viral Infection

Table 1. Viruses that use the cytoskeleton

Virus family Examples	Genome	Morphology Diameter	Cytosolic structure	Mode of entry	Site of replication
Rhabdoviridae Rabies virus Mokola virus	ss (−) RNA ~12 kb	Envelope 75×180 nm^2	Internal ribo-nucleoprotein core	Endocytosis Fusion with endosomal membrane	Cytosol
Picornaviridae Poliovirus	ss (+) RNA ~8 kb	No envelope 30 nm	Genome	Fusion pore at plasma or endosomal membrane	Cytosol, on small vesicles
Herpesviridae HSV1 PRV	dsDNA 120–230 kb	Envelope 150–200 nm	Icosahedral capsid	Fusion at plasma membrane	Nucleus
Retroviridae HIV-1	ss (+) RNA dimer 2×7–13 kb	Envelope 80–130 nm	Internal nucleocapsid, reverse transcription complex	Fusion at plasma membrane, endocytosis possible	Cytosolic reverse transcription, nuclear transcription
Adenoviridae Ad2, Ad5	dsDNA 36–38 kb	No envelope 70–100 nm	Icosahedral capsid	Endocytosis, lysis	Nucleus
Parvoviridae CPV AAV	ssDNA (+) or (−) 5 kb	No envelope 18–26 nm	Icosahedral capsid ???	Endocytosis, fusion pore?	Nucleus
Poxviridae Vaccinia virus	dsDNA 130–300 kb	Envelope 2 infectious forms: IMV 200×300 nm^2	Complex cores, IMV, IEV	Unclear, different for IMV or EEV	Perinuclear viral factories
Baculoviridae AcMNPV	dsDNA 90–160 kb	Envelope 30–60×300 nm^2	Rod-shaped nucleocapsid	Endocytosis, membrane fusion	Nucleus

ds, double stranded; ss, single stranded

mechanisms for directed transport (Luby-Phelps 2000). This is particularly evident in the case of neurotropic viruses that travel long distances in the axon during retrograde or anterograde transport (Smith and Enquist 2002). For example, it would take a herpes virus capsid 231 years to diffuse 10 mm in the axonal cytoplasm (Sodeik 2000).

Viruses use two alternative strategies for intracellular transport. One is to hijack cytoplasmic membrane traffic. During cell entry many viruses pass through the endocytic pathway to the cell centre (see the chapter by Sieczkarski and Whittaker, this volume). After budding, virions travel inside vesicles derived from the endoplasmic reticulum and the Golgi apparatus to the plasma membrane for viral egress (see the chapter by Maggioni and Braakman, this volume). Alternatively, viral components interact directly with the cytoskeletal transport machinery. It is helpful to distinguish between cytoplasmic and cytosolic transport. The first is a general term referring to both transport inside a membrane and cytosolic transport. The latter involves direct interactions between cytosolic viral components and the cytoskeleton (Sodeik 2000).

We begin with an overview of the organisation and function of the cytoskeleton and then discuss selected viruses whose cytoplasmic transport is either well characterised in molecular terms, such as herpesvirus and adenovirus, or mechanistically unique, as for vaccinia virus and baculovirus (Table 1).

2
The Cytoskeleton

The host cytoskeleton forms a three-dimensional network which is regulated by many accessory proteins and defines the cell's shape as well as its internal organisation. It mediates physical robustness, cell–cell contact, cell crawling, cell division, organelle or RNA transport during interphase and chromosome movement during mitosis or meiosis. Cytoskeletal filaments are assembled by non-covalent binding from protein subunits: Monomeric actin polymerises into microfilaments with a diameter of 5–6 nm, dimeric tubulin assembles hollow tubes of 25 nm and proteins such as keratin or vimentin form 10-nm intermediate filaments.

2.1
Intermediate Filaments

The intermediate filament proteins show a characteristic tripartite domain organization. N- and C-terminal globular head and tail domains of varying size and sequence flank a centrally located rod domain which mediates lateral interactions during assembly (Coulombe et al. 2001). The rope-like intermediate filaments span the entire cytosol and provide mechanical strength against shear forces.

Intermediate filaments do not seem to play any role in intracellular transport because they have no polarity and no motor proteins have been identified which use them as tracks.

During the late phase of many viral infections, intermediate filaments are rearranged or disassembled by viral proteases (Chen et al. 1993; Ferreira et al. 1994; Luftig and Lupo 1994; Brunet et al. 2000). If the intermediate filaments are disassembled, microinjected beads can diffuse more freely (Luby-Phelps 2000). One could assume that the same holds true for viral particles, and that viral modification of the intermediate filaments facilitates egress.

The formation of aggresomes also leads to the reorganisation of intermediate filaments into a perinuclear cage in a cellular attempt to sequester misfolded and unassembled protein subunits (Garcia-Mata et al. 1999; Kopito 2000). African Swine Fever virus might actually exploit this pathway in order to concentrate viral structural proteins for assembly in a viral factory (Heath et al. 2001).

2.2
Actin Filaments

The most abundant protein of many cells, actin, exists in monomeric form as globular actin or G-actin and in filamentous form called F-actin or microfilaments. Polymerisation proceeds by the reversible endwise addition of actin subunits that stimulates hydrolysis of bound ATP. Each filament is polarised, and its ends have distinct biochemical properties. The fast-growing end is called the barbed or plus-end, and the slow-growing end the pointed or minus-end (Welch and Mullins 2002). In cells, polymerisation occurs primarily at the plus-end and must be tightly regulated, because the turnover of actin subunits in vivo is 100–200 times faster than with pure actin (Zigmond 1993).

Actin filaments are flexible structures which are organised into a variety of linear bundles and arrays, two-dimensional networks or three-dimensional gels. Numerous G- and F-actin binding proteins regulate the monomer pool, the formation of filaments and higher-order networks as well as filament depolymerisation for monomer recycling. For example, binding of the Arp2/3 complex to the lateral side of actin filaments leads to a dendritic pattern of actin filament nucleation (Welch and Mullins 2002).

Besides actin filaments dispersed throughout the entire cell, all cells contain a cortical network of actin filaments, generally oriented with their plus-ends facing towards the plasma membrane. This actin cortex is the key player in various cell motility processes which are based on filament assembly and the action of myosins (Welch and Mullins 2002; Kieke and Titus 2003).

However, the actin cortex beneath the plasma membrane can also be an obstacle for virus entry or budding (Marsh and Bron 1997). This may be the reason why many viruses enter cells via endosomes which easily traverse the actin cortex (see the chapter by Sieczkarski and Whittaker, this volume). Certain viruses might even fuse with either the plasma membrane or an endosome depending on the cell type they are infecting (Miller and Hutt-Fletcher 1992; Nicola et al. 2003).

The cortical actin might also be a barrier for virus assembly at the plasma membrane. The cortical actin might be locally depolymerised, or the force of actin polymerisation could be used to drive the budding process (Cudmore et al. 1997; Garoff et al. 1998). Recent evidence suggests that a shell of cortical actin also surrounds many membrane organelles. Thus viruses could pick up actin and actin-binding proteins during budding as has been shown for human immunodeficiency virus (HIV) (Ott et al. 1996; Liu et al. 1999; Wilk et al. 1999).

Regulated binding to cytosolic actin filaments can control the subcellular localisation of a protein; an example is the nucleoprotein of influenza virus (Digard et al. 1999). Despite many reports on nuclear actin only recent evidence supports a role of nuclear actin in chromatin remodelling, splicing, nuclear import and export (reviewed in Rando et al. 2000; Pederson and Aebi 2002). Many viruses use actin during replication, but with the exception of nucleocapsid assembly of some baculoviruses (Kasman and Volkman 2000) and the export of unspliced genomic HIV-1 RNAs (Kimura et al. 2000; Hofmann et al. 2001), these activities all occur within the cytoplasm.

2.3 Microtubules

Microtubules (MTs) are long, hollow cylinders assembled from heterodimers of α- and β-tubulin and MT-associated proteins (MAPs). The head-to-tail association of tubulin into 13 protofilaments leads to the formation of MTs, which are further stabilised by lateral interactions between adjacent protofilaments. MTs are, like actin filaments, polar structures with a dynamic fast-growing plus-end and a less dynamic minus-end (Desai and Mitchison 1997; Downing 2000). The overall subcellular MT organisation is highly polarised. MT dynamics vary considerably between different regions of a cell, during the cell cycle and throughout differentiation. Thus MT turnover is tightly controlled in space and over time.

The plus-ends of MTs are the primary sites of MT growth and shortening, resulting in a phenomenon called dynamic instability (Mitchison and Kirschner 1984). After subunit addition at the end, the GTP bound to β-tubulin is slowly hydrolysed. The consecutive loss of this GTP-tubulin cap leads to catastrophic MT depolymerisation. GTP-tubulin is more likely to be added to a GTP-tubulin cap than to a MT rapidly depolymerising. Plus-end binding proteins such as CLIP-170, EB1 or APC stabilise a region of about 1 μm at the plus-end of MTs and thus enable them to interact with the plasma membrane, organelles or the kinetochores of chromosomes. This search-and-capture mechanism prevents depolymerisation, and the MT stabilisation can even reorient the MTOC and thus the entire MT network (Howard and Hyman 2003).

In most animal cells, minus-ends are attached to a MT-organising centre (MTOC) called a centrosome and typically located close to the cell nucleus whereas the plus-ends are pointing towards the cell periphery and plasma membrane. However, several cell types contain a substantial number of non-centrosomal MTs (Keating and Borisy 1999). Even at steady state—when a population of MTs has reached constant mass—individual MTs may be polymerising or depolymerising, and such dynamics modulate intracellular transport (Keating and Borisy 1999; Giannakakou et al. 2002). Polarised epithelial cells are characterised by an apical and a basal MT web and longitudinally arranged MTs which point with their minus-ends to the apical and with their plus-ends to the basal membrane (Topp et al. 1996). Neuronal MTs in axons are also longitudinally arranged with their minus-ends pointing to the soma and their plus-ends to the axon terminal, whereas most dendrites have MTs of mixed polarity (Goldstein and Yang 2000; Hirokawa and Takemura 2003).

Many MAPs such as tau contain a basic domain which interacts with MTs via the acidic C-terminal domains of tubulin which are exposed on the MT surface (Downing 2000). Also, viral proteins with a basic domain such as many RNA- or DNA-binding proteins have the potential to interact with the MT surface (Elliott and O'Hare 1997; Ploubidou et al. 2000; Mallardo et al. 2001).

MTs are involved in long-distance transport as highlighted by the transport requirements through neuronal dendrites and axons. Moreover, the reversible association to MTs and motors regulates the subcellular localisation of important signalling proteins and transcription factors (Ziegelbauer et al. 2001; Giannakakou et al. 2002; Schnapp 2003).

3
Molecular Motors

Many proteins interact with tubulin and actin to regulate their structural organisation. A subset of those, the motor proteins, use conformational changes powered by ATP hydrolysis to transport cargo along the filaments. If motors are fixed on a surface such as a membrane, they move the filaments relative to that surface. If instead the filament is fixed, motors transport cargo along that filament.

Myosins mediate translocation along actin filaments, whereas dynein and kinesins move on MTs. It has been suggested that in animal cells long-range transport occurs along MTs and short-range transport along actin filaments (reviewed in Brown 1999; Verhey 2003). A viral particle could be transported either by directly recruiting a motor or by attaching to a cargo of a given motor. Moreover, polymerising actin filaments can push around different cargo such as endocytic vesicles, intracellular bacteria or vaccinia virus (Frischknecht and Way 2001).

3.1
Myosins

All myosins share a common N-terminal motor domain of about 500 amino acids that binds to actin filaments (www.mrc-lmb.cam.ac.uk/myosin/myosin.html). A flexible neck region connects the motor domain to a tail region of variable size which determines the specific function of each myosin and the specificity of cargo binding. Myosins have been grouped into 18 classes according to homologous sequences in the tail domain (Kieke and Titus 2003).

Most myosins walk towards the plus-ends of actin filaments, with the exception of myosin VI and possibly myosin IXb, which are minus-end-directed motors (Kieke and Titus 2003). The myosin IIs are often called 'conventional' myosins and include the best-studied example of skeletal muscle myosin. Cytoplasmic myosin II consists of two heavy chains and two pairs of light chains. Myosin II generates contractile forces during cytokinesis, maintains cortical tension and is crucial for cell motility. Myosin I, V and VI are involved in membrane traffic; moreover, myosin I plays a role in cell motility and signal transduction (Kieke and Titus 2003).

3.2
Kinesins

Kinesins transport membranes, cytoskeletal filaments, viral particles, mRNAs and proteins along MTs and are engaged in organising the mitotic spindle as well as chromosome translocation (Hirokawa and Takemura 2003). The defining criterion for a kinesin is a MT-binding motor domain of 320 amino acids, which is folded into a core with an overall structure similar to myosins and G proteins (Schliwa and Woehlke 2003; www.proweb.org/kinesin//index.html). The motor domain is linked to structural and regulatory domains, which attach to other cofactors, adaptor proteins or interaction modules (Karcher et al. 2002; Schnapp 2003). The kinesins are sorted into three groups according to the position of the motor domain at the N-terminus, an internal position, or the C-terminus, hence the names Kin-N, Kin-I and Kin-C kinesins, and based on homology into 14 classes (Hirokawa and Takemura 2003).

N-type kinesins, which include conventional kinesin (kinesin-1, KIF5), move towards MT plus-ends. Kinesin-1 is a heterodimer of two heavy chains (KHC) that each contain a motor domain, an α-helical stalk responsible for dimerization and a C-terminal tail domain that binds two light chains (KLC). The KLCs can dock onto adaptors or cargo receptors and link kinesin to the cargo (Hirokawa and Takemura 2003). Kinesin-1 transports various cargos such as axonal vesicles, mitochondria, lysosomes, endocytic vesicles, tubulin oligomers, intermediate filament proteins or mRNA complexes (Hirokawa and Takemura 2003). Kinesin-1 is also responsible for the transport of vaccinia virus (VV) to the plasma membrane (Rietdorf et al. 2001).

The respective cargos of the N-type kinesins KIF1A and KIF1B are synaptic vesicles and mitochondria (Nangaku et al. 1994; Okada et al. 1995). The N-type heterotrimeric kinesin with the KIF3 heavy chains mediates axonal transport of fodrin-containing vesicles, and in epithelial cells KIF3 interacts with APC, a MT plus-end binding protein (Takeda et al. 2000; Jimbo et al. 2002). Kin-C kinesins move towards the MT minus-ends (Hirokawa and Takemura 2003). Most C-type kinesins act during mitosis, but some are implicated in membrane transport in neuronal dendrites and polarised epithelial cells (Hanlon et al. 1997; Saito et al. 1997; Noda et al. 2001; Xu et al. 2002). Kin-I kinesins most likely do not work as motors but destabilise MTs (Desai et al. 1999; Hunter et al. 2003).

3.3
Cytoplasmic Dynein

Axonemal and cytoplasmic dyneins transport cargo towards MTs minus-ends. Axonemal dyneins provide the driving force for the beating movement of cilia and flagella (King 2003). Cytoplasmic dynein plays an essential role during mitosis and is responsible for the perinuclear localisation of several organelles around the MTOC as well as retrograde organelle transport in axons (Karki and Holzbaur 1999). Cytoplasmic dynein is also required for the transport of non-membranous cargo such as NuMA, aggresomes, viral capsids, neurofilaments and pericentrin particles (Merdes and Cleveland 1997; Garcia-Mata et al. 1999; Suomalainen et al. 1999; Shah et al. 2000; Sharp et al. 2000; Young et al. 2000; Döhner et al. 2002). It is even involved in the transport of mRNAs and proteins regulating transcription (Galigniana et al. 2001; Giannakakou et al. 2002; Tekotte and Davis 2002).

Cytoplasmic dynein is a 20 S protein complex in the shape of a Y and consists of two dynein heavy chains (DHCs; 530 kDa), two intermediate chains (DICs; 70–80 kDa), four light intermediate chains (DLICs; 50–60 kDa), and three families of light chains (DLCs, 7–14 kDa, see below; Karki and Holzbaur 1999; King 2003). DHCs are members of the AAA family of ATPases and mediate MT binding, ATP hydrolysis and force generation (King 2000). The DICs belong to the WD-repeat protein family and are located at the base of the dynein motor. DLICs, DICs and DLCs have been implicated in motor regulation and cargo binding (King 2003). The subunits differ in their subcellular localisation and tissue expression, suggesting that cytoplasmic dynein exists in many distinct isoforms with unique subunit composition (Nurminsky et al. 1998; Tai et al. 2001).

For most, if not all, minus-end-directed transport processes cytoplasmic dynein requires a cofactor called dynactin (Karki and Holzbaur

1999). Dynactin is another large 20 S protein complex consisting of 11 different subunits (Holleran et al. 1998; Eckley et al. 1999). The p150^{Glued} subunit of dynactin binds directly to the DICs. Interestingly, dynactin was recently reported to also play a role in melanosome transport catalysed by heterotrimeric kinesin (Deacon et al. 2003).

3.3.1
Binding Partners of Dynein Light Chains

Many DLC binding partners have been identified in yeast two-hybrid screens, sometimes supported by GST-pull down or co-immunoprecipitation assays after transient overexpression of the proteins. However, it has often been difficult to confirm these interactions with endogenous proteins under physiological conditions, and to prove that they lead to the recruitment of a motor. Based on sequence homology, DLCs have been grouped into the LC7/Roadblock, LC8/PIN and Tctex families (King 2003).

Members of the LC7 family bind to DICs and play a role in intracellular transport and mitosis in *Drosophila* (Bowman et al. 1999; Susalka et al. 2002). A mammalian member, mLC7, interacts with transforming growth factor-β (Tang et al. 2002).

DLCs of the LC8/PIN family bind to neuronal nitric oxide synthase and are therefore also called PIN for 'protein inhibitor of nitric oxide synthase' (Jaffrey and Snyder 1996). Further interaction partners are IκBα, Bim of the Bcl-2 family, postsynaptic density-95/guanylate kinase domain-associated protein, nuclear respiratory factor, the erect wing gene product in *Drosophila*, swallow protein, DIC and, interestingly, the 3'-UTR of parathyroid hormone RNA (King 2003). LC8/PIN is also a subunit of the actin motor myosin V (Espindola et al. 2000). The interaction often involves a conserved K/R-XTQT or G-I/V-QVD sequence motif in the DLC binding partner (Lo et al. 2001; Rodriguez-Crespo et al. 2001; Martinez-Moreno et al. 2003).

LC8/PIN also interacts with the phosphoprotein of rabies and Mokola virus (Raux et al. 2000; Jacob et al. 2000). Rabies virus spreads via neurons and travels long distances from the site of entry to the neuronal cell body. Axonal transport of rabies virus requires intact MTs, but it is unclear whether vesicles containing virions or cytosolic capsids are transported (Ceccaldi et al. 1989; Tsiang et al. 1991). Rabies viruses deleted for the LC8 binding motif no longer incorporate LC8, but pathogenesis is only impaired in already attenuated strains, and only in very young mice (Mebatsion 2001; Poisson et al. 2001). LC8/PIN also binds to p54 of

African swine fever virus, whose infection is inhibited if dynein is blocked (Alonso et al. 2001; Heath et al. 2001). Yeast two-hybrid screens revealed an interaction between HIV integrase and DYN2, a putative LC8-like yeast dynein light chain of 92 amino acids (Richard de Soultrait et al. 2002). Several additional potential viral interaction partners for LC8/PIN such as HSV1 helicase, adenovirus protease, vaccinia virus (VV) polymerase and, surprisingly, the extracellular domain of respiratory syncytial virus attachment protein have been identified by screening libraries of overlapping dodecapeptides in ligand blots (Martinez-Moreno et al. 2003).

DLCs of the Tctex class bind to Doc2, p59^{fyn} kinase, rhodopsin, FIP1, Trk receptors, DIC, the cytosolic tail of CD5 and CD155, often via the binding motif K/R-K/R-XX-K/R (King 2003). CD155 is the receptor for poliovirus that invades the CNS by haematogenous or neural spread (Mueller et al. 2002). After binding to the receptor the poliovirus genome is released into the cytosol through a membrane pore. It is unclear whether pore formation occurs already at the plasma membrane or after delivery to an endosome (Hogle 2002). If axonal transport occurs inside an endocytic membrane, the binding of CD155 to Tctex-1 and thus possibly dynein could mediate retrograde transport of virus-containing vesicles to the cell body of the motor neuron (Mueller et al. 2002).

4
Experimental Approaches

4.1
The Tool Kit

There are several reversible pharmacological drugs that target cytoskeletal proteins. The response of different cell types to these drugs varies greatly, and therefore each inhibitor is tested initially with different concentrations and incubation times (Jordan and Wilson 1999).

Cytochalasin and latrunculin depolymerise actin filaments, whereas jasplakinolide promotes actin polymerisation (Cooper 1987; Bubb et al. 1994; Ayscough 1998). Colchicine, colcemide, vincristine, vinblastine and nocodazole (podophyllotoxin) depolymerise MTs, whereas taxol (paclitaxel) stabilises the MT network (Jordan and Wilson 1999). Because all these are not competitive inhibitors but interfere with the dynamic equilibrium between subunits and filaments, the cells need to be treated for quite some time, until an effect on the steady-state distribution of the cytoskeleton is manifested (Cooper 1987; Jordan and Wilson 1999).

Butanedione monoxime (BDM) inhibits ATP hydrolysis of myosin II, V, VI and possibly I (Herrmann et al. 1992; Cramer and Mitchison 1995). However, higher concentrations may also target other ATPases (Schlichter et al. 1992; Mojon et al. 1993; Phillips and Altschuld 1996). KT5926, wortmannin and ML-7 block the activity of myosin light chain kinase, and thus the only known downstream target, the non-muscle myosin II (Nakanishi et al. 1990; Nakanishi et al. 1992; Ruchhoeft and Harris 1997). Adociasulfate blocks kinesins (Sakowicz et al. 1998), and dyneins are inhibited by erythro-9-[3-(2-hydroxynonyl)]adenine that also affects adenosine deaminase and cGMP-stimulated phosphodiesterase (Penningroth 1986; Mery et al. 1995).

Microinjecting function-blocking antibodies, adding anti-sense or small interfering RNAs or overexpressing a dominant-negative protein can inhibit many motors. For example, excess dynamitin blocks dynactin and overexpression of the cargo-binding domain of KLC inhibits conventional kinesin (Echeverri et al. 1996; Burkhardt et al. 1997; Valetti et al. 1999; Rietdorf et al. 2001).

Testing many inhibitors aiming at alternative targets by different molecular mechanisms provides a guide for analysing the role of the cytoskeleton in virus infection. To determine which viral proteins are involved in cytoplasmic transport, viral mutants with defined genotypes have been generated.

4.2
Video Microscopy

Intracellular movement of host and viral particles is best analysed by video or digital time-lapse microscopy (Lippincott-Schwartz et al. 2000; Zhang et al. 2002). This technique adds another dimension to the analysis of fixed cells, because not only the steady-state distribution of viral components but also the dynamics of their transport is analysed. Because of their small size and limited contrast, viral particles with the exception of vaccinia virus need to be tagged with a fluorescent molecule for tracking in live cells.

Direct chemical coupling to small fluorescent dyes was used to analyse cytoplasmic transport of non-enveloped viruses (Georgi et al. 1990; Leopold et al. 1998; Suomalainen et al. 1999). Because many viruses undergo substantial disassembly steps during entry (Greber et al. 1993), one needs to identify the protein to which the dye has been attached and, if possible, analyse trafficking of differently tagged viral structures.

To analyse the infection of enveloped viruses, the fluorescent tag must be linked to a structure which after membrane fusion remains associated with the genome during its cytosolic passage. Green fluorescent protein (GFP)-tagged capsid or core proteins have been used to track pseudorabies virus and HIV (Smith et al. 2001; McDonald et al. 2002). Because all viral particles can be taken up by endocytosis, even if that is not their physiological entry port (see the chapter by Sieczkarski and Whittaker, this volume), experiments using fluorescently labelled virions require care to distinguish cytosolic particles from those in endosomes (Suomalainen et al. 1999; McDonald et al. 2002; Sodeik 2002). Likewise, when analysing viral egress, one needs to distinguish GFP-tagged virions from GFP-fusion protein in the biosynthetic pathway or in the cytosol (Rietdorf et al. 2001). The recent improvements in the design of fluorescent proteins allow dual-colour imaging of different viral particles in live cells (Zhang et al. 2002).

Attaching multiple fluorescent molecules to a virus or subviral particle could alter virus assembly and subsequently virus entry. Fluorescent dyes are usually quite hydrophobic, and GFP is a protein of 27 kDa resulting in a rather large mutation of any viral protein. Therefore, it is crucial to test with several assays that a GFP-tagged virus behaves similarly to the unlabelled virus. Among the physiological cargo of dynein are aggregated protein complexes that are packed into aggresomes at the MTOC (Garcia-Mata et al. 1999; Kopito 2000). Thus the cells might send many non-functional viral proteins along MTs to the MTOC for refolding or final degradation (Sodeik 2002).

Many of these issues can now be overcome by single-molecule imaging techniques (Seisenberger et al. 2001). However, they require specialised equipment to visualise the emission with a low signal-to-noise ratio. Most results so far have been obtained with a high ratio of dye to particle. With the use of fluorescent proteins with improved excitation and emission spectra and digital cameras with increasing sensitivity, movies of viral cytoplasmic transport at physiological conditions can now be recorded.

4.3
Virological Assays, Cytoskeletal Cooperation, and Viral Modifications

Besides suitable tools to analyse the cytoskeleton, many experiments have been developed to study individual phases of a viral life cycle. In addition to blocking cytoplasmic transport, the inhibitors used could also influence viral infection in other ways.

For example, because a cell's membrane traffic relies heavily on actin and MT transport, important host factors, such as viral receptors, could mis-localise, thus resulting in a reduced or increased virus binding, internalisation or other changes in the viral life cycle (Döhner et al. 2002). If the readout for cytoplasmic transport during virus entry is nuclear viral gene expression, a potential role of nuclear actin or myosin in viral transcription should be considered (Pederson and Aebi 2002). The host cytoskeleton can also directly influence viral transcription as shown for paramyxoviruses or reverse transcription and viral budding as in the case of HIV (Sasaki et al. 1995; Bukrinskaya et al. 1998; Gupta et al. 1998).

Moreover, the three types of cytoskeletal filaments are interdependent entities which operate together in cells (Fuchs and Yang 1999; Goode et al. 2000). The last years have shown a close interaction between MT plus-ends and the actin cortex, and several proteins provide physical links between different filament types (Coulombe et al. 2000; Allan and Näthke 2001; Leung et al. 2002). Thus a prolonged incubation with any of the inhibitors could lead to pleiotropic effects. For example, targeting MTs with depolymerising drugs or with antibodies to stable MTs also results in the collapse of intermediate filaments to a perinuclear region (Gurland and Gundersen 1995).

Many cells contract, and in the extreme case completely round up, if actin filaments are disassembled, because their substrate adhesion via focal contacts is weakened (Ayscough 1998). This reduces the transport distances, for example from the plasma membrane to the nucleus, and cargo could reach a particular binding site such as the nuclear pore in the absence of the appropriate filament system by diffusion rather than by active transport as under physiological conditions. Moreover, although cytoskeletal filaments provide tracks for cytoplasmic viral transport, they also restrict the space that is open for translocation (Luby-Phelps 2000). Thus depolymerising cytoskeletal filaments removes both transport trails and steric barriers.

For these reasons, it is sometimes difficult to determine the precise functional role of the cytoskeleton during virus infection. In addition, late in infection the cytoskeletal filaments are often rearranged, complicating the analysis of their role during assembly and egress (Avitabile et al. 1995; Brunet et al. 2000; Dreschers et al. 2001). However, by taking these caveats and limitations into consideration, we have learnt an amazing amount about the molecular cross-talk between viruses and the host cytoskeleton.

5
Neurotropic Herpesvirus and Cytoplasmic Transport

5.1
Microtubule Transport During Entry

Neurotropic viruses are dependent on efficient transport because they travel long distances during pathogenesis. Prominent examples are the alphaherpesviruses such as human herpes simplex virus (HSV). After initial replication in exposed mucosal epithelia, progeny virions enter local nerve endings of peripheral neurons and the capsid without the envelope is retrogradely transported to the nucleus located in a peripheral nerve ganglion. Here, HSV1 establishes a latent infection that on stress can be reactivated to a lytic infection of the neuron. Newly synthesised virus is then released from the nerve endings and re-infects the peripheral epithelial tissue (Roizman and Knipe 2001; Smith and Enquist 2002). MTs are required in epithelial and neuronal cells as well as during entry and egress (Kristensson et al. 1986; Topp et al. 1994, 1996; Avitabile et al. 1995; Sodeik et al. 1997; Miranda-Saksena et al. 2000; Kotsakis et al. 2001; Mabit et al. 2002).

The fusion of the HSV1 envelope with the plasma membrane releases into the cytosol a capsid and about 20 different proteins of the tegument, a protein layer encasing the capsid. The capsid, associated with a subset of tegument proteins, is transported along MTs to the MTOC, localised in the cell centre, and further to the nucleus (Kristensson et al. 1986; Lycke et al. 1988; Sodeik et al. 1997; Mabit et al. 2002). The docking of the capsid at the nuclear pore induces the translocation of the viral genome into the nucleoplasm (Ojala et al. 2000), where viral replication, transcription and later during infection capsid assembly take place (Fig. 1A).

Incoming HSV1 capsids co-localise with cytoplasmic dynein and dynactin in epithelial cells (Sodeik et al. 1997; Döhner et al. 2002). Electron micrographs of detergent-extracted, infected cells reveal structures resembling the Y-shaped cytoplasmic dynein at the capsid vertices (Sodeik et al. 1997), where the tegument has a more ordered structure and the tegument protein VP1-3 might be located (Zhou et al. 1999). Moreover, inhibiting dynein function by the overexpression of dynamitin reduces transport of HSV1 capsids to the nucleus and early viral gene expression without having any effect on virus binding or internalisation (Döhner et al. 2002). Whether and how capsids move further from the MTOC to the nuclear pore complex is unclear (Sodeik 2002).

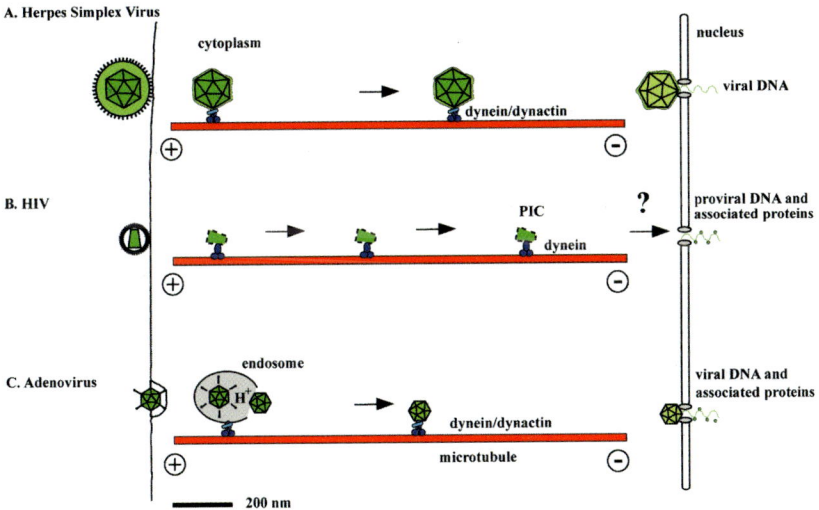

Fig. 1A–C. Cytosolic transport during virus entry. **A** After fusion of the envelope of herpes simplex virus type 1 (HSV1) with the plasma membrane, the capsid (*dark green*) and the tegument proteins (*green*) are released into the cytosol. The capsid with a subset of tegument proteins associated binds to microtubules (*red*) and is transported by dynein and dynactin (*blue*) to the MT-organising centre where the MT minus-ends are located. After capsid (*light green*) binding to the nuclear pore complex, the viral DNA is injected into the nucleoplasm. **B** Human immunodeficiency virus type-1 (HIV-1) enters cells by fusion with the plasma membrane. During passage through the cytosol, the viral RNA genome is reverse transcribed into DNA in a structure called reverse transcription complex or preintegration complex (*PIC*; *green*). This structure co-localises with microtubules (*red*) and requires dynein (*blue*) for its transport to the nucleus. The proviral DNA is imported into the nucleus and integrated into a host chromosome. **C** Adenovirus of the subgroup C is internalised by receptor-mediated endocytosis and induces lysis of the endosome (*grey*) to access the cytosol. The cytosolic capsid (*green*) binds to microtubules (*red*) and is transported by dynein and dynactin (*blue*) to the cell centre. During entry, the capsids are stepwise disassembled (indicated by the different shades of *green*). At the nuclear pores, the viral DNA and associated proteins are released from the capsid (*light green*) and imported into the nucleoplasm

In vivo analysis of the entry into epithelial cells by time-lapse digital fluorescence microscopy showed that GFP-tagged capsids moved along MTs towards *and* away from the nucleus with maximal speeds of 1.1 µm/s (Döhner, Büttner, Wolfstein, Schmidt and Sodeik, in preparation). Incoming pseudorabies virus, another alphaherpesvirus, moves at rates averaging 1.3 µm/s. The transport is saltatory and bi-directional, but in neuronal processes with a retrograde bias towards the cell body (Smith and Enquist 2002).

Alphaherpesvirus capsids must possess a viral receptor either for dynein or for dynactin, which allows them to engage the host MT system for efficient transport from the plasma membrane to the host nucleus, a particularly long distance after infection of neurons via a synapse. In unpolarised epithelial cells capsids show minus- and plus-ended directed transport during entry, whereas in cultured neurons the retrograde transport towards the nucleus predominates.

5.2
Microtubule Transport During Egress

Herpesvirus capsids assemble in the nucleus and after genome packaging acquire a primary envelope by budding at the inner nuclear membrane. The traditional view was that from then on the virus particle remained within the secretory pathway and was released from infected cells by fusion of a vesicle containing the enveloped virion with the plasma membrane (reviewed in Roizman and Knipe 2001; Mettenleiter 2002). In this scenario, all tegument proteins were added to the capsids already in the nucleus, and cytoplasmic transport was mediated by the interaction of membrane vesicles, possibly modified by the addition of viral proteins, with the host cytoskeleton.

Several studies in the last decade suggested that many if not all herpesviruses use an alternative pathway. Accordingly, transiently enveloped viral particles leave the periplasmic space by fusion with the outer nuclear envelope or the membrane of the endoplasmic reticulum which is continuous with the nuclear envelope (Roizman and Knipe 2001; Mettenleiter 2002). By this procedure, capsids traverse the nuclear envelope and reach the cytosol. The naked cytosolic capsids can recruit tegument proteins and engage the host cytoskeleton to travel to the site of secondary envelopment. Newly synthesised HSV1 particles tagged with GFP-VP11/12 move along MTs over distances of up to 49 µm with average velocities of 2 µm/s (Willard 2002).

In unpolarised epithelial cells, secondary envelopment most likely occurs at the *trans*-Golgi network (TGN) or an endosome (Skepper et al. 2001; Mettenleiter 2002). In non-infected cultured cells, these organelles are clustered around the MTOC in close proximity to the nucleus. Thus, at least early during assembly, again dynein possibly assisted by dynactin could transport the capsids to the desired location. However, later during infection, the MTs are marginalised in bundles to the periphery

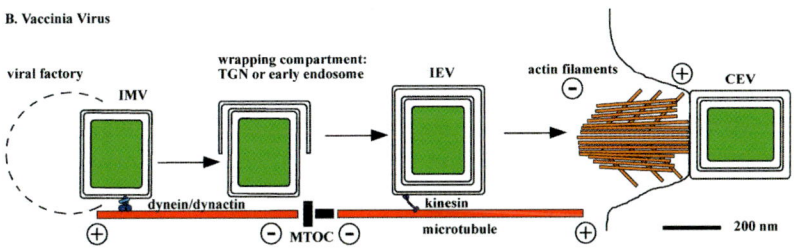

Fig. 2A, B. Cytosolic transport during virus egress. **A** After nuclear HSV1 capsid assembly and genome packaging, the capsid (*dark green*) acquires a primary envelope by budding at the inner nuclear membrane. This primary envelope is lost by subsequent fusion with the outer nuclear membrane, and the capsid is released into the cytosol. Most if not all tegument proteins (*light green*) are acquired in the cytosol or during the second envelopment. In neurons, capsids with associated tegument and membrane vesicles (*grey*) containing viral envelope proteins are transported along microtubules (*red*) to the presynapse. Axonal transport is mediated by conventional kinesin (*dark blue*). Presumably, secondary envelopment occurs in the nerve endings. **B** The intracellular mature form (IMV) of vaccinia virus containing the core (*green*) is assembled in viral factories located in the cytoplasm. The IMV is transported to the MTOC region along MTs by dynein (*blue*) and dynactin (light blue). At the MTOC, the IMV is enwrapped by a membrane cisterna derived from the *trans*-Golgi network or an early endosome, resulting in the formation of the intracellular enveloped virus (IEV). The IEV requires MTs and conventional kinesin (*dark blue*) for its transport to the plasma membrane. Shortly before or after fusion of the outer IEV membrane with the plasma membrane, polymerisation of actin filaments (*orange*) is induced. The cell-associated virus (CEV) is pushed on top of a long microvillus into the medium or a neighbouring cell

of the cytoplasm, and there is no longer an obvious MTOC (Avitabile et al. 1995; Elliott and O'Hare 1998, 1999; Kotsakis et al. 2001). The reorganisation of the MT network most likely also changes the steady-state distribution of the membrane organelles. The localisation of MT plus- or minus-ends has not been determined late in infection, and therefore it is difficult to predict which MT motors would then catalyse capsid transport to the organelle of secondary envelopment.

In a typical neuronal axon, the MT plus-ends are uniformly oriented towards the synapses and the minus-ends to the cell body (Goldstein and Yang 2000). There are no reasons to believe that this organisation is changed during a herpesvirus infection. Progeny virus spreads via the axons, and the traditional model predicted that axonal transport of fully assembled virions occurs inside vesicles. However almost 10 years ago Cunningham and collaborators demonstrated that in human primary sensory neurons HSV1 naked capsids were closely associated with axonal MTs (Penfold et al. 1994). This study suggested that in neurons the secondary envelopment might take place in the presynapse (Fig. 2A).

Immune labelling showed that the glycoproteins are transported in vesicles, separately and with different kinetics from the capsids (Holland et al. 1999; Miranda-Saksena et al. 2000; Ohara et al. 2000). In pseudorabies virus, the type II integral membrane protein US9 is required for targeting of all tested viral membrane proteins to the axon, but not for axonal transport of capsids and some tegument proteins (Tomishima and Enquist 2001). Immunoelectron microscopy showed that progeny axonal cytosolic capsids co-localise with conventional kinesin (Diefenbach et al. 2002). As during virus entry, GFP-tagged pseudorabies virus is transported in both directions in axons late in infection. But during egress the anterograde transport predominates. Average plus-end-directed velocities are 2 µm/s and minus-end directed velocities 1.3 µm/s (Smith et al. 2001).

Thus, during axonal transport, progeny capsids recruit a plus-end MT motor, most likely conventional kinesin. In epithelial cells a minus-end-directed MT motor could ensure, at least early during assembly, transport to the organelle of secondary envelopment.

5.3
Biochemical Analysis of Viral Capsid Transport

Several assays have been developed to characterise the host factors as well as the viral receptors involved in MT transport. GFP-labelled capsids were isolated after detergent extraction of purified virions (Bearer et al. 2000; Wolfstein, Nagel, Döhner, Allan and Sodeik, in preparation). Bearer et al. (2000) injected such capsids into the giant axon of the squid *Loligo paelei*, a classic system to analyse fast MT-mediated transport. HSV1 capsids labelled with co-packaged VP16-GFP that had been expressed from a complementing cell line moved in a continuous fashion in retrograde direction with average velocities of 2.2 µm/s. This unidirectional, non-saltatory movement is different to the motility of GFP-tagged herpes particles after infection of vertebrate cells (Smith et al. 2001; Willard 2002; Döhner, Büttner, Wolfstein, Schmidt and Sodeik, in preparation).

Wolfstein et al. (in preparation) reconstituted capsid motility in vitro with Cy3-labeled MTs, VP26GFP-tagged capsids, an ATP-regenerating system and a cytosolic protein fraction. In vitro transport was also saltatory, with long continuous runs of more than 10 µm in one direction, and the capsids often changed MTs. In this assay capsid transport along MTs depended on their tegument composition. It could be inhibited by the addition of recombinant dynamitin, suggesting that the motility was catalysed by dynein or heterotrimeric kinesin (Wolfstein, Nagel, Döhner, Allan and Sodeik, in preparation). Such biochemical assays can now be used to characterise the molecular interactions between viral and host factors.

5.4
Interaction of Herpesvirus with Cytoskeletal Proteins

VP22, one of the major tegument proteins of HSV1, and GFP-VP22 co-localise with MTs after transfection, whereas in infected cells VP22 does not show a prominent enrichment on MTs (Elliott and O'Hare 1997, 1999; Kotsakis et al. 2001). The transient restructuring of the MT network during virus infection coincides with the translocation of VP22 from the cytoplasm to the nucleus (Kotsakis et al. 2001). Transient expression of VP22 or infection with HSV1 leads to a stabilisation and hyperacetylation of MTs (Elliott and O'Hare 1998, 1999). A basic VP22 fragment contains the MT binding domain (Martin et al. 2002).

In addition, VP22 interacts with non-muscle myosin II, and the myosin inhibitor BDM reduces virus yields (van Leeuwen et al. 2002). Because VP22 is released into the cytosol on HSV1 fusion with the plasma membrane, this interaction could also facilitate the penetration of the actin cortex by the incoming viral capsid.

The first herpesvirus protein reported to interact with the DIC subunit of cytoplasmic dynein in GST pull down assays was the UL34 of

HSV1, with a predicted type II integral membrane topology (Ye et al. 2000; Reynolds et al. 2002). UL34 of HSV and pseudorabies virus is targeted to the inner nuclear membrane, where during budding it is incorporated into the primary envelope of periplasmic virions (Klupp et al. 2000; Reynolds et al. 2002). In contrast to the inner nuclear membrane, cytoplasmic membranes and extracellular virions do not contain UL34; thus it cannot be involved in viral cytoplasmic transport.

The heavy chain of conventional kinesin interacts with the HSV1 tegument protein US11. Moreover, US11 and the major capsid protein VP5 were reported to be transported with similar kinetics into the axons of dissociated rat neurons (Diefenbach et al. 2002).

6
Retrovirus Entry and Budding

HIV, the most prominent representative of retroviruses, enters cells primarily by membrane fusion at the plasma membrane (Greene and Peterlin 2002), although some strains can also infect cells via endocytic uptake (Fackler and Peterlin 2000). During passage through the cytosol, the viral RNA genome is reverse transcribed into DNA in a structure named the reverse transcription complex (RTC) or pre-integration complex (PIC). PIC is targeted to the nuclear pore complex and imported into the nucleus, and the HIV genome is integrated into a chromosome (Greene and Peterlin 2002; see the chapter by Byland and Marsh, this volume, and Fig. 1B).

Cytochalasin prevents the co-clustering of the HIV receptors CD4 and CXCR4 on gp120 binding and thus somehow inhibits efficient HIV fusion with the plasma membrane (Iyengar et al. 1998). On cell fractionation RTCs are localised in a cytoskeletal fraction, and reverse transcription requires intact actin filaments (Bukrinskaya et al. 1998).

Intracellular trafficking of HIV can be visualised with a GFP-Vpr fusion protein which is packaged into virions (McDonald et al. 2002). By live cell imaging GFP-Vpr-labelled subviral particles co-localise with MTs, move in curvilinear paths in the cytoplasm and accumulate around the MTOC as early as 2 h after infection. HIV transport is completely blocked in the presence of both nocodazole and latrunculin, but not by either one alone. The incoming PIC of human foamy virus, another retrovirus, also accumulates around the MTOC by a MT depending transport (Saib et al. 1997).

When dynein is inhibited by a microinjected antibody, the relative transport of RTCs to the nucleus is reduced (McDonald et al. 2002). Interestingly, in this situation the RTCs accumulate in the cell periphery, similar to what has been described for HSV capsids after overexpression of dynamitin (Döhner et al. 2002). Moreover, yeast two-hybrid assays showed an interaction of HIV integrase with the yeast homologue of dynein LC8 (Richard de Soultrait et al. 2002).

Most likely, HIV subviral particles that are assembled in the cytoplasm also use active transport to reach the plasma membrane, where retrovirus budding often takes place (Greene and Peterlin 2002). Both MTs and the binding motive for the host factor hnRNP A2 present in genomic HIV RNA are involved in trafficking of RNA granules (Mouland et al. 2001). The human T cell leukaemia virus, a retrovirus distantly related to HIV, requires a cell-cell junction and MTs for efficient transmission. T cell adhesion induces a rapid reorganisation of the MT network towards the junction that becomes a kind of virological synapse for budding (Igakura et al. 2003).

Filamentous actin and myosin seem to play important roles during HIV budding at the plasma membrane (Sasaki et al. 1995). The interaction of the HIV negative effector Nef with Vav, the host guanine nucleotide exchange factor of Cdc42 and Rac—two small GTPases regulating the actin cytoskeleton—could modify the actin cortex before virus budding (Fackler et al. 1999). Nef is also packaged into virions (10–100 molecules/particle) and has an important role in viral entry early after virus fusion but before the completion of reverse transcription (Schaeffer et al. 2001; Forshey and Aiken 2003). It is tempting to speculate that Nef, possibly in concert with other proteins of the RTC, could again remodel the actin cortex to facilitate access to the MT plus-ends for further transport.

The kinesin motor KIF4A is abundantly expressed in neurons and transports vesicles containing the cell adhesion protein L1 (Peretti et al. 2000). Yeast two-hybrid assays and chromatography showed that Gag polyproteins and matrix protein of murine leukaemia virus, HIV and SIV bind to KIF4 (Kim et al. 1998; Tang et al. 1999). Such an interaction could mediate a plus-end-directed transport of the RTC during entry, as well as the transport of RNA granules to the plasma membrane for assembly.

7
Adenovirus Entry and Microtubules

The cell entry of adenovirus is characterised by the stepwise disassembly of incoming virions (Greber et al. 1993; Fig. 1C). Adenoviruses are internalised by receptor-mediated endocytosis and induce the lysis of endosomal membranes to gain access to the cytosol (Shenk 2001). Ad2 and Ad5 virions of the subgroup C attach to MTs in vitro and in vivo (Luftig and Weihing 1975; Miles et al. 1980). If cells have been treated with vincristine to induce MT paracrystals, the virions display projections emanating from the vertices that are shorter and thicker than the fibre, and thus might represent bound host factors (Dales and Chardonnet 1973). The capsids use dynein for transport along MTs towards the MTOC because depolymerising the MT network, overexpressing dynamitin and microinjection of function-blocking dynein antibodies all inhibit Ad2 and Ad5 infection (Suomalainen et al. 1999; Leopold et al. 2000; Mabit et al. 2002).

As with herpesvirus, it is unclear whether and how viral capsids which have reached the MTOC are forwarded to the nuclear pores. Dynamic MTs are not required for nuclear targeting of incoming Ad particles. In fact, suppressing MT dynamics enhances the nuclear targeting efficiency of Ad2 (Giannakakou et al. 2002; Mabit et al. 2002). At the nuclear pores the viral DNA is injected into the nucleoplasm (Greber et al. 1993; Trotman et al. 2001), where viral replication, transcription and capsid assembly take place.

Analysing Ad2 particles that were visualised by the covalent attachment of fluorescent dyes, Suomalainen et al. (1999) described two types of motilities. Several transport events last for several seconds with an average velocity of 0.3–0.5 µm/s, depending on the cell type, and peak velocities of up to 2.6 µm/s which are directed exclusively towards the MTOC. Other particles rapidly alter direction towards or away from the nucleus with speeds of up to 0.5 µm/s. As with many other viruses, the incoming Ad2 particles induce a complex signalling cascade, which in this case activates protein kinase A and p38/MAPK pathways to boost minus-end-directed MT capsid transport (Suomalainen et al. 2001). In contrast to Ad2, Leopold et al. (1998, 2000) showed that labelled Ad5 particles rarely switch directions but also move along curvilinear tracks with an average speed of 2.2 µm/s. In the absence of a MT network, both groups detected only short-range random movements (Suomalainen et al. 1999; Leopold et al. 2000).

Glotzer et al. (2001) labelled Ad5 particles by co-packaging a GFP-labelled protein that bound to specific sequences engineered into the viral genome. Such particles moved both towards and away from the nucleus and, surprisingly, with comparable peak velocities of 1 µm/s and average velocities of 0.32 µm/s or 0.26 µm/s, irrespective of the presence or absence of MTs. Moreover, viral gene expression is not reduced if MTs have been depolymerised before infection (Glotzer et al. 2001). To explain these conflicting results on cytosolic adenovirus capsid transport, the parallel analysis of differently labelled particles with an identical experimental set-up and time resolution will be necessary (for a discussion, see Mabit et al. 2002).

The adenoviral protein Ad E3-14.7K, an inhibitor of TNFα-induced cell death, interacts with a small GTPase called FIP1, which in turn interacts with TCTEL1, a human homologue of the DLC Tctex-1 (Lukashok et al. 2000). Because Ad E3-14.7K is not a structural protein, it cannot have a role in cytosolic transport of Ad particles. However, many cellular proteins require MT-mediated transport to reach their final destination (Giannakakou et al. 2000), and the interaction of Ad E3-14.7K with a FIP1-TCTEL1 complex could aid in its cytosolic trafficking.

Altogether these studies suggest that adenovirus capsids possess a viral receptor either for dynein or for dynactin, which allows them to use MTs for efficient transport to the host nucleus after exit from the endosome. The disruption of the intermediate filament network facilitates viral egress (Chen et al. 1993), and progeny adenoviruses are released by cell lysis.

8
Parvovirus Entry and Microtubules

Parvoviruses enter cells by receptor-mediated endocytosis and seem to cross endosomal membranes at different stages of the endocytic pathway. Canine parvovirus (CPV) is detected in clathrin-coated vesicles, recycling endosomes, late endosomes and finally lysosomes, and CPV-containing vesicles are often associated with MTs. The transport from early to late endosomes as well as that of CPV-containing vesicles to the perinuclear area requires intact MTs and cytoplasmic dynein (Aniento et al. 1993; Suikkanen et al. 2002). Injected anti-capsid antibodies block CPV infection, suggesting that capsids enter the cytosol during infection. Moreover, nocodazole and a microinjected anti-dynein antibody block the nuclear accumulation of the injected capsids, suggesting that cytosolic CPV are transported by dynein along MTs to the nucleus (Suikkanen et al. 2003).

Adeno-associated virus (AAV) particles also move in a nocodazole- and cytochalasin B-sensitive manner to the nucleus. However, it remains to be determined whether virions are transported inside vesicles or freely in the cytosol (Sanlioglu et al. 2000). Real-time imaging of viral particles labelled with a single fluorescent molecule shows diffusion of free virions, diffusion of virus within endosomes and directed MT-dependent motion with velocities of 1.8–3.7 µm/s (Seisenberger et al. 2001).

9
Poxviruses—Multiple Cargos for Microtubules and Actin Tails

9.1
Cytosolic Vaccinia Virus Cores

The most complicated virus in terms of entry and assembly is vaccinia virus (VV), a poxvirus. One of its unique features is the formation of two infectious viral particles. Assembly commences in viral factories leading to the formation of the intracellular mature virus (IMV; Sodeik et al. 1993), a large enveloped brick-shaped particle. Relative to the IMV, the extracellular enveloped virus (EEV) has one additional membrane, the so-called envelope (Smith et al. 2002; Sodeik and Krijnse-Locker 2002).

Because both forms lead to infection, they must differ in how the DNA genome is uncoated and released into the cytosol where viral replication takes place. There are conflicting reports on whether both EEV and IMV, or only IMV, require actin filaments for entry and whether both or only IMV enter directly at the plasma membrane rather than by endocytosis (Payne and Norrby 1978; Doms et al. 1990; Vanderplasschen et al. 1998). In a recent study, the cells responded only to IMV but not EEV with the formation of long cell surface protrusions which were induced by a signalling cascade involving the actin-binding protein ezrin and the small GTPase Rac (Krijnse Locker et al. 2000). Accordingly, the IMV but not the EEV requires actin dynamics and filopodia for entry at the plasma membrane.

Although VV entry is still poorly understood, at the end of this process cores lacking any membrane are delivered into the cytosol. Early VV mRNAs are first transcribed inside these cores and then accumulate in large granular structures which recruit ribosomes for translation and are some distance away from the cores (Mallardo et al. 2001). Both the cores and the early mRNAs are associated with MTs, and nocodazole reduces mRNA synthesis (Mallardo et al. 2001). The in vitro binding of isolated cores to MTs requires the DNA- and RNA-binding protein L4R and the core protein A10L, which both co-localise with stabilised MTs in infected cells (Ploubidou et al. 2000; Mallardo et al. 2001).

Next, the incoming genome is uncoated, released from the core and amplified in specialised replication sites that are surrounded by membranes of the endoplasmic reticulum (Tolonen et al. 2001). After synthesis of the structural proteins from the late mRNAs, IMVs are formed within cytoplasmic viral factories from non-infectious precursors called crescents and immature virions (Fig. 2B).

9.2
Particle Transport During Virus Egress

Overexpression of dynamitin, or the deletion of the IMV protein A27L, inhibits the minus-end-directed MT transport of IMV particles from the viral factories to the MTOC (Ploubidou et al. 2000; Sanderson et al. 2000; Fig. 2B). At the MTOC, the IMV is wrapped in a membrane cisterna derived from the -TGN or an early endosome and intracellular enveloped virus (IEV) with two additional membranes relative to the IMV is formed (Smith et al. 2002). If cells are infected in the presence of nocodazole, no IEVs are made and there is a threefold reduction in the amount of IMVs synthesized (Ploubidou et al. 2000).

IEVs move to the plasma membrane in a saltatory and directional manner with an average speed of 1–2 μm/s along MTs and even switch from one MT to another (Geada et al. 2001; Hollinshead et al. 2001; Rietdorf et al. 2001; Ward and Moss 2001). IEVs co-localise with kinesin-1, and the overexpression of the cargo-binding domain of KLC inhibits their transport (Rietdorf et al. 2001). Besides kinesin, the integral membrane protein A36R and the peripheral membrane protein F12L are necessary for IEV transport to the plasma membrane (Rietdorf et al. 2001; van Eijl et al. 2002).

Shortly before or after the fusion of the outer membrane of IEV with the plasma membrane, the polymerisation of actin filaments is induced (Hollinshead et al. 2001). The cell-associated virus (CEV) that remains attached to the plasma membrane for quite some time is pushed on top of a long microvillus into the medium or a neighbouring cell at a speed of 2.8 μm/s (Cudmore et al. 1995). This process facilitates VV spread in a cell monolayer and in the tissue. The IEV protein A36R, which has an unusual long cytosolic tail and no obvious mammalian homologues, is the vaccinia actin tail nucleator. Tyr phosphorylation of A36R by a member of the Src-kinase family leads to the recruitment of the adaptor proteins Nck and Grb2, the WASP-interacting protein (WIP) as well as N-WASP (Wiskott-Aldrich syndrome protein). WASP activates the Arp2/3 complex and thus motility based on actin polymerisation (Frischknecht et al. 1999; Moreau et al. 2000; Scaplehorn et al. 2002). Another interesting rearrangement of the host actin cytoskeleton is the induction of cell migration that depends on early VV gene expression and the formation of cellular projections up to 160 μm in length that requires late VV gene expression (Sanderson et al. 1998).

Considering how many different types of large VV particles have to be shuffled around during entry and assembly, it is probably not surprising that there is considerable cross-talk between VV proteins and both cytosolic transport systems, MTs and microfilaments.

10
Actin Remodelling During Baculovirus Infection

Baculoviruses such as *Autographa californica M* nuclear polyhedrosis virus (AcMNPV) catalyse several distinct rearrangements of the actin cytoskeleton during different stages of the viral infection.

After virus uptake by endocytosis and acid-induced endosomal escape, the incoming cytosolic capsids induce prominent transient actin cables which often have a single capsid at one end and seem to be involved in capsid transport to the nucleus (Charlton and Volkman 1993; van Loo et al. 2001). Isolated AcMNPV capsids can nucleate actin polymerisation in vitro, and competition data with cytochalasin suggest that these capsids bind to the minus-end of the filaments (Lanier and Volkman 1998). Two AcMNPV proteins, p39 and p78/83, can bind to actin directly (Lanier and Volkman 1998). Moreover, p78/83 contains regions of homology to WASP which bind monomeric actin and the Arp2/3 complex which nucleates assembly of new actin filaments (Machesky et al. 2001). This suggests that p78/83 could function as a viral WASP homolog in inducing actin filament assembly.

Interestingly, in contrast to other viruses, depolymerising the MTs with nocodazole speeds up the onset of baculovirus gene expression, at least in transduction experiments using mammalian cells, whereas infection in the presence of cytochalasin reduces the infection efficiency (van

Loo et al. 2001). The myosin inhibitor BDM inhibits capsid transport to the nucleus, suggesting that a myosin could also be involved in capsid transport (Lanier and Volkman 1998).

The expression of the early viral protein Arif-1 (actin rearrangement-inducing factor 1) is responsible for second modification of the actin cytoskeleton, the appearance of actin aggregates localised at the plasma membrane (Roncarati and Knebel-Mörsdorf 1997). The amino acid sequence of Arif-1 predicts a N-terminal signal peptide and three transmembrane regions of about 20 amino acids each, and, moreover, Arif-1 seems to be phosphorylated at tyrosine residues (Dreschers et al. 2001). Arif-1 is not required for infection of cultured cells but might play a role during pathogenesis in insects.

The later phases of baculovirus infection are characterized by the appearance of F-actin inside the nucleus in a ring close to the nuclear membrane which is essential for nucleocapsid morphogenesis (Ohkawa and Volkman 1999). Six gene products of AcMNPV are required for the recruitment of G-actin to the nucleus (Ohkawa et al. 2002).

11
Perspectives

There is probably not a single virus for which the cytoskeleton does not play important roles during entry, replication, assembly or egress. With the emerging technologies in digital time-lapse microscopy, the further characterisation of viral transport will improve our understanding of the underlying host mechanisms.

Deciphering the molecular interactions between viruses and the cytoskeleton could provide important new targets for the development of antiviral therapy. Although the cytoskeletal filaments and associated motors play major roles in the cellular metabolism, viruses might interact with these structures via unique sites different from the cellular interaction partners. Other potential targets might be specific host proteins which viruses use to hook onto the cytoplasmic transport machinery.

In our quest to develop efficient vectors for therapeutic gene expression, it has become apparent that the transport of naked DNA to the nucleus, where transcription usually takes place, is a major barrier for the transfection of non-dividing cells, and hence for the application of human non-viral gene therapy vectors in vivo. However, several viruses have found effective solutions to this problem, namely successful negotiation with the host cytoskeleton to ensure efficient cytoplasmic transportation to the nuclear pores of the nuclear envelope. The future design of viral vectors will include these specific nuclear targeting mechanisms while deleting any factors responsible for cell toxicity and the induction of the immune response.

Acknowledgements We thank Rudi Bauerfeind (Cell Biology, MHH) and Elena Korenbaum (Biophysical Chemistry, MHH) for many helpful suggestions on the manuscript and Oliver Fackler (Virology, University of Heidelberg) for insightful discussions on HIV cell entry.

References

Allan, V, Näthke, IS (2001) Catch and pull a microtubule: getting a grasp on the cortex. Nat Cell Biol 3: E226–8

Alonso, C, Miskin, J, Hernaez, B, Fernandez-Zapatero, P, Soto, L, Canto, C, Rodriguez-Crespo, I, Dixon, L, Escribano, JM (2001) African swine fever virus protein p54 interacts with the microtubular motor complex through direct binding to light-chain dynein. J Virol 75:9819–27

Aniento, F, Emans, N, Griffiths, G, Gruenberg, J (1993) Cytoplasmic dynein-dependent vesicular transport from early to late endosomes. J Cell Biol 123:1373–87

Avitabile, E, Di Gaeta, S, Torrisi, MR, Ward, PL, Roizman, B, Campadelli-Fiume, G (1995) Redistribution of microtubules and Golgi apparatus in herpes simplex virus-infected cells and their role in viral exocytosis. J Virol 69:7472–82

Ayscough, K (1998) Use of latrunculin-A, an actin monomer-binding drug. Methods Enzymol 298:18–25

Bearer, EL, Breakefield, XO, Schuback, D, Reese, TS, LaVail, JH (2000) Retrograde axonal transport of herpes simplex virus: evidence for a single mechanism and a role for tegument. Proc Natl Acad Sci U S A 97:8146–50

Bowman, AB, Patel-King, RS, Benashski, SE, McCaffery, JM, Goldstein, LS, King, SM (1999) *Drosophila* roadblock and *Chlamydomonas* LC7: a conserved family of dynein-associated proteins involved in axonal transport, flagellar motility, and mitosis. J Cell Biol 146:165–80

Brown, SS (1999) Cooperation between microtubule- and actin-based motor proteins. Annu Rev Cell Dev Biol 15:63–80

Brunet, JP, Jourdan, N, Cotte-Laffitte, J, Linxe, C, Geniteau-Legendre, M, Servin, A, Quero, AM (2000) Rotavirus infection induces cytoskeleton disorganization in human intestinal epithelial cells: implication of an increase in intracellular calcium concentration. J Virol 74:10801–6

Bubb, MR, Senderowicz, AM, Sausville, EA, Duncan, KL, Korn, ED (1994) Jasplakinolide, a cytotoxic natural product, induces actin polymerization and competitively inhibits the binding of phalloidin to F-actin. J Biol Chem 269:14869–71

Bukrinskaya, A, Brichacek, B, Mann, A, Stevenson, M (1998) Establishment of a functional human immunodeficiency virus type 1 (HIV-1) reverse transcription complex involves the cytoskeleton. J Exp Med 188:2113–25

Burkhardt, JK, Echeverri, CJ, Nilsson, T, Vallee, RB (1997) Overexpression of the dynamitin (p50) subunit of the dynactin complex disrupts dynein-dependent maintenance of membrane organelle distribution. J Cell Biol 139:469-84

Ceccaldi, PE, Gillet, JP, Tsiang, H (1989) Inhibition of the transport of rabies virus in the central nervous system. J Neuropathol Exp Neurol 48:620-30

Charlton, CA, Volkman, LE (1993) Penetration of *Autographa californica* nuclear polyhedrosis virus nucleocapsids into IPLB Sf 21 cells induces actin cable formation. Virology 197:245-54

Chen, PH, Ornelles, DA, Shenk, T (1993) The adenovirus L3 23-kilodalton proteinase cleaves the amino-terminal head domain from cytokeratin 18 and disrupts the cytokeratin network of HeLa cells. J Virol 67:3507-14

Cooper, JA (1987) Effects of cytochalasin and phalloidin on actin. J Cell Biol 105:1473-8

Coulombe, PA, Bousquet, O, Ma, L, Yamada, S, Wirtz, D (2000) The 'ins' and 'outs' of intermediate filament organization. Trends Cell Biol 10:420-8

Coulombe, PA, Ma, L, Yamada, S, Wawersik, M (2001) Intermediate filaments at a glance. J Cell Sci 114:4345-7

Cramer, LP, Mitchison, TJ (1995) Myosin is involved in postmitotic cell spreading. J Cell Biol 131:179-89

Cudmore, S, Cossart, P, Griffiths, G, Way, M (1995) Actin-based motility of vaccinia virus. Nature 378:636-8

Cudmore, S, Reckmann, I, Way, M (1997) Viral manipulations of the actin cytoskeleton. Trends Microbiol 5:142-8

Dales, S, Chardonnet, Y (1973) Early events in the interaction of adenoviruses with HeLa cells. IV. Association with microtubules and the nuclear pore complex during vectorial movement of the inoculum. Virology 56:465-83

Deacon, SW, Serpinskaya, AS, Vaughan, PS, Fanarraga, ML, Vernos, I, Vaughan, KT, Gelfand, VI (2003) Dynactin is required for bidirectional organelle transport. J Cell Biol 160:297-301

Desai, A, Mitchison, TJ (1997) Microtubule polymerization dynamics. Annu Rev Cell Dev Biol 13:83-117

Desai, A, Verma, S, Mitchison, TJ, Walczak, CE (1999) Kin I kinesins are microtubule-destabilizing enzymes. Cell 96:69-78

Diefenbach, RJ, Miranda-Saksena, M, Diefenbach, E, Holland, DJ, Boadle, RA, Armati, PJ, Cunningham, AL (2002) Herpes simplex virus tegument protein US11 interacts with conventional kinesin heavy chain. J Virol 76:3282-3291

Digard, P, Elton, D, Bishop, K, Medcalf, E, Weeds, A, Pope, B (1999) Modulation of nuclear localization of the influenza virus nucleoprotein through interaction with actin filaments. J Virol 73:2222-31

Döhner, K, Wolfstein, A, Prank, U, Echeverri, C, Dujardin, D, Vallee, R, Sodeik, B (2002) Function of dynein and dynactin in herpes simplex virus capsid transport. Mol Biol Cell 13:2795-809

Doms, RW, Blumenthal, R, Moss, B (1990) Fusion of intra- and extracellular forms of vaccinia virus with the cell membrane. J Virol 64:4884-92

Downing, KH (2000) Structural basis for the interaction of tubulin with proteins and drugs that affect microtubule dynamics. Annu Rev Cell Dev Biol 16:89-111

Dreschers, S, Roncarati, R, Knebel-Mörsdorf, D (2001) Actin rearrangement-inducing factor of baculoviruses is tyrosine phosphorylated and colocalizes to F-actin at the plasma membrane. J Virol 75:3771–8

Echeverri, CJ, Paschal, BM, Vaughan, KT, Vallee, RB (1996) Molecular characterization of the 50-kD subunit of dynactin reveals function for the complex in chromosome alignment and spindle organization during mitosis. J Cell Biol 132:617–33

Eckley, DM, Gill, SR, Melkonian, KA, Bingham, JB, Goodson, HV, Heuser, JE, Schroer, TA (1999) Analysis of dynactin subcomplexes reveals a novel actin-related protein associated with the arp1 minifilament pointed end. J Cell Biol 147:307–20

Elliott, G, O'Hare, P (1997) Intercellular trafficking and protein delivery by a herpesvirus structural protein. Cell 88:223–33

Elliott, G, O'Hare, P (1998) Herpes simplex virus type 1 tegument protein VP22 induces the stabilization and hyperacetylation of microtubules. J Virol 72:6448–55

Elliott, G, O'Hare, P (1999) Live-cell analysis of a green fluorescent protein-tagged herpes simplex virus infection. J Virol 73:4110–9

Espindola, FS, Suter, DM, Partata, LB, Cao, T, Wolenski, JS, Cheney, RE, King, SM, Mooseker, MS (2000) The light chain composition of chicken brain myosin-Va: calmodulin, myosin-II essential light chains, and 8-kDa dynein light chain/PIN. Cell Motil Cytoskeleton 47:269–81

Fackler, OT, Luo, W, Geyer, M, Alberts, AS, Peterlin, BM (1999) Activation of Vav by Nef induces cytoskeletal rearrangements and downstream effector functions. Mol Cell 3:729–39

Fackler, OT, Peterlin, BM (2000) Endocytic entry of HIV-1. Curr Biol 10:1005–8

Ferreira, LR, Moussatche, N, Moura Neto, V (1994) Rearrangement of intermediate filament network of BHK-21 cells infected with vaccinia virus. Arch Virol 138:273–85

Forshey, BM, Aiken, C (2003) Disassembly of human immunodeficiency virus type 1 cores in vitro reveals association of Nef with the subviral ribonucleoprotein complex. J Virol 77:4409–14

Frischknecht, F, Moreau, V, Rottger, S, Gonfloni, S, Reckmann, I, Superti-Furga, G, Way, M (1999) Actin-based motility of vaccinia virus mimics receptor tyrosine kinase signalling. Nature 401:926–9

Frischknecht, F, Way, M (2001) Surfing pathogens and the lessons learned for actin polymerization. Trends Cell Biol 11:30–38

Fuchs, E, Yang, Y (1999) Crossroads on cytoskeletal highways. Cell 98:547–50

Galigniana, MD, Radanyi, C, Renoir, JM, Housley, PR, Pratt, WB (2001) Evidence that the peptidylprolyl isomerase domain of the hsp90-binding immunophilin FKBP52 is involved in both dynein interaction and glucocorticoid receptor movement to the nucleus. J Biol Chem 276:14884–9

Garcia-Mata, R, Bebok, Z, Sorscher, EJ, Sztul, ES (1999) Characterization and dynamics of aggresome formation by a cytosolic GFP-chimera. J Cell Biol 146:1239–54

Garoff, H, Hewson, R, Opstelten, DJ (1998) Virus maturation by budding. Microbiol Mol Biol Rev 62:1171–90

Geada, MM, Galindo, I, Lorenzo, MM, Perdiguero, B, Blasco, R (2001) Movements of vaccinia virus intracellular enveloped virions with GFP tagged to the F13L envelope protein. J Gen Virol 82:2747–60

Georgi, A, Mottola-Hartshorn, C, Warner, A, Fields, B, Chen, LB (1990) Detection of individual fluorescently labeled reovirions in living cells. Proc Natl Acad Sci U S A 87:6579–83

Giannakakou, P, Nakano, M, Nicolaou, KC, O'Brate, A, Yu, J, Blagosklonny, MV, Greber, UF, Fojo, T (2002) Enhanced microtubule-dependent trafficking and p53 nuclear accumulation by suppression of microtubule dynamics. Proc Natl Acad Sci U S A 99:10855–60

Giannakakou, P, Sackett, DL, Ward, Y, Webster, KR, Blagosklonny, MV, Fojo, T (2000) p53 is associated with cellular microtubules and is transported to the nucleus by dynein. Nat Cell Biol 2:709–17

Glotzer, JB, Michou, AI, Baker, A, Saltik, M, Cotten, M (2001) Microtubule-independent motility and nuclear targeting of adenoviruses with fluorescently labeled genomes. J Virol 75:2421–34

Goldstein, LS, Yang, Z (2000) Microtubule-based transport systems in neurons: the roles of kinesins and dyneins. Annu Rev Neurosci 23:39–71

Goode, BL, Drubin, DG, Barnes, G (2000) Functional cooperation between the microtubule and actin cytoskeletons. Curr Opin Cell Biol 12:63–71

Greber, UF, Willetts, M, Webster, P, Helenius, A (1993) Stepwise dismantling of adenovirus 2 during entry into cells. Cell 75:477–86

Greene, WC, Peterlin, BM (2002) Charting HIV's remarkable voyage through the cell: Basic science as a passport to future therapy. Nat Med 8:673–80

Gupta, S, De, BP, Drazba, JA, Banerjee, AK (1998) Involvement of actin microfilaments in the replication of human parainfluenza virus type 3. J Virol 72:2655–62

Gurland, G, Gundersen, GG (1995) Stable, detyrosinated microtubules function to localize vimentin intermediate filaments in fibroblasts. J Cell Biol 131:1275–90

Hanlon, DW, Yang, Z, Goldstein, LS (1997) Characterization of KIFC2, a neuronal kinesin superfamily member in mouse. Neuron 18:439–51

Heath, CM, Windsor, M, Wileman, T (2001) Aggresomes resemble sites specialized for virus assembly. J Cell Biol 153:449–55

Herrmann, C, Wray, J, Travers, F, Barman, T (1992) Effect of 2,3-butanedione monoxime on myosin and myofibrillar ATPases. An example of an uncompetitive inhibitor. Biochemistry 31:12227–32

Hirokawa, N, Takemura, R. (2003). Kinesin superfamily proteins. In Molecular Motors (M. Schliwa, ed.), pp. 79–109. Wiley-VCH Verlag, Weinheim

Hofmann, W, Reichart, B, Ewald, A, Muller, E, Schmitt, I, Stauber, RH, Lottspeich, F, Jockusch, BM, Scheer, U, Hauber, J, Dabauvalle, MC (2001) Cofactor requirements for nuclear export of Rev response element (RRE)- and constitutive transport element (CTE)-containing retroviral RNAs. An unexpected role for actin. J Cell Biol 152:895–910

Hogle, JM (2002) Poliovirus cell entry: common structural themes in viral cell entry pathways. Annu Rev Microbiol 56:677–702

Holland, DJ, Miranda-Saksena, M, Boadle, RA, Armati, P, Cunningham, AL (1999) Anterograde transport of herpes simplex virus proteins in axons of peripheral human fetal neurons: an immunoelectron microscopy study. J Virol 73:8503–11

Holleran, EA, Karki, S, Holzbaur, EL (1998) The role of the dynactin complex in intracellular motility. Int Rev Cytol 182:69–109

Hollinshead, M, Rodger, G, Van Eijl, H, Law, M, Hollinshead, R, Vaux, DJ, Smith, GL (2001) Vaccinia virus utilizes microtubules for movement to the cell surface. J Cell Biol 154:389–402

Howard, J, Hyman, AA (2003) Dynamics and mechanics of the microtubule plus end. Nature 422:753–8

Hunter, AW, Caplow, M, Coy, DL, Hancock, WO, Diez, S, Wordeman, L, Howard, J (2003) The kinesin-related protein MCAK is a microtubule depolymerase that forms an ATP-hydrolyzing complex at microtubule ends. Mol Cell 11:445–57

Igakura, T, Stinchcombe, JC, Goon, PK, Taylor, GP, Weber, JN, Griffiths, GM, Tanaka, Y, Osame, M, Bangham, CR (2003) Spread of HTLV-I between lymphocytes by virus-induced polarization of the cytoskeleton. Science 299:1713–6

Iyengar, S, Hildreth, JE, Schwartz, DH (1998) Actin-dependent receptor colocalization required for human immunodeficiency virus entry into host cells. J Virol 72:5251–5

Jacob, Y, Badrane, H, Ceccaldi, PE, Tordo, N (2000) Cytoplasmic dynein LC8 interacts with lyssavirus phosphoprotein. J Virol 74:10217–22

Jaffrey, SR, Snyder, SH (1996) PIN: an associated protein inhibitor of neuronal nitric oxide synthase. Science 274:774–7

Jimbo, T, Kawasaki, Y, Koyama, R, Sato, R, Takada, S, Haraguchi, K, Akiyama, T (2002) Identification of a link between the tumour suppressor APC and the kinesin superfamily. Nat Cell Biol 4:323–7

Jordan, MA, Wilson, L (1999) The use and action of drugs in analyzing mitosis. Methods Cell Biol 61:267–95

Karcher, RL, Deacon, SW, Gelfand, VI (2002) Motor-cargo interactions: the key to transport specificity. Trends Cell Biol 12:21–7

Karki, S, Holzbaur, EL (1999) Cytoplasmic dynein and dynactin in cell division and intracellular transport. Curr Opin Cell Biol 11:45–53

Kasman, LM, Volkman, LE (2000) Filamentous actin is required for lepidopteran nucleopolyhedrovirus progeny production. J Gen Virol 81:1881–8

Keating, TJ, Borisy, GG (1999) Centrosomal and non-centrosomal microtubules. Biol Cell 91:321–9

Kieke, MC, Titus, MA. (2003). The myosin superfamily: An overview. *In* Molecular Motors (M. Schliwa, ed.), pp. 3–44. Wiley-VCH Verlag, Weinheim

Kim, W, Tang, Y, Okada, Y, Torrey, TA, Chattopadhyay, SK, Pfleiderer, M, Falkner, FG, Dorner, F, Choi, W, Hirokawa, N, Morse, HC, 3rd (1998) Binding of murine leukemia virus Gag polyproteins to KIF4, a microtubule-based motor protein. J Virol 72:6898–901

Kimura, T, Hashimoto, I, Yamamoto, A, Nishikawa, M, Fujisawa, JI (2000) Rev-dependent association of the intron-containing HIV-1 gag mRNA with the nuclear actin bundles and the inhibition of its nucleocytoplasmic transport by latrunculin-B. Genes Cells 5:289–307

King, SM (2000) AAA domains and organization of the dynein motor unit. J Cell Sci 113:2521–6

King, SM. (2003). Dynein motors: Structure, mechanochemistry and regulation. *In* Molecular Motors (M. Schliwa, ed.), pp. 45–78. Wiley-VCH Verlag, Weinheim

Klupp, BG, Granzow, H, Mettenleiter, TC (2000) Primary envelopment of pseudorabies virus at the nuclear membrane requires the UL34 gene product. J Virol 74:10063-73

Kopito, RR (2000) Aggresomes, inclusion bodies and protein aggregation. Trends Cell Biol 10:524-30

Kotsakis, A, Pomeranz, LE, Blouin, A, Blaho, JA (2001) Microtubule reorganization during herpes simplex virus type 1 infection facilitates the nuclear localization of VP22, a major virion tegument protein. J Virol 75:8697-711

Krijnse Locker, J, Kuehn, A, Schleich, S, Rutter, G, Hohenberg, H, Wepf, R, Griffiths, G (2000) Entry of the two infectious forms of vaccinia virus at the plasma membrane is signaling-dependent for the IMV but not the EEV. Mol Biol Cell 11:2497-511

Kristensson, K, Lycke, E, Roytta, M, Svennerholm, B, Vahlne, A (1986) Neuritic transport of herpes simplex virus in rat sensory neurons in vitro. Effects of substances interacting with microtubular function and axonal flow [nocodazole, taxol and erythro-9-3-(2-hydroxynonyl)adenine]. J Gen Virol 67:2023-8

Lanier, LM, Volkman, LE (1998) Actin binding and nucleation by *Autographa californica M* nucleopolyhedrovirus. Virology 243:167-77

Leopold, PL, Ferris, B, Grinberg, I, Worgall, S, Hackett, NR, Crystal, RG (1998) Fluorescent virions: dynamic tracking of the pathway of adenoviral gene transfer vectors in living cells. Hum Gene Ther 9:367-78

Leopold, PL, Kreitzer, G, Miyazawa, N, Rempel, S, Pfister, KK, Rodriguez-Boulan, E, Crystal, RG (2000) Dynein- and microtubule-mediated translocation of adenovirus serotype 5 occurs after endosomal lysis. Hum Gene Ther 11:151-65

Leung, CL, Green, KJ, Liem, RK (2002) Plakins: a family of versatile cytolinker proteins. Trends Cell Biol 12:37-45

Lippincott-Schwartz, J, Roberts, TH, Hirschberg, K (2000) Secretory protein trafficking and organelle dynamics in living cells. Annu Rev Cell Dev Biol 16:557-89

Liu, B, Dai, R, Tian, CJ, Dawson, L, Gorelick, R, Yu, XF (1999) Interaction of the human immunodeficiency virus type 1 nucleocapsid with actin. J Virol 73:2901-8

Lo, KW, Naisbitt, S, Fan, JS, Sheng, M, Zhang, M (2001) The 8-kDa dynein light chain binds to its targets via a conserved (K/R)XTQT motif. J Biol Chem 276:14059-66

Luby-Phelps, K (2000) Cytoarchitecture and physical properties of cytoplasm: volume, viscosity, diffusion, intracellular surface area. Int Rev Cytol 192:189-221

Luftig, RB, Lupo, LD (1994) Viral interactions with the host-cell cytoskeleton: the role of retroviral proteases. Trends Microbiol 2:178-82

Luftig, RB, Weihing, RR (1975) Adenovirus binds to rat brain microtubules in vitro. J Virol 16:696-706

Lukashok, SA, Tarassishin, L, Li, Y, Horwitz, MS (2000) An adenovirus inhibitor of tumor necrosis factor alpha-induced apoptosis complexes with dynein and a small GTPase. J Virol 74:4705-9

Lycke, E, Hamark, B, Johansson, M, Krotochwil, A, Lycke, J, Svennerholm, B (1988) Herpes simplex virus infection of the human sensory neuron. An electron microscopy study. Arch Virol 101:87-104

Mabit, H, Nakano, MY, Prank, U, Saam, B, Döhner, K, Sodeik, B, Greber, UF (2002) Intact microtubules support adenovirus and herpes simplex virus infections. J Virol 76:9962–71

Machesky, LM, Insall, RH, Volkman, LE (2001) WASP homology sequences in baculoviruses. Trends Cell Biol 11:286–7

Mallardo, M, Schleich, S, Krijnse Locker, J (2001) Microtubule-dependent organization of vaccinia virus core-derived early mRNAs into distinct cytoplasmic structures. Mol Biol Cell 12:3875–91

Marsh, M, Bron, R (1997) SFV infection in CHO cells: cell-type specific restrictions to productive virus entry at the cell surface. J Cell Sci 110:95–103

Martin, A, O'Hare, P, McLauchlan, J, Elliott, G (2002) Herpes simplex virus tegument protein VP22 contains overlapping domains for cytoplasmic localization, microtubule interaction, and chromatin binding. J Virol 76:4961–70

Martinez-Moreno, M, Navarro-Lerida, I, Roncal, F, Albar, JP, Alonso, C, Gavilanes, F, Rodriguez-Crespo, I (2003) Recognition of novel viral sequences that associate with the dynein light chain LC8 identified through a pepscan technique. FEBS Lett 544:262–267

McDonald, D, Vodicka, MA, Lucero, G, Svitkina, TM, Borisy, GG, Emerman, M, Hope, TJ (2002) Visualization of the intracellular behavior of HIV in living cells. J Cell Biol 159:441–52

Mebatsion, T (2001) Extensive attenuation of rabies virus by simultaneously modifying the dynein light chain binding site in the P protein and replacing Arg333 in the G protein. J Virol 75:11496–502

Merdes, A, Cleveland, DW (1997) Pathways of spindle pole formation: different mechanisms; conserved components. J Cell Biol 138:953–6

Mery, PF, Pavoine, C, Pecker, F, Fischmeister, R (1995) Erythro-9-(2-hydroxy-3-nonyl)adenine inhibits cyclic GMP-stimulated phosphodiesterase in isolated cardiac myocytes. Mol Pharmacol 48:121–30

Mettenleiter, TC (2002) Herpesvirus assembly and egress. J Virol 76:1537–47

Miles, BD, Luftig, RB, Weatherbee, JA, Weihing, RR, Weber, J (1980) Quantitation of the interaction between adenovirus types 2 and 5 and microtubules inside infected cells. Virology 105:265–9

Miller, N, Hutt-Fletcher, LM (1992) Epstein-Barr virus enters B cells and epithelial cells by different routes. J Virol 66:3409–14

Miranda-Saksena, M, Armati, P, Boadle, RA, Holland, DJ, Cunningham, AL (2000) Anterograde transport of herpes simplex virus type 1 in cultured, dissociated human and rat dorsal root ganglion neurons. J Virol 74:1827–39

Mitchison, T, Kirschner, M (1984) Dynamic instability of microtubule growth. Nature 312:237–42

Mojon, D, Zhang, W, Oetliker, H (1993) Inhibition by 2,3-butanedione-monoxime of mitochondrial ADP-dependent respiration and muscle contraction. Biochem Mol Biol Int 31:501–7

Moreau, V, Frischknecht, F, Reckmann, I, Vincentelli, R, Rabut, G, Stewart, D, Way, M (2000) A complex of N-WASP and WIP integrates signalling cascades that lead to actin polymerization. Nat Cell Biol 2:441–8

Mouland, AJ, Xu, H, Cui, H, Krueger, W, Munro, TP, Prasol, M, Mercier, J, Rekosh, D, Smith, R, Barbarese, E, Cohen, EA, Carson, JH (2001) RNA trafficking signals in human immunodeficiency virus type 1. Mol Cell Biol 21:2133–43

Mueller, S, Cao, X, Welker, R, Wimmer, E (2002) Interaction of the poliovirus receptor CD155 with the dynein light chain Tctex-1 and its implication for poliovirus pathogenesis. J Biol Chem 277:7897–904

Nakanishi, S, Kakita, S, Takahashi, I, Kawahara, K, Tsukuda, E, Sano, T, Yamada, K, Yoshida, M, Kase, H, Matsuda, Y, et al. (1992) Wortmannin, a microbial product inhibitor of myosin light chain kinase. J Biol Chem 267:2157–63

Nakanishi, S, Yamada, K, Iwahashi, K, Kuroda, K, Kase, H (1990) KT5926, a potent and selective inhibitor of myosin light chain kinase. Mol Pharmacol 37:482–8

Nangaku, M, Sato-Yoshitake, R, Okada, Y, Noda, Y, Takemura, R, Yamazaki, H, Hirokawa, N (1994) KIF1B, a novel microtubule plus end-directed monomeric motor protein for transport of mitochondria. Cell 79:1209–20

Nicola, AV, McEvoy, AM, Straus, SE (2003) Roles for endocytosis and low pH in herpes simplex virus entry into HeLa and Chinese hamster ovary cells. J Virol 77:5324–32

Noda, Y, Okada, Y, Saito, N, Setou, M, Xu, Y, Zhang, Z, Hirokawa, N (2001) KIFC3, a microtubule minus end-directed motor for the apical transport of annexin XII-Ib-associated Triton-insoluble membranes. J Cell Biol 155:77–88

Nurminsky, DI, Nurminskaya, MV, Benevolenskaya, EV, Shevelyov, YY, Hartl, DL, Gvozdev, VA (1998) Cytoplasmic dynein intermediate-chain isoforms with different targeting properties created by tissue-specific alternative splicing. Mol Cell Biol 18:6816–25

Ohara, PT, Chin, MS, LaVail, JH (2000) The spread of herpes simplex virus type 1 from trigeminal neurons to the murine cornea: an immunoelectron microscopy study. J Virol 74:4776–86

Ohkawa, T, Rowe, AR, Volkman, LE (2002) Identification of six *Autographa californica* multicapsid nucleopolyhedrovirus early genes that mediate nuclear localization of G-actin. J Virol 76:12281–9

Ohkawa, T, Volkman, LE (1999) Nuclear F-actin is required for AcMNPV nucleocapsid morphogenesis. Virology 264:1-4

Ojala, PM, Sodeik, B, Ebersold, MW, Kutay, U, Helenius, A (2000) Herpes simplex virus type 1 entry into host cells: reconstitution of capsid binding and uncoating at the nuclear pore complex in vitro. Mol Cell Biol 20:4922–31

Okada, Y, Yamazaki, H, Sekine-Aizawa, Y, Hirokawa, N (1995) The neuron-specific kinesin superfamily protein KIF1A is a unique monomeric motor for anterograde axonal transport of synaptic vesicle precursors. Cell 81:769–80

Ott, DE, Coren, LV, Kane, BP, Busch, LK, Johnson, DG, Sowder, RC, 2nd, Chertova, EN, Arthur, LO, Henderson, LE (1996) Cytoskeletal proteins inside human immunodeficiency virus type 1 virions. J Virol 70:7734–43

Payne, LG, Norrby, E (1978) Adsorption and penetration of enveloped and naked vaccinia virus particles. J Virol 27:19–27

Pederson, T, Aebi, U (2002) Actin in the nucleus: what form and what for? J Struct Biol 140:3–9

Penfold, ME, Armati, P, Cunningham, AL (1994) Axonal transport of herpes simplex virions to epidermal cells: evidence for a specialized mode of virus transport and assembly. Proc Natl Acad Sci U S A 91:6529–33

Penningroth, SM (1986) Erythro-9-[3-(2-hydroxynonyl)]adenine and vanadate as probes for microtubule-based cytoskeletal mechanochemistry. Methods Enzymol 134:477–87

Peretti, D, Peris, L, Rosso, S, Quiroga, S, Caceres, A (2000) Evidence for the involvement of KIF4 in the anterograde transport of L1-containing vesicles. J Cell Biol 149:141–52

Phillips, RM, Altschuld, RA (1996) 2,3-Butanedione 2-monoxime (BDM) induces calcium release from canine cardiac sarcoplasmic reticulum. Biochem Biophys Res Commun 229:154–7

Ploubidou, A, Moreau, V, Ashman, K, Reckmann, I, Gonzalez, C, Way, M (2000) Vaccinia virus infection disrupts microtubule organization and centrosome function. EMBO J 19:3932–44

Ploubidou, A, Way, M (2001) Viral transport and the cytoskeleton. Curr Opin Cell Biol 13:97–105

Poisson, N, Real, E, Gaudin, Y, Vaney, MC, King, S, Jacob, Y, Tordo, N, Blondel, D (2001) Molecular basis for the interaction between rabies virus phosphoprotein P and the dynein light chain LC8: dissociation of dynein-binding properties and transcriptional functionality of P. J Gen Virol 82:2691–6

Rando, OJ, Zhao, K, Crabtree, GR (2000) Searching for a function for nuclear actin. Trends Cell Biol 10:92–7

Raux, H, Flamand, A, Blondel, D (2000) Interaction of the rabies virus P protein with the LC8 dynein light chain. J Virol 74:10212–6

Reynolds, AE, Wills, EG, Roller, RJ, Ryckman, BJ, Baines, JD (2002) Ultrastructural localization of the herpes simplex virus type 1 UL31, UL34, and US3 proteins suggests specific roles in primary envelopment and egress of nucleocapsids. J Virol 76:8939–52

Richard de Soultrait, V, Caumont, A, Durrens, P, Calmels, C, Parissi, V, Recordon, P, Bon, E, Desjobert, C, Tarrago-Litvak, L, Fournier, M (2002) HIV-1 integrase interacts with yeast microtubule-associated proteins. Biochim Biophys Acta 1575:40–8

Rietdorf, J, Ploubidou, A, Reckmann, I, Holmstrom, A, Frischknecht, F, Zettl, M, Zimmermann, T, Way, M (2001) Kinesin-dependent movement on microtubules precedes actin-based motility of vaccinia virus. Nat Cell Biol 3:992–1000

Rodriguez-Crespo, I, Yelamos, B, Roncal, F, Albar, JP, Ortiz de Montellano, PR, Gavilanes, F (2001) Identification of novel cellular proteins that bind to the LC8 dynein light chain using a pepscan technique. FEBS Lett 503:135–41

Roizman, B, Knipe, DM. (2001). Herpes simplex viruses and their replication. *In* Fundamental Virology (D. M. Knipe and P. M. Howley, eds.), pp. 1123–1183. Lippincott Williams & Wilkins, Philadelphia

Roncarati, R, Knebel-Mörsdorf, D (1997) Identification of the early actin-rearrangement-inducing factor gene, arif-1, from *Autographa californica* multicapsid nuclear polyhedrosis virus. J Virol 71:7933–41

Ruchhoeft, ML, Harris, WA (1997) Myosin functions in *Xenopus* retinal ganglion cell growth cone motility in vivo. J Neurobiol 32:567–78

Saib, A, Puvion-Dutilleul, F, Schmid, M, Peries, J, de The, H (1997) Nuclear targeting of incoming human foamy virus Gag proteins involves a centriolar step. J Virol 71:1155–61

Saito, N, Okada, Y, Noda, Y, Kinoshita, Y, Kondo, S, Hirokawa, N (1997) KIFC2 is a novel neuron-specific C-terminal type kinesin superfamily motor for dendritic transport of multivesicular body-like organelles. Neuron 18:425–38

Sakowicz, R, Berdelis, MS, Ray, K, Blackburn, CL, Hopmann, C, Faulkner, DJ, Goldstein, LS (1998) A marine natural product inhibitor of kinesin motors. Science 280:292–5

Sanderson, CM, Hollinshead, M, Smith, GL (2000) The vaccinia virus A27L protein is needed for the microtubule-dependent transport of intracellular mature virus particles. J Gen Virol 81 Pt 1:47–58

Sanderson, CM, Way, M, Smith, GL (1998) Virus-induced cell motility. J Virol 72:1235–43

Sanlioglu, S, Benson, PK, Yang, J, Atkinson, EM, Reynolds, T, Engelhardt, JF (2000) Endocytosis and nuclear trafficking of adeno-associated virus type 2 are controlled by rac1 and phosphatidylinositol-3 kinase activation. J Virol 74:9184–96

Sasaki, H, Nakamura, M, Ohno, T, Matsuda, Y, Yuda, Y, Nonomura, Y (1995) Myosin-actin interaction plays an important role in human immunodeficiency virus type 1 release from host cells. Proc Natl Acad Sci U S A 92:2026–30

Scaplehorn, N, Holmstrom, A, Moreau, V, Frischknecht, F, Reckmann, I, Way, M (2002) Grb2 and Nck act cooperatively to promote actin-based motility of vaccinia virus. Curr Biol 12:740–5

Schaeffer, E, Geleziunas, R, Greene, WC (2001) Human immunodeficiency virus type 1 Nef functions at the level of virus entry by enhancing cytoplasmic delivery of virions. J Virol 75:2993–3000

Schlichter, LC, Pahapill, PA, Chung, I (1992) Dual action of 2,3-butanedione monoxime (BDM) on K^+ current in human T lymphocytes. J Pharmacol Exp Ther 261:438–46

Schliwa, M, Woehlke, G (2003) Molecular motors. Nature 422:759–65

Schnapp, BJ (2003) Trafficking of signaling modules by kinesin motors. J Cell Sci 116:2125–35

Seisenberger, G, Ried, MU, Endress, T, Buning, H, Hallek, M, Brauchle, C (2001) Real-time single-molecule imaging of the infection pathway of an adeno-associated virus. Science 294:1929–32

Shah, JV, Flanagan, LA, Janmey, PA, Leterrier, JF (2000) Bidirectional translocation of neurofilaments along microtubules mediated in part by dynein/dynactin. Mol Biol Cell 11:3495–508

Sharp, DJ, Rogers, GC, Scholey, JM (2000) Cytoplasmic dynein is required for poleward chromosome movement during mitosis in *Drosophila* embryos. Nat Cell Biol 2:922–30

Shenk, T. (2001). Adenoviridae: The viruses and their replication. *In* Fundamental Virology (D. M. Knipe and P. M. Howley, eds.), pp. 1053–1088. Lippincott Williams & Wilkins, Philadelphia

Skepper, JN, Whiteley, A, Browne, H, Minson, A (2001) Herpes simplex virus nucleocapsids mature to progeny virions by an envelopment→deenvelopment→reen-\toreenvelopmentpathway.JVirol75 : 5697 − −702

Smith, GA, Enquist, LW (2002) Break ins and break outs: Viral interactions with the cytoskeleton of mammalian cells. Annu Rev Cell Dev Biol 18:135–61

Smith, GA, Gross, SP, Enquist, LW (2001) Herpesviruses use bidirectional fast-axonal transport to spread in sensory neurons. Proc Natl Acad Sci U S A 98:3466–70

Smith, GL, Vanderplasschen, A, Law, M (2002) The formation and function of extracellular enveloped vaccinia virus. J Gen Virol 83:2915–31

Sodeik, B (2000) Mechanisms of viral transport in the cytoplasm. Trends Microbiol 8:465–72

Sodeik, B (2002) Unchain my heart, baby let me go—the entry and intracellular transport of HIV. J Cell Biol 159:393–5

Sodeik, B, Doms, RW, Ericsson, M, Hiller, G, Machamer, CE, van 't Hof, W, van Meer, G, Moss, B, Griffiths, G (1993) Assembly of vaccinia virus: role of the intermediate compartment between the endoplasmic reticulum and the Golgi stacks. J Cell Biol 121:521–41

Sodeik, B, Ebersold, MW, Helenius, A (1997) Microtubule-mediated transport of incoming herpes simplex virus 1 capsids to the nucleus. J Cell Biol 136:1007–21

Sodeik, B, Krijnse-Locker, J (2002) Assembly of vaccinia virus revisited: de novo membrane synthesis or acquisition from the host? Trends Microbiol 10:15–24

Suikkanen, S, Aaltonen, T, Nevalainen, M, Välilehto, O, Lindholm, L, Vuento, M, Vihinen-Ranta, M (2003) Exploitation of microtubule cytoskeleton and dynein during parvoviral traffic towards the nucleus. J Virol 77:10270–9

Suikkanen, S, Saajarvi, K, Hirsimaki, J, Valilehto, O, Reunanen, H, Vihinen-Ranta, M, Vuento, M (2002) Role of recycling endosomes and lysosomes in dynein-dependent entry of canine parvovirus. J Virol 76:4401–11

Suomalainen, M, Nakano, MY, Boucke, K, Keller, S, Greber, UF (2001) Adenovirus-activated PKA and p38/MAPK pathways boost microtubule-mediated nuclear targeting of virus. EMBO J 20:1310–9

Suomalainen, M, Nakano, MY, Keller, S, Boucke, K, Stidwill, RP, Greber, UF (1999) Microtubule-dependent plus- and minus end-directed motilities are competing processes for nuclear targeting of adenovirus. J Cell Biol 144:657–72

Susalka, SJ, Nikulina, K, Salata, MW, Vaughan, PS, King, SM, Vaughan, KT, Pfister, KK (2002) The roadblock light chain binds a novel region of the cytoplasmic dynein intermediate chain. J Biol Chem 277:32939–46

Tai, AW, Chuang, JZ, Sung, CH (2001) Cytoplasmic dynein regulation by subunit heterogeneity and its role in apical transport. J Cell Biol 153:1499–510

Takeda, S, Yamazaki, H, Seog, DH, Kanai, Y, Terada, S, Hirokawa, N (2000) Kinesin superfamily protein 3 (KIF3) motor transports fodrin-associating vesicles important for neurite building. J Cell Biol 148:1255–65

Tang, Q, Staub, CM, Gao, G, Jin, Q, Wang, Z, Ding, W, Aurigemma, RE, Mulder, KM (2002) A novel transforming growth factor-beta receptor-interacting protein that is also a light chain of the motor protein dynein. Mol Biol Cell 13:4484–96

Tang, Y, Winkler, U, Freed, EO, Torrey, TA, Kim, W, Li, H, Goff, SP, Morse, HC, 3rd (1999) Cellular motor protein KIF-4 associates with retroviral Gag. J Virol 73:10508–13

Tekotte, H, Davis, I (2002) Intracellular mRNA localization: motors move messages. Trends Genet 18:636–42

Tolonen, N, Doglio, L, Schleich, S, Krijnse Locker, J (2001) Vaccinia virus DNA replication occurs in endoplasmic reticulum-enclosed cytoplasmic mini-nuclei. Mol Biol Cell 12:2031–46

Tomishima, MJ, Enquist, LW (2001) A conserved α-herpesvirus protein necessary for axonal localization of viral membrane proteins. J Cell Biol 154:741–752

Topp, KS, Bisla, K, Saks, ND, Lavail, JH (1996) Centripetal transport of herpes simplex virus in human retinal pigment epithelial cells in vitro. Neuroscience 71:1133–44

Topp, KS, Meade, LB, LaVail, JH (1994) Microtubule polarity in the peripheral processes of trigeminal ganglion cells: relevance for the retrograde transport of herpes simplex virus. J Neurosci 14:318–25

Trotman, LC, Mosberger, N, Fornerod, M, Stidwill, RP, Greber, UF (2001) Import of adenovirus DNA involves the nuclear pore complex receptor CAN/Nup214 and histone H1. Nat Cell Biol 3:1092–100

Tsiang, H, Ceccaldi, PE, Lycke, E (1991) Rabies virus infection and transport in human sensory dorsal root ganglia neurons. J Gen Virol 72:1191–4

Valetti, C, Wetzel, DM, Schrader, M, Hasbani, MJ, Gill, SR, Kreis, TE, Schroer, TA (1999) Role of dynactin in endocytic traffic: effects of dynamitin overexpression and colocalization with CLIP-170. Mol Biol Cell 10:4107–20

van Eijl, H, Hollinshead, M, Rodger, G, Zhang, WH, Smith, GL (2002) The vaccinia virus F12L protein is associated with intracellular enveloped virus particles and is required for their egress to the cell surface. J Gen Virol 83:195–207

van Leeuwen, H, Elliott, G, O'Hare, P (2002) Evidence of a role for nonmuscle myosin II in herpes simplex virus type 1 egress. J Virol 76:3471–81

van Loo, ND, Fortunati, E, Ehlert, E, Rabelink, M, Grosveld, F, Scholte, BJ (2001) Baculovirus infection of nondividing mammalian cells: mechanisms of entry and nuclear transport of capsids. J Virol 75:961–70

Vanderplasschen, A, Hollinshead, M, Smith, GL (1998) Intracellular and extracellular vaccinia virions enter cells by different mechanisms. J Gen Virol 79:877–87

Verhey, KJ. (2003). Motors and membrane trafficking. In Molecular Motors (M. Schliwa, ed.), pp. 377–409. Wiley-VCH Verlag, Weinheim

Ward, BM, Moss, B (2001) Vaccinia virus intracellular movement is associated with microtubules and independent of actin tails. J Virol 75:11651–63

Welch, MD, Mullins, RD (2002) Cellular control of actin nucleation. Annu Rev Cell Dev Biol 18:247–88

Wilk, T, Gowen, B, Fuller, SD (1999) Actin associates with the nucleocapsid domain of the human immunodeficiency virus Gag polyprotein. J Virol 73:1931–40

Willard, M (2002) Rapid directional translocations in virus replication. J Virol 76:5220–5232

Xu, Y, Takeda, S, Nakata, T, Noda, Y, Tanaka, Y, Hirokawa, N (2002) Role of KIFC3 motor protein in Golgi positioning and integration. J Cell Biol 158:293–303

Ye, GJ, Vaughan, KT, Vallee, RB, Roizman, B (2000) The herpes simplex virus 1 U(L)34 protein interacts with a cytoplasmic dynein intermediate chain and targets nuclear membrane. J Virol 74:1355–63

Young, A, Dictenberg, JB, Purohit, A, Tuft, R, Doxsey, SJ (2000) Cytoplasmic dynein-mediated assembly of pericentrin and gamma tubulin onto centrosomes. Mol Biol Cell 11:2047–56

Zhang, J, Campbell, RE, Ting, AY, Tsien, RY (2002) Creating new fluorescent probes for cell biology. Nat Rev Mol Cell Biol 3:906-18

Zhou, ZH, Chen, DH, Jakana, J, Rixon, FJ, Chiu, W (1999) Visualization of tegument-capsid interactions and DNA in intact herpes simplex virus type 1 virions. J Virol 73:3210-8

Ziegelbauer, J, Shan, B, Yager, D, Larabell, C, Hoffmann, B, Tjian, R (2001) Transcription factor MIZ-1 is regulated via microtubule association. Mol Cell 8:339-49

Zigmond, SH (1993) Recent quantitative studies of actin filament turnover during cell locomotion. Cell Motil Cytoskeleton 25:309-16

Nuclear Import in Viral Infections

U. F. Greber[1] (✉) · M. Fornerod[2]

[1] Zoologisches Institut der Universität Zürich, Winterthurerstrasse 190,
8057 Zürich, Switzerland
ufgreber@zool.unizh.ch
[2] The Netherlands Cancer Institute H4, Plesmanlaan 121, 1066 CX, Amsterdam,
The Netherlands

1	Introduction	110
2	Nuclear Pore Complexes, Guardians of the Nucleus	111
3	Soluble Import Receptors Control Nuclear Access	113
4	Models of Nuclear Pore Complex Translocation	114
5	Ran-Independent Transport	116
6	Nuclear Import of Viral Genomes	117
6.1	Parvoviruses	118
6.2	Hepadnaviruses, DNA Retroviruses	119
6.3	Retrotransposons	120
6.4	Lentiviruses	121
6.5	Adenoviruses	123
6.6	Herpesviruses	124
6.7	Papovaviruses	125
6.8	Negative-Sense RNA Viruses	127
7	Perspectives	128
	References	129

Abstract The separation of transcription in the nucleus and translation in the cytoplasm requires nucleo-cytoplasmic exchange of proteins and RNAs. Viruses have evolved strategies to capitalize on the nucleo-cytoplasmic trafficking machinery of the cell. Here, we first discuss the principal mechanisms of receptor-mediated nuclear import of proteinaceous cargo through the nuclear pore complex, the gate keeper of the cell nucleus. We then focus on viral strategies leading to nuclear import of genomes and subgenomic particles. Nucleo-cytoplasmic transport is directly important for those viruses that are replicating in the nucleus, such as DNA tumor viruses and RNA viruses, including parvoviruses, the DNA retroviruses hepadnaviruses, RNA-retrotransposons and retroviruses, adenoviruses, herpesviruses, papovaviruses, and particular negative-sense RNA viruses, such as the orthomyxovirus influenza virus. The viral strategies of nuclear import turn out to be surprisingly diverse. Their investigation continues to give insight into how nucleic acids pass in and out of the nucleus.

1
Introduction

Viruses interfere with all aspects of cell physiology. This can lead to lytic infections releasing large amounts of progeny particles from the infected cells, persistent infections producing moderate amounts of progeny, or latent infections without apparent particle production. All of these possible outcomes can ensure viral survival and spread. How viruses interact with their hosts has largely been selected for by the pressure of the immune system and the underlying cell biological machineries. Normally, host cells are protected from infections by multiple barriers during entry, replication, and viral release. Major hurdles for the incoming viruses are the extracellular matrix and the accessibility of receptors, the plasma membrane, the cytoplasm, and the nuclear membrane. An effective strategy to overcome the extracellular matrix is to tightly bind to a receptor in the plasma membrane (Wimmer 1994). Interestingly, some viruses have chosen to use a receptor that is cryptic, that is, not readily accessible from the apical lumen of polarized cells where incoming respiratory or enteric viruses initiate their infections (Spear 2002; Stehle and Dermody 2003). Examples of viral receptors localized in cell junctions include the coxsackievirus B adenovirus (Ad) receptor (CAR) serving as a high-affinity binding site for most Ad serotypes, the nectin-1 receptor of alpha herpesviruses, and the junctional adhesion molecule (JAM) of reoviruses. For those viruses the extracellular matrix is a particular challenge. One way to overcome it is to use the opportunity of entry sites where the integrity of the epithelial barrier is disrupted or otherwise nonpolarized cells are facing the lumen. Although it is not entirely clear why this viral strategy has been so successful, it was recently suggested for Ads that newly synthesized fiber proteins can disrupt cell-cell contacts, loosen up the epithelial integrity, and aid viral release from infected basal cells into the apical lumen (Walters et al. 2002). This may be potentiated by the ability of an additional viral protein, penton base, to abduce fiber from the infected to the surrounding noninfected cells (Trotman et al. 2003). Thus CAR may serve two different viral functions, as an exit and an entry receptor. Once an incoming virus has bound to a cell receptor, it may catch a ride in endocytic vesicles shuttling from the plasma membrane to the interior of the cell (see the chapter by Sieczkarski and Whittaker, this volume). Other viruses have evolved strategies to penetrate the plasma membrane or endosomes after receiving appropriate signals from the cell (see the chapter Earp et al., this volume). Alone, the delivery of the viral nucleic acid in the cytosol is in most cases not sufficient for replicating the genome or maintaining it in a stable state within the host. Regulated cytoplasmic trafficking and nuclear deposition of an encapsidated genome is in many cases important to ensure efficient access to the replication site (see the chapter by Piguet, this volume). Viral replication and particle production can lead to vertical transmissions, for example, in the case of papillomaviruses spreading from the infected epithelium of the mother to the newborn. Alternatively, horizontal transmissions can disseminate the viral particles throughout a population. This requires efficient entry, genome production, and packaging processes as well as effective viral egress and suppression of host antiviral defense reactions.

Clearly, all of these steps have evolved through an intimate relationship of the virus with its host, and in many instances they depend on the cell nucleus (for recent reviews on nuclear import of incoming viruses, see Greber and Fassati 2003; Whittaker 2003). In this review, we summarize the cornerstones of nuclear import of incoming viral genomes and highlight some of the better-studied postentry strategies of viruses co-opting the cellular mechanisms of nucleo-cytoplasmic exchange.

2
Nuclear Pore Complexes, Guardians of the Nucleus

The nuclear envelope is the limiting feature of the nucleus harboring the large majority of the genomic information of the cell. It contains randomly positioned nuclear pore complexes (NPCs) controlling the exchange of cytosolic and nuclear macromolecules (Nakielny and Dreyfuss 1999). NPCs are large protein structures of approximately 125 MDa, about 25 times the mass of a ribosome (Reichelt et al. 1990). They have an eightfold symmetry if viewed in projection from the cytoplasmic or the nuclear side (reviewed in Stoffler et al. 1999; Wente 2000). Ultrastructural analyses using transmission electron microscopy (TEM), scanning EM, and atomic force microscopy have shown that the vertebrate NPC is a ~65-nm-deep channel, decorated on each side with eight filaments attached to a coaxial ring that project away from the nuclear envelope (reviewed in Allen et al. 2000; Fahrenkrog et al. 2001). The nature of a central plug is controversial. It can either be an integral part of the NPC or cargo in transit (for a recent discussion, see Suntharalingam and Wente 2003). The ~50–100-nm filaments of the nuclear face join together at their distal ends to form a so-called nuclear basket or fish trap.

The vertebrate NPC is built of 30–40 different proteins, termed nucleoporins (Cronshaw et al. 2002; Fahrenkrog et al. 2001; Rout et al. 2000; Ryan and Wente 2000). Nucleoporins are proteins whose steady-state distribution is predominantly in the NPC and that are somehow involved in building the NPC or serving the NPC's transport function (Vasu and Forbes 2001). A motif characteristic for nucleoporins is the FG-repeat consisting of a large number of phenylalanine-glycine dipeptides, spaced by 4–20 intervening amino acids. These FG-repeat regions have been shown to interact with transport receptors of the karyopherin/importin type, as well as with the unrelated mRNA export receptor NFX1/TAP complexed to p15 (see below). In vertebrates, FG-repeat nucleoporins have been localized to different positions of the NPC structure, such as the nuclear face (Nup153), the cytoplasmic face (CAN/Nup214), the cytoplasmic filaments (RanBP2/Nup358), and the central channel (p62, Pom121). The exact localization of most other vertebrate nucleoporins remains to be determined. The vertebrate NPCs are relatively static in the plane of the nuclear membrane (Daigle et al. 2001), presumably because they are bound to the lamina network at the inner nuclear membrane. Nonetheless, evidence is emerging to indicate that individual nucleoporins can be dynamic and shuttle between intranuclear or cytoplasmic pools. For example, the NPCs of the single cell algae *Chlamydomonas reinhardtii* were observed to move to the posterior pole on deflagellation of the cell (Colon-Ramos et al. 2003). Increased levels of β-tubulin accumulate in the cytoplasm proximal to the rearranged NPCs near the site where the new flagella will be built, suggesting a link between cytoplasmic transcript localization and nuclear envelope architecture, in particular NPC localization. Interestingly, non-NPC proteins are found to transiently associate with NPCs, for example, proteins involved in cell cycle control and in protein modification with the small ubiquitin modifier (SUMO) peptide that mediates protein-protein interactions. In vertebrate cells, RanBP2/Nup358 has an E3 ligase activity that transfers SUMO to acceptor proteins (Pichler et al. 2002). The Ran:GTPase activating protein (RanGAP1) is sumoylated and docks at the NPC. Additionally, it was found that the SUMO conjugating enzyme Ubc9 can associate with RanGAP1 and RanBP2/Nup358 and the SUMO protease SENP2 localizes to the NPC by interacting with Nup153 (Hang and Dasso 2002; Zhang et al. 2002). It is possible that the SUMO machinery at the NPC modifies particular cargoes in transit between the nucleus and cytoplasm (Greber and Carafoli 2002; Rodriguez et al. 2001). Yet another level of NPC regulation may be imposed by phosphorylation. It was found that in vitro reconstituted nuclear protein import is inhibited by the phosphatase inhibitor okadaic acid and partially reversed by kinase inhibitors without affecting the association of import substrates with the NPC, suggesting that phosphorylation of NPC components inhibits nuclear import (Kehlenbach and Gerace 2000). These results, although preliminary, suggest that activation of specific cellular signaling pathways can influence the number and perhaps the size of the molecules that can cross the NPC. It emerges that the NPC integrates modifier and transport functions of high capacity and high selectivity (Gorlich and Kutay 1999). Thus the NPC seems to be an ideal target of viruses aiming to subvert cellular control functions.

3
Soluble Import Receptors Control Nuclear Access

Efficient macromolecular trafficking across the NPC is both signal- and receptor dependent. Nuclear localization (NLS) or nuclear export (NES) signals on cargo proteins are recognized by transport receptors of the karyopherin/importin β family, of which there are currently more than 20 members in vertebrates (reviewed by Fornerod and Ohno 2002; Gorlich and Kutay 1999; Strom and Weis 2001). Some of them are expressed in a tissue- and developmentally restricted manner and can also serve functions other than those required for nuclear transport (see, e.g., Geles et al. 2002). The prevalent function of importin and exportin superfamily proteins in interphase cells is to mediate nucleo-cytoplasmic transport of proteins, either free or bound to nucleic acids. Importins are characterized by their ability to bind cargo and the small GTPase Ran loaded with GTP (Ran:GTP), and they interact with the NPC. Although most importins and exportins are rather specialized, recognizing a restricted set of cargoes, the import receptor importin β transports a wide variety of cargo, including viral proteins. Importin β can cooperate with the adaptor importin α and with other importin family members, and it serves to import proteins bearing classic SV40-type basic NLS and also uridine-rich small nuclear ribonucleoproteins (snRNPs) (reviewed by Kuersten et al. 2001). Import complexes are allowed to form in the cytoplasm, where the GTPase activating enzyme RanGAP1 restricts the levels of Ran:GTP and the nuclear transport factor (NTF) 2 transports Ran:GDP into the nucleus. Inside the nucleus, Ran:GTP is replenished by the chromatin-bound factor RCC1, exchanging GTP for GDP in the nucleotide binding pocket of Ran. Ran:GTP dissociates the importin-cargo complex. The cargo is distributed to subnuclear sites, and the receptors are recycled to the cytoplasm.

4
Models of Nuclear Pore Complex Translocation

NPCs mediate two types of translocation, passive diffusion of small substances and receptor-dependent translocations of signal-bearing cargoes. The size exclusions of both transport modes are kinetically determined. That is, within a time frame of a few hours the largest cargoes seen to diffuse through the NPC have been about 9 nm in diameter (reviewed in Gerace and Burke 1988). The largest cargoes of facilitated transport are 39-nm gold particles coated with the nucleophilic protein nucleoplasmin and the transport factors importin α and β (Pante and Kann 2002). Furthermore, the rates of substance flux through an individual NPC are considerable, namely, in the order of 1,000 cargo molecules per second (Ribbeck and Gorlich 2001; Smith et al. 2002).

One model of translocation proposes that increasing affinities between nucleoporins and transport receptors along the distance of the NPC determine the direction of translocation (Bednenko et al. 2003). This model is supported by the strong interaction of the import receptor importin β to the FG repeat region of Nup153 located at the nuclear baskets (Shah et al. 1998) and the strong interaction of the export receptor CRM1/exportin1 to the FG repeat region of the nucleoporin CAN/Nup214 located at the cytoplasmic filaments (Askjaer et al. 1999; Kehlenbach et al. 1999). Nup153 and CAN/Nup214 could therefore represent the terminal and highest-affinity binding sites of an import and export reaction, respectively (Ben-Efraim and Gerace 2001). Several lines of evidence argue against the affinity gradient being the sole force of directional transport within the NPC. First, directionality can be reversed by the reversal of the Ran:GTP gradient across the nuclear envelope, indicating that the relative binding strengths of nucleoporins with receptors are not necessary for directionality of transport (Nachury and Weis 1999). This argues that the Ran:GTP gradient and the associated stabilities of cargo-transport factors are sufficient to achieve vectorial translocation of cargo. Second, the translocation would be driven by binding to the highest-affinity site and release by binding of nuclear Ran:GTP (import) or Ran:GTP hydrolysis (export). However, nuclear import can occur in the absence of GTP hydrolysis on Ran (Schwoebel et al. 1998), and single rounds of import or export can occur in the absence of Ran

(Englmeier et al. 1999; Huber et al. 2002; Ribbeck et al. 1999). In addition, the export and reimport rates of a complex consisting of the export factor CRM1, NES-containing cargo, and Ran:GTP in *Xenopus* oocytes are similar (Becskei and Mattaj 2003), further suggesting that the affinity gradient model may only partly account for the observed transport directionalities.

On the basis of kinetic experiments a second model has proposed that the FG repeats of nucleoporins form a hydrophobic meshwork within the central NPC channel that is restrictive for most proteins because of their hydrophilic surfaces (Ribbeck and Gorlich 2001). The model predicts that transport receptors immerse into this meshwork by virtue of their rather hydrophobic surface and diffuse through the NPC channel. Supporting evidence for this model shows that import receptors bind to hydrophobic interaction resins and the permeability barrier of the NPC can be compromised by the addition of hexanediole, which is thought to weaken or break up the meshwork (Ribbeck and Gorlich 2002). This hypothesis can potentially explain the nature of the diffusion barrier through the central channel of the NPC, even though it cannot explain the specific role of some nucleoporins for specific transport pathways, for example, Nup82 for mRNA export (Grandi et al. 1995; Hurwitz and Blobel 1995) and Nup153 for NLS-protein import (Walther et al. 2001).

In a third model, called the "virtual gate" (Rout et al. 2000), filamentous nucleoporins have been suggested to play a major role in protein translocation. In this scenario, the random Brownian movement of the cytoplasmic filaments would clear away noncargo proteins from the entrance of the NPC channel. Importin-cargo complexes on the other hand would be concentrated by binding to the cytoplasmic filaments and subsequently diffuse through the central translocation channel into the nucleus, where they would be disassembled by Ran:GTP (Rout et al. 2000). This model is attractive, but it still awaits experimental support. Nucleoporin FG repeats are likely to be in an unstructured conformation and therefore indeed could be considered as microfilaments (Denning et al. 2003). Interestingly, however, the removal of the cytoplasmic filaments of the NPC does not reduce NLS import or increase the passive permeability of the NPC in vitro (Walther et al. 2002). Together, all three models agree that the interactions of transport factor and nucleoporin FG regions are key for translocation through the NPC, but none of the models alone can yet explain all the observed features of nucleo-cytoplasmic transport. Of particular interest here is the observation that entry of large cargo into the NPC alone is apparently not sufficient for translocation (Lyman et al. 2002) and that the translocation efficiency correlates with the number of bound transport factors (Ribbeck and Gorlich 2002). This is particularly interesting in the light of a recent biophysical analysis of the NPC by atomic force microscopy suggesting that transport receptors with a relatively weak affinity to the FG repeats of the nucleoporins lead to a dilation of the pore whereas high-affinity receptors seem to promote an expansion of the pore in the vertical direction (Jaggi et al. 2003). This all suggests that diffusion alone, particularly of large cargo, may not be sufficient for NPC translocation.

5
Ran-Independent Transport

Importins and exportins do not require the Ran system to enter or exit the nucleus, presumably because of their interactions with nucleoporin FG repeats (Kose et al. 1997, 1999; Nakielny and Dreyfuss 1998; Ribbeck et al. 1999). Importins and exportins consist of approximately 20 repeats of the so-called HEAT or Armadillo type (Andrade et al. 2001). Interestingly, β-catenin and importin α, which have a similar overall structure, are also able to move in and out of the nucleus independent of Ran (Henderson and Fagotto 2002; Miyamoto et al. 2002). β-Catenin is a component of cell-cell adhesion junctions binding directly to cadherins and linking to the actin cytoskeleton via α-catenin. The biological activity of β-catenin is controlled by the wnt signaling pathway and in particular by the adenomatous polyposis coli protein, a predominantly cytoplasmic factor that binds and thereby facilitates β-catenin degradation. Failure of β-catenin degradation and concomitant nuclear accumulation of β-catenin are associated with colon cancer. β-Catenin contains no recognizable NLS, and it was found to be imported into nuclei of permeabilized cells in the absence of transport factors and Ran (Fagotto et al. 1998; Yokoya et al. 1999). Whether β-catenin has any shuttling function like the importin factors is not known.

Another example of importin-independent transport is the MAP kinase ERK. ERKs play a key role in transducing extracellular signals into physiological cell responses during proliferation, differentiation, and early embryonic development. There are at least two pathways known for ERK nuclear import, Ran-dependent transport of dimeric phosphorylated ERK (Khokhlatchev et al. 1998) and passive diffusion of monomeric ERK (Adachi et al. 1999). In digitonin-permeabilized cells, GFP-fused ERK accumulates in the nucleus independently of phosphorylation and in the absence of soluble factors and ATP (Matsubayashi et al.

2001). This is inhibited by wheat germ agglutinin or by an excess of importin β at reduced temperature, suggesting that nuclear accumulation of GFP-ERK requires NPCs. ERK was found to bind the carboxy-terminal FG repeats of CAN/Nup214 in vitro, suggesting that it could interact directly with the NPC. This implies that the importin-independent translocation depends on specific interactions of ERK and the NPC, distinct from the interactions during importin-dependent translocation of dimeric ERK (Shibayama et al. 2002). It will be interesting to see how these distinct nuclear import pathways of ERK connect to particular signal transduction pathways.

Interestingly, nuclear localization of certain large substrates, namely the spliceosomal U snRNP particles has been reported to be Ran independent (Huber et al. 2002; Rollenhagen et al. 2003). It is mediated by the adaptor protein snurportin, which binds to the m3G-cap structure at the 5' end of the U RNAs and, in addition, has an importin β binding domain (IBB). Importin β and snurportin were sufficient for a single round of import of U1 snRNPs in digitonin-permeabilized mammalian cells and did not require Ran and trinucleotides, such as ATP or GTP. When the IBB domain of snurportin was exchanged with the IBB domain of importin α, nuclear import of U1 snRNPs was dependent on Ran. This suggests that different IBB domains influence interactions of importin β with nucleoporins of the NPC and determine whether Ran is required to dissociate these interactions. Interestingly, the size of the cargo did not seem to be a major factor impeding translocation in these experiments, because the U5 snRNP with a molecular mass larger than one million daltons was also imported in a Ran- and energy-independent manner when native snurportin was present.

6
Nuclear Import of Viral Genomes

Viruses that replicate in the nucleus of nondividing cells with an intact nuclear membrane transfer their genomes through the NPC into the nucleus. The strategies to target the NPC differ among viruses but invariably involve specific pathways of signaling, endocytosis, access to the cytosol, and cytoplasmic transport (for reviews, see Greber 2002; Meier and Greber 2003; Ploubidou and Way 2001; Poranen et al. 2002; Sodeik 2000). Nuclear import of incoming viral genomes also depends on viral uncoating and in some cases involves an increase of capsid affinity for the NPC (for reviews, see Greber et al. 1994; Whittaker et al. 2000). In addition, certain viruses and viruslike particles (VLPs) undergo endogenous replication and can shuttle their genomes in and out of the nucleus without going through a complete virus assembly process.

6.1
Parvoviruses

The nonenveloped parvoviruses are among the smallest virus particles found both intracellularly and extracellularly. They consist of a simple icosahedral capsid of about 26 nm bearing a single-strand DNA genome of about 5,000 bases. Parvoviruses internalize by receptor-mediated endocytosis, arrive in the cytosol, interact with the NPC, and deliver their single-strand DNA into the nucleus, where the synthesis of the viral second-strand DNA occurs (reviewed in Greber 2002; Sieczkarski and Whittaker 2002). For example, the canine parvoviruses are small enough to pass through the NPC without prior dissociation. However, microinjected cytosolic particles have not been found to be imported in significant amounts into the nucleus (Weichert et al. 1998). In infected cells, the amino-terminal sequence of the capsid protein VP1 is exposed by an unknown stimulus, and this was found to correlate with an increased nuclear localization of capsids or capsid fragments (Vihinen-Ranta et al. 2002). Accordingly, genetic deletion studies with the minute virus of mice (MVM) have shown that the amino-terminal basic clusters of VP1 are required for the onset of infection, suggesting that importin α and β are involved in VP1 import (Lombardo et al. 2002). It is not clear, however, whether additional NLSs on other capsid proteins need to be exposed for nuclear import of the viral genome or whether the capsid disassembles before DNA import (for further discussion, see Greber and Fassati 2003).

Similarly, it is unclear how the adeno-associated viruses (AAVs) uncoat and import their ssDNA genomes. One report states that the AAV serotype 2 entered the nuclei in intact form (Bartlett et al. 2000). AAV2 was also reported to be imported into purified nuclei from cultured cells in a nonsaturable manner, independent of temperature and NPC inhibitors (Hansen et al. 2001). Nuclei of parallel samples excluded the 150-kDa dextran but not the 4-kDa dextran, suggesting that they were size selective and competent to retain small dextran. Another study, however, suggested that fluorescent AAV2 rapidly entered HeLa cells and remained in a perinuclear area up to 24 h after infection (Xiao et al. 2002). Viral uncoating was thought to happen before or during nuclear entry of viral DNA and intact AAV particles were difficult to detect inside the nuclei. In the presence of coinfecting Ad5, however, viral capsid fluorescence was found within the nuclei and viral DNA cofractionated with enriched nuclei. This was independent of the NPC inhibitor thapsigargin (Greber and Gerace 1995), which blocked nuclear access of small dextrans and AAV replication (Xiao et al. 2002). The data suggested that in the presence of Ad5, AAV2 particles could enter the nucleus. The underlying mechanisms of the proposed NPC-independent nuclear translocation are, however, unknown, and the nuclear factors triggering capsid disassembly remain to be identified.

6.2
Hepadnaviruses, DNA Retroviruses

Hepadnaviruses, such as hepatitis B virus (HBV), are another family of small viruses that could translocate through the NPC without disassembly, just based on capsid size. Hepadnaviruses are enveloped viruses with an icosahedral capsid of up to 36 nm and a partly double-strand DNA genome that arises through reverse transcription from a pregenomic positive-sense RNA (Ganem and Schneider 2001). Human HBVs are important pathogens, causing acute and chronic hepatitis and eventually liver cancer. Their entry is difficult to study in cell cultures for reasons that are not precisely known. Recombinant hepatitis B virus capsids produced in *E. coli* were found to associate with the NPCs of permeabilized cells in the presence of importin α and β, provided that the capsids were phosphorylated by protein kinase C before incubation with the permeabilized cells (Kann et al. 1999). PKC somehow exposed an NLS in the carboxy terminus of the capsid proteins, and this seemed to be important for the interaction with the NPC. A homologous NLS is present in the duck hepatitis B virus (dHBV) capsid protein, but it is positioned more centrally in the protein and not near potential phosphorylation sites (Mabit et al. 2001). The dHBV NLS was essential for NPC association of the nucleocapsid and for virus production as shown with a knock-out mutant, confirming the important role of this NLS in the viral life cycle. High-resolution EM in *Xenopus* oocytes subsequently showed that recombinant HBV capsids that were phosphorylated with PKC were localized at and within the NPCs but not in the nucleoplasm, suggesting that the recombinant capsids were arrested in the NPC nuclear baskets (Pante and Kann 2002). Because the baskets are known to be dynamic and to have a high turnover (Daigle et al. 2001; Jarnik and Aebi 1991; Nakielny et al. 1999), they might impose a diffusional barrier. Alternatively, some NPC proteins might remain bound to these capsids through multiple direct and indirect interactions. A recent study suggests that mature HBV capsids isolated from purified virions disassemble and release their DNA within the nucleus of permeabilized cells (Rabe et al. 2003). Surprisingly, the presence of the soluble factors importin α and β was sufficient to release a significant amount of DNA into the nucleus of these permeabilized cells. Interestingly, immature viral capsids derived from virion-producing hepatoma cell lines were found to be phosphorylated and bound to the NPC much like the *E. coli*-expressed capsids. However, the immature capsids did not uncoat their nucleic acids which were present as a mixture of RNA and DNA intermediates. This revealed two critical steps for nuclear import of HBV DNA genomes. The first step is the exposure of a capsid NLS, making the particle competent to bind importins and associate with the NPC, and the second step is the release of the reverse-transcribed double-strand DNA genome from the capsid. Analogous to the disassembly program of the species C Ads and herpes simplex virus type 1 (HSV1) (Greber 1998; Greber and Fassati 2003), HBV may contain a built-in priming mechanism that allows a disassembly-competent capsid to associate with the critical cellular structure, the NPC, and there receive the final trigger for DNA release. The nature of the final cue triggering disassembly of HBV capsids is still unknown.

6.3
Retrotransposons

Retrotransposons containing long terminal repeats (LTR) are found in many eukaryotic cells. They are closely related to retroviruses and encode Gag, reverse transcriptase (RT), and integrase (IN) proteins (Boeke and Stoye, 1997). The absence of envelope (ENV) genes suggests that they do not switch their host cells as frequently as the ENV-containing retroviruses. Nonetheless, retrotransposons shuttle between the cytoplasm and the nucleus. Their Gag proteins assemble into coats, and this is required for reverse transcription of the mRNA genome. IN integrates the DNA genome into the host chromosomes. The maturation of the viral particles leads to Gag processing by the viral protease (PR), yielding three major products, matrix (MA), capsid (CA), and nucleocapsid (NC). For example, the retrotransposon Ty1 and Ty2 of yeast *Saccharomyces cerevisiae*, the Tf1 and Tf2 elements of *Schizosaccharomyces pombe*, and the copia elements of *Drosophila melanogaster* form VLPs. Tf1 and Tf2 are model systems to study retroviruses (Levin et al. 1990), and the expression of Tf1 mRNA leads to high levels of transposition. A

Gag-like protein has been shown to cosediment with a complex of RT, IN, and viral cDNA (Levin et al. 1993). This Gag-like protein has an NLS required for nuclear import and transposition (Dang and Levin 2000). Nuclear import depends on the FXFG-type NPC protein Nup124, which is directly targeted by Tf1 (Balasundaram et al. 1999; Dang and Levin 2000). Yeast cells mutated in Nup124 block the nuclear import of Tf1 protein and also cDNA. This import defect appears to be specific, because overall protein or RNA import was not affected in this mutant. It is not clear whether Nup124 has a functional homolog in other eukaryotes. Nup124 shows the highest resemblance to *S. cerevisiae* Nup1p, a nucleoporin located at the nuclear basket, but Nup1 is essential. A nuclear targeting signal in integrase is involved in nuclear transport of Ty3, a retrotransposon of *S. cerevisiae* (Lin et al. 2001). It is not clear, though, whether VLPs disassemble to let the retrotransposon genome cross the NPC.

6.4
Lentiviruses

Like retrotransposons, the genomes of retroviruses are unspliced mRNA that is packaged into a nucleocapsid. Unlike retrotransposons, the nucleocapsid is enwrapped by a lipid membrane. The viral replication cycle involves a double-strand DNA derived from the genomic mRNA by reverse transcription in the cytoplasm in the so-called reverse transcription complex (RTC), giving rise to the preintegration complex (PIC), followed by integration of the proviral DNA into the host chromosomes (Cooper et al. 1995; for a recent perspective, see Trono 2003). Whereas oncoretroviruses require dividing cells for infection, lentiviruses, such as the human immunodeficiency virus (HIV), infect both dividing and nondividing cells. The lentivirus PIC traffics through the cytoplasm and is imported into the nucleus through the NPC (for recent reviews, see Greber and Fassati 2003; Piller et al. 2003). Lentiviral DNA could be found inside the nucleus of CD4 receptor- and CCR5 coreceptor-expressing HeLa cells (Bell et al. 2001). Large amounts of viral DNA were not integrated into chromosomes, supporting the notion that these cells had efficiently imported PIC. PIC is a pleiomorphic structure of double-strand DNA and several viral proteins, including RT, IN, MA, and the auxiliary protein Vpr as well as other unidentified cellular proteins (Miller et al. 1997). RT is, however, found to be shed from the reverse transcription complex soon after entry (Fassati and Goff 2001). Lentiviral PIC also contains a particular structural DNA element, a polypurine tract serving as a second site of plus-strand DNA synthesis. Viral mutants bearing deletions of this DNA flap structure have a slightly reduced ability to access the nucleus (Zennou et al. 2000). Although IN and MA contain basic NLSs that function in the context of the isolated proteins, their roles in PIC import are controversial (this has been reviewed elsewhere; see, e.g., Goff 2001; Greber and Fassati 2003; Piller et al. 2003).

The small accessory protein Vpr of 14 kDa plays several interesting roles. It appears to enhance HIV infection in macrophages, and, interestingly, it interacts with the NPC (Fouchier et al. 1998; Popov et al. 1998; Vodicka et al. 1998). However, Vpr is not essential for viral infectivity in all nondividing cells; for example, it is not needed in T cells (Eckstein et al. 2001). The role of Vpr in nuclear import of PIC has been controversial (see, e.g., Bouyac-Bertoia et al. 2001). Nonetheless, Vpr uses several nuclear import pathways that are Ran- and importin independent (Jenkins et al. 1998). Moreover, it shuttles between the nucleus and the cytoplasm through a CRM1-dependent NES (Sherman et al. 2001). Furthermore, yeast two-hybrid interactions, in situ localizations, and coimmunoprecipitations have shown that Vpr associates with the NPC protein hCG1, the mammalian homolog of the yeast nucleoporin Rip1p/Nup42 (Le Rouzic et al. 2002). hCG1 was originally identified in yeast two-hybrid screens binding to the human export factor TAP and to CRM1 binding the HIV-1 Rev protein (Farjot et al. 1999; Katahira et al. 1999; Wiegand et al. 2002). The yeast homolog Rip1p is involved in the export of mRNAs encoding heat shock proteins (Saavedra et al. 1997), and a carboxy-terminal truncation of Rip1p can be rescued by the carboxy terminus of hCG1, suggesting that Rip1p and hCG1 are performing homologous functions (Strahm et al. 1999). Vpr interacts with FG repeats of several nucleoporins in vitro, but it does not use the FG repeats of hCG1 and rather binds a nonconserved domain in the amino-terminal portion of hCG1 (Le Rouzic et al. 2002). In contrast, the hCG1 FG repeats interact with the transport factors TAP and CRM1, implying that Vpr might not interfere with hCG1-dependent export pathways. The functional implications of the Vpr interactions with hCG1 are, however, not known. It is also unknown whether Vpr disturbs normal NPC transport functions and how it is involved in the formation of herniations and transient perforations of the nuclear envelope and cell cycle arrest in the G_2 phase (de Noronha et al. 2001). Nevertheless, it is conceivable that, depending on the cell type and the viral isolate, the association of Vpr with the NPC is a key event supporting PIC import and infection.

Another class of key factors of RTC nuclear import are likely to be the importins. A recent analysis of transport factors of RTC import in primary macrophages points toward importin 7, which is normally involved in histone H1 import (Fassati et al. 2003). Using in vitro nuclear import assays in permeabilized cells and small interfering RNAs for importin 7, these authors showed that importin 7 stimulated nuclear import of RTC in a Ran-dependent manner and was required in HIV infection. It was preferentially involved in IN import, but IN interacted with other importins, including importin β. The inhibition of the importin β pathway by an excess of the importin β binding domain had only moderate effects on RTC import, suggesting that importin β is not sufficient for RTC import (Gallay et al. 1997). Consistently, mutations of basic NLSs in IN did not affect nuclear import of PIC and had little effect on IN import into HeLa cells (Petit et al. 2000). However, IN contains another nonbasic NLS located in the middle of the protein. This NLS is required for accumulation of viral DNA in the nucleus and for infection but does not affect the catalytic activity of IN (Bouyac-Bertoia et al. 2001). It will now be important to develop a coherent picture of how the different players of PIC import are coordinated and how their action relates to the conformation of the PIC and its precursor, the RTC. Interestingly, some consensus of RTC structure and function now seems to emerge based on microscopy and biochemical analyses. It is suggested that after viral fusion with the plasma membrane the incoming lentiviral core remains intact until it is delivered into the cytosol (Fassati and Goff 2001; McDonald et al. 2002; Nermut and Fassati 2003). Reverse transcription and additional cues then induce capsid uncoating, and this might in turn be essential for the completion of the reverse transcription process. Intriguingly, host restrictions targeting the p24 Gag component CA might help inhibiting viral uncoating (Besnier et al. 2002), and thus viral uncoating and subsequent import can be points of host antiviral defense (Turelli et al. 2001). Resolving the molecular mechanisms underlying these blocks promises to reveal more of the intricate relationship of retroviruses with their hosts.

6.5
Adenoviruses

Adenoviruses (Ads) are large nonenveloped DNA viruses with a capsid diameter of about 90 nm (Burnett 1997). Their entry pathways are largely determined by the fiber proteins that bind particular cell surface receptors. Ad particles navigate their genome through the cytoplasm along microtubules, much like the HSV1 capsids (Mabit et al. 2002). In contrast to retroviral capsids, the Ad capsids dock to the NPC of both infected cells and purified nuclei (Chardonnet and Dales 1970; Morgan et al. 1969; Wisnivesky et al. 1999). The species C Ads dock to a nucleoporin, CAN/Nup214 of the cytoplasmic side of the NPC in the absence of cytosolic factors (Trotman et al. 2001). In all likelihood, docking occurs through the major capsid protein hexon. However, docking is not sufficient for DNA release from the capsid, which is crucial for infection (Cotten and Weber 1995; Greber et al. 1997, 1996). To uncoat, the NPC-docked Ad2 particle binds the mobile linker histone H1 that exits the nucleus for short periods (Trotman et al. 2001). This binding is thought to occur at the acidic polyglutamate domains of hexon located at the outside of the capsid (Rux and Burnett 2000). These acidic stretches are conserved in the species C Ads but not in other species. Binding of H1 is, however, still not sufficient for Ad2 disassembly, and the histone H1 import factors must be present (Trotman et al. 2001). Additional cytosolic factors such as Hsp70 have been implicated as well (Saphire et al. 2000). It is unknown how these molecular requirements relate to the earlier notion that cells treated with *p*-hydroxymercuric benzoate (PHMB) block viral uncoating and inhibit a cellular ATPase activity present in isolated nuclei (Chardonnet and Dales 1972; Dales and Chardonnet 1973). It should be noted here that the Ad2 capsids that are docked to the NPCs have previously gone through a limited disassembly program, removing the fibers and destabilizing minor capsid components (Greber et al. 1993; Nakano et al. 2000). Thus it is presumably the simultaneous binding of the carboxy-terminal domain of CAN/Nup214, the histone H1, and the H1 import factors importin β and importin 7 on the weakened capsid that finally triggers capsid disassembly, allowing the release of the viral DNA. How Ad serotypes of species other than species C dock to the NPC and uncoat their genomes is unknown.

6.6
Herpesviruses

Most of the information about nuclear import of herpesviruses comes from studies with herpes simplex virus type 1 (HSV1) and pseudorabies virus (PRV), belonging to the alpha herpesviruses subfamily. After fusion of the viral envelope with the plasma membrane, the 120-nm large HSV1 and PRV nucleocapsids are transported through the cytoplasm and are targeted to the NPCs similar to Ad capsids, as indicated by EM of cultured cells (Granzow et al. 1997; Lycke et al. 1988; Roizman and Knipe 2001; Sodeik et al. 1997). Remarkably, the capsid seems to be positioned in a particular orientation with respect to the NPC, perhaps placing the portal protein channel, through which the viral DNA was packaged, proximal to the NPC mouth (Newcomb et al. 2001). The docked capsids uncoat and inject their linear double-strand DNA genome through the nuclear pores into the nucleoplasm and often leave an empty capsid behind at the NPC (for a recent review, see Greber and Fassati 2003). The precise composition of the NPC-docked capsids is not known, but the capsids comprise at least four different proteins and additional factors from the tegument—a mass of protein and mRNA located between the nucleocapsid and the envelope in intact virions. Interestingly, the HSV-1 mutant tsB7 appears to be defective at capsid disassembly and/or DNA release if grown at the restrictive temperature but this mutant still docks at the NPC (Knipe and Smith 1986). Although tsB7 bears multiple mutations, this may mean that tegument proteins are part of a viral checkpoint for DNA release. Indeed, capsids containing the major tegument proteins VP1/2, VP13/14, and VP22 were found to bind to isolated nuclei when recombinant importin β was present (Ojala et al. 2000). Capsid binding to nuclei was inhibited by an excess of Ran:GTP, implying that importin β bound to capsids. The addition of cytosol and ATP together with shifting the temperature up to 37°C triggered genome uncoating, that is, the viral DNA became sensitive to added deoxyribonuclease. Although the cytosolic factors required for genome uncoating are unknown, the removal of the tegument proteins VP1/2, VP13/14, and VP22 by proteolytic digestions before nuclear binding inhibited capsid binding to nuclei and DNA release in vitro. Studies of viral mutants, furthermore, suggested that the full-length VP22 is not needed for a single round of HSV1 infection, including viral entry, replication, and assembly, but that it may be required for spreading virus particles in vivo (reviewed in Mettenleiter 2002). This implies that VP22 may not be a key factor of HSV capsid docking to the NPC. It will be interesting to see whether HSV capsids release their DNA at a specialized region in a capsid vertex, for example, the portal of entry, which is known to be important for the packaging of the genome into the newly synthesized capsids (for a discussion, see Fuller 2003).

6.7
Papovaviruses

The papilloma and polyoma virus subfamilies (papovavirus family) comprise more than 80 different human and animal viruses (Cole 1996; Howley 1996). They are nonenveloped icosahedral double-strand DNA viruses about 60 nm in diameter. Depending on the virus and the cell type, their entry occurs by clathrin-mediated endocytosis, caveolar endocytosis, or noncaveolar cholesterol-dependent endocytosis (Ashok and Atwood 2003; Bousarghin et al. 2003; Day et al. 2003; Fausch et al. 2003; Pelkmans et al. 2001; Pho et al. 2000; Selinka et al. 2002). In some cases entry also involves coreceptors (for a recent example, see Caruso et al. 2003).

One of the better-studied polyoma viruses is the simian vacuolating virus 40 (SV40). SV40 entry occurs via cholesterol-rich membrane domains that are enriched in signaling molecules (reviewed by Pelkmans and Helenius 2002). In caveosomes, SV40 is sorted for transport to the smooth ER from where it somehow transfers its DNA into the nucleus. This seems to involve a cytosolic phase, because neutralizing anti-SV40 antibodies injected into the cytoplasm inhibited the infection (for reviews of the older literature, see Greber and Kasamatsu 1996; Kasamatsu and Nakanishi 1998). There is recent evidence to suggest that SV40 exposes the NLS of the minor VP2/3 protein and this NLS is recognized by importin α (Nakanishi et al. 2002). Some of the key questions that remain to be solved include Where do these conformational changes occur in the cell? and What is the composition of the nucleoprotein complex that is delivered into the nucleus?

Papillomaviruses are more difficult to study because the infectious units are relatively rare in isolated virions and the viruses replicate in terminally differentiated keratinocytes (Howley and Lowy 2001). The high-risk human papillomaviruses (including HPVs 16, 18, 31, 33, 45) are frequently found in invasive cervical carcinomas, whereas the low-risk HPVs (types 6 and 11) are associated with rather benign condylomata acuminata (zur Hausen 2002). Nuclear import of HPVs has been studied with viruslike particles (VLPs) and the capsid proteins L1 and L2. The L1 protein is the major capsid component and forms VLPs in the absence of L2, but incorporation of L2 seems to be important for infectivity (Florin et al. 2002). Intact L1 capsids of HPV11 were found to be excluded from the nucleus, but L1 capsomeres were imported into the nuclei of digitonin-permeabilized HeLa cells (Merle et al. 1999). In this system, docking of HPV11 VLPs was inhibited by antibodies against importin $\alpha 2$ and $\beta 1$. These results suggested that capsid disassembly was required for HPV11 L1 nuclear import and that the classic importins and additional factors were involved. Likewise, the L1 protein of the high-risk HPV16 was imported into the nucleus of digitonin-permeabilized HeLa cells in a complex with importin $\alpha 2$ and $\beta 1$ in a Ran-dependent manner (Nelson et al. 2002). Interestingly, it was observed in this study that L1 capsomeres also interacted with importin $\beta 2$, which is normally involved in the import of M9-containing proteins, such as hnRNP A1. The related L1 of the low-risk HPV11 also interacted with importin $\beta 2$ and importin $\beta 3$ and thereby inhibited the nuclear import of hnRNP A1 (Nelson et al. 2003). Whether this has any physiological significance during virus assembly or release is unknown.

6.8
Negative-Sense RNA Viruses

Influenza virus is an orthomyxovirus with a segmented negative-sense RNA genome that requires the nucleus for replication (reviewed in Lamb and Krug 2001). It contains a lipid-protein envelope and eight different genomic RNPs containing the RNA, nucleoprotein (NP), and the heterotrimeric RNA-dependent RNA polymerase. The RNPs are tethered to the envelope through the matrix protein M. Although the length of the RNPs varies from about 20 to 100 nm, their diameter is probably smaller than 25 nm, that is, it would pass through the NPC without major uncoating reactions. To a large extent, viral disassembly happens in late endosomes, where the low pH triggers conformational changes of viral hemagglutinin, which then acts to catalyze fusion of the viral envelope with the limiting endosomal membrane (for a recent review, see Colman and Lawrence 2003). In addition, endosomal protons dissociate the interactions of the matrix protein M1 and the RNPs within the virion, which is required for NP nuclear import in cultured cells (reviewed in Lamb et al. 1994; Whittaker et al. 2000). Import of RNPs has also been investigated in digitonin-permeabilized cells. It was found that RNA import required the coating of the RNA with NP and the presence of importin α (i.e., the npi-1 and npi-3 proteins), importin β and Ran (O'Neill et al. 1995; Wang et al. 1997). The NLS on NP is nontypical, namely, SxGTKRSYxxM for npi-1 and TKRSxxxM for npi-3, different from the basic SV40 large T-type NLS. Additional predominantly basic NLSs have been identified in the NP protein, but it is not clear whether, in the context of the RNP, these NLSs bind the nucleic acid or import factors.

Thogoto virus, a tick-borne orthomyxovirus with six genomic segments of negative polarity and a structure much like influenza virus, provides an interesting example where nuclear trafficking of the incoming viral genome is intercepted by an innate host defense mechanism based on the interferon-induced large GTPases of the Mx family (for a recent review, see Haller and Kochs 2002). Mx proteins were shown to bind the viral RNPs, and this inhibited RNP nuclear import (Kochs and Haller 1999). Yet another example of Mx protein-mediated interference with cytoplasmic trafficking of viral RNPs has been reported with bunyaviruses, namely, La Crosse virus (LACV), an arbovirus causing pediatric encephalitis (Kochs et al. 2002). LACV replicates in the cytoplasm, but in MxA expressing cells the LACV RNPs were sequestered into cytoplasmic fibrillar structures and rendered unavailable for replication. It will be interesting to see whether the cytoplasmically sequestered genomes are targeted for degradation or whether the cell enters a self destruction program before producing progeny virus.

7
Perspectives

The ongoing molecular analysis of viral import into the nucleus has revealed that incoming viruses undergo several conformational changes that allow them to destabilize the protective capsid and in some cases increase their nuclear affinity. In the case of DNA and RNA retroviruses, some of these changes are linked to genome maturation processes in the cytoplasm, rendering the capsids competent for cytoplasmic transport and interactions with the nuclear pore complexes. Viral nucleoproteins interact with the NPC both directly and through cellular import factors. Direct attachment to the NPC is particularly effective in the case of viruses that have a relatively hydrophobic surface such as the Ads or the parvoviruses. These direct interactions can serve to position the particle such that it becomes competent for releasing the nucleic acid. Alternatively, direct interactions can lead to capsid translocation through the pore, resulting in infection after intranuclear genome uncoating. Important challenges for future research include the identification of the nuclear, cytoplasmic, and viral factors required for genome uncoating, identification of the factors involved in NPC translocation of the nucleoprotein complexes, and tackling the subnuclear events immediately after nuclear import. We expect that research on nuclear import of incoming viral particles will continue to reveal basic cellular mechanisms, which might include new aspects of the innate host antiviral defense, similar to antiviral signaling processes.

Acknowledgements Research in the lab of UFG is supported by funds from the Swiss National Science Foundation, the Cancer League of the Canton of Zürich, and the Canton of Zürich. MF is supported by a grant from The Netherlands Cancer Institute.

References

Adachi, M, Fukuda, M, and Nishida, E (1999) Two co-existing mechanisms for nuclear import of MAP kinase: passive diffusion of a monomer and active transport of a dimer. EMBO J 18:5347–58

Allen, TD, Cronshaw, JM, Bagley, S, Kiseleva, E, and Goldberg, MW (2000) The nuclear pore complex: mediator of translocation between nucleus and cytoplasm. J Cell Sci 113:1651–9

Andrade, F, Bull, HG, Thornberry, NA, Ketner, GW, Casciola-Rosen, LA, and Rosen, A (2001) Adenovirus L4-100K assembly protein is a granzyme B substrate that potently inhibits granzyme B-mediated cell death. Immunity 14:751–761

Ashok, A, and Atwood, WJ (2003) Contrasting roles of endosomal pH and the cytoskeleton in infection of human glial cells by JC virus and simian virus 40. J Virol 77:1347–56

Askjaer, P, Bachi, A, Wilm, M, Bischoff, FR, Weeks, DL, Ogniewski, V, Ohno, M, Niehrs, C, Kjems, J, Mattaj, IW, and Fornerod, M (1999) RanGTP-regulated interactions of CRM1 with nucleoporins and a shuttling DEAD-box helicase. Mol Cell Biol 19:6276–85

Balasundaram, D, Benedik, MJ, Morphew, M, Dang, VD, and Levin, HL (1999) Nup124p is a nuclear pore factor of *Schizosaccharomyces pombe* that is important for nuclear import and activity of retrotransposon Tf1. Mol Cell Biol 19:5768–84

Bartlett, JS, Wilcher, R, and Samulski, RJ (2000) Infectious entry pathway of adeno-associated virus and adeno-associated virus vectors. J Virol 74:2777–2785

Becskei, A, and Mattaj, IW (2003) The strategy for coupling the RanGTP gradient to nuclear protein export. Proc Natl Acad Sci U S A 100:1717–22

Bednenko, J, Cingolani, G, and Gerace, L (2003) Nucleocytoplasmic transport: navigating the channel. Traffic 4:127–35

Bell, P, Montaner, LJ, and Maul, GG (2001) Accumulation and intranuclear distribution of unintegrated human immunodeficiency virus type 1 DNA. J Virol 75:7683–91

Ben-Efraim, I, and Gerace, L (2001) Gradient of increasing affinity of importin beta for nucleoporins along the pathway of nuclear import. J Cell Biol 152:411–8

Besnier, C, Takeuchi, Y, and Towers, G (2002) Restriction of lentivirus in monkeys. Proc Natl Acad Sci U S A 99:11920–5

Boeke, JD, and Stoye, J.P. (1997) Retrotransposons, endogenous retroviruses and the evolution of retroelements. In: JM Coffin, SH Hughes and HE Varmus eds) Retroviruses. Cold Spring Harbor Press, Cold Spring Harbor, NY, pp. 343–435

Bousarghin, L, Touze, A, Sizaret, PY, and Coursaget, P (2003) Human papillomavirus types 16, 31, and 58 use different endocytosis pathways to enter cells. J Virol 77:3846–50

Bouyac-Bertoia, M, Dvorin, JD, Fouchier, RA, Jenkins, Y, Meyer, BE, Wu, LI, Emerman, M, and Malim, MH (2001) HIV-1 infection requires a functional integrase NLS. Mol Cell 7:1025–35

Burnett, RM (1997) The structure of adenovirus. *In*: W Chiu, RM Burnett and RL Garcea eds) Structural biology of viruses. Oxford Press, Oxford, pp. 209–238

Caruso, M, Belloni, L, Sthandier, O, Amati, P, and Garcia, MI (2003) Alpha4beta1 integrin acts as a cell receptor for murine polyomavirus at the postattachment level. J Virol 77:3913–21

Chardonnet, Y, and Dales, S (1970) Early events in the interaction of adenoviruses with HeLa cells. I. Penetration of type 5 and intracellular release of the DNA genome. Virology 40:462–477

Chardonnet, Y, and Dales, S (1972) Early events in the interaction of adenoviruses with HeLa cells: III. Relationship between an ATPase activity in nuclear envelopes and transfer of core material: a hypothesis. Virology 48:342–359

Cole, CN (1996) Polyomavirinae: The viruses and their replication. *In:*BN Fields, DM Knipe and PM Howley eds) Fundamental Virology. Lippincott-Raven, New York, pp. 917–945

Colman, PM, and Lawrence, MC (2003) The structural biology of type I viral membrane fusion. Nat Rev Mol Cell Biol 4:309–19

Colon-Ramos, DA, Salisbury, JL, Sanders, MA, Shenoy, SM, Singer, RH, and Garcia-Blanco, MA (2003) Asymmetric distribution of nuclear pore complexes and the cytoplasmic localization of beta2-tubulin mRNA in *Chlamydomonas reinhardtii*. Dev Cell 4:941–52

Cooper, GM, Greenberg-Temin, R, and Sugden, B (1995) The DNA provirus, Howard Temin's scientific legacy, American Society for Microbiology, Washington,

Cotten, M, and Weber, JM (1995) The adenovirus protease is required for virus entry into host cells. Virology 213:494–502

Cronshaw, JM, Krutchinsky, AN, Zhang, W, Chait, BT, and Matunis, MJ (2002) Proteomic analysis of the mammalian nuclear pore complex. J Cell Biol 158:915–27

Daigle, N, Beaudouin, J, Hartnell, L, Imreh, G, Hallberg, E, Lippincott-Schwartz, J, and Ellenberg, J (2001) Nuclear pore complexes form immobile networks and have a very low turnover in live mammalian cells. J Cell Biol 154:71–84

Dales, S, and Chardonnet, Y (1973) Early events in the interaction of adenoviruses with HeLa cells. IV. Association with microtubules and the nuclear pore complex during vectorial movement of the inoculum. Virol. 56:465–483

Dang, VD, and Levin, HL (2000) Nuclear import of the retrotransposon Tf1 is governed by a nuclear localization signal that possesses a unique requirement for the FXFG nuclear pore factor Nup124p. Mol Cell Biol 20:7798–812

Day, PM, Lowy, DR, and Schiller, JT (2003) Papillomaviruses infect cells via a clathrin-dependent pathway. Virology 307:1–11

de Noronha, CM, Sherman, MP, Lin, HW, Cavrois, MV, Moir, RD, Goldman, RD, and Greene, WC (2001) Dynamic disruptions in nuclear envelope architecture and integrity induced by HIV-1 Vpr. Science 294:1105–8

Denning, DP, Patel, SS, Uversky, V, Fink, AL, and Rexach, M (2003) Disorder in the nuclear pore complex: the FG repeat regions of nucleoporins are natively unfolded. Proc Natl Acad Sci U S A 100:2450–5

Eckstein, DA, Sherman, MP, Penn, ML, Chin, PS, De Noronha, CM, Greene, WC, and Goldsmith, MA (2001) HIV-1 Vpr enhances viral burden by facilitating infection of tissue macrophages but not nondividing CD4+ T cells. J Exp Med 194:1407–19

Englmeier, L, Olivo, JC, and Mattaj, IW (1999) Receptor-mediated substrate translocation through the nuclear pore complex without nucleotide triphosphate hydrolysis. Curr Biol 9:30–41

Fagotto, F, Gluck, U, and Gumbiner, BM (1998) Nuclear localization signal-independent and importin/karyopherin-independent nuclear import of beta-catenin. Curr Biol 8:181–90

Fahrenkrog, B, Stoffler, D, and Aebi, U (2001) Nuclear pore complex architecture and functional dynamics. Curr Top Microbiol Immunol 259:95–117

Farjot, G, Sergeant, A, and Mikaelian, I (1999) A new nucleoporin-like protein interacts with both HIV-1 Rev nuclear export signal and CRM-1. J Biol Chem 274: 17309–17

Fassati, A, and Goff, SP (2001) Characterization of intracellular reverse transcription complexes of human immunodeficiency virus type 1. J Virol 75:3626–35

Fassati, A, Gorlich, D, Harrison, I, Zaytseva, L, and Mingot, JM (2003) Nuclear import of HIV-1 intracellular reverse transcription complexes is mediated by importin 7. EMBO J 22:3675–3685

Fausch, SC, Da Silva, DM, and Kast, WM (2003) Differential uptake and cross-presentation of human papillomavirus virus-like particles by dendritic cells and Langerhans cells. Cancer Res 63:3478–82

Florin, L, Schafer, F, Sotlar, K, Streeck, RE, and Sapp, M (2002) Reorganization of nuclear domain 10 induced by papillomavirus capsid protein l2. Virology 295:97–107

Fornerod, M, and Ohno, M (2002) Exportin-mediated nuclear export of proteins and ribonucleoproteins. Results Probl Cell Differ 35:67–91

Fouchier, RA, Meyer, BE, Simon, JH, Fischer, U, Albright, AV, Gonzalez-Scarano, F, and Malim, MH (1998) Interaction of the human immunodeficiency virus type 1 Vpr protein with the nuclear pore complex. J Virol 72:6004–13

Fuller, S (2003) Snapshots of viral maturation. Nat Struct Biol 10:322–3

Gallay, P, Hope, T, Chin, D, and Trono, D (1997) HIV-1 infection of nondividing cells through the recognition of integrase by the importin/karyopherin pathway. Proc Natl Acad Sci U S A 94:9825–30

Ganem, D, and Schneider, R.J. (2001) Hepadnaviridae: the viruses and their replication. In: BN Fields, DM Knipe and PM Howley eds) Fundamental Virology. Lippincott Williams & Wilkins, Philadelphia, pp. 1285–1331

Geles, KG, Johnson, JJ, Jong, S, and Adam, SA (2002) A role for *Caenorhabditis elegans* importin IMA-2 in germ line and embryonic mitosis. Mol Biol Cell 13:3138–47

Gerace, L, and Burke, B (1988) Functional organization of the nuclear envelope. Annu Rev Cell Biol 4:335–374

Goff, SP (2001) Intracellular trafficking of retroviral genomes during the early phase of infection: viral exploitation of cellular pathways. J Gene Med 3:517–28

Gorlich, D, and Kutay, U (1999) Transport between the cell nucleus and the cytoplasm. Annu Rev Cell Dev Biol 15:607–60

Grandi, P, Schlaich, N, Tekotte, H, and Hurt, EC (1995) Functional interaction of Nic96p with a core nucleoporin complex consisting of Nsp1p, Nup49p and a novel protein Nup57p. EMBO J 14:76–87

Granzow, H, Weiland, F, Jons, A, Klupp, BG, Karger, A, and Mettenleiter, TC (1997) Ultrastructural analysis of the replication cycle of pseudorabies virus in cell culture: a reassessment. J Virol 71:2072–82

Greber, UF (1998) Virus assembly and disassembly: the adenovirus cysteine protease as a trigger factor. Rev Med Virol 8:213–222

Greber, UF (2002) Signalling in viral entry. Cell Mol Life Sci 59:608–626

Greber, UF, and Carafoli, E (2002) Signalling takes control of nucleo-cytoplasmic transport. EMBO Rep 3:410–414

Greber, UF, and Fassati, A (2003) Nuclear import of viral DNA genomes. Traffic 4:136–43

Greber, UF, and Gerace, L (1995) Depletion of calcium from the lumen of the endoplasmic reticulum reversibly inhibits passive diffusion and signal-mediated transport into the nucleus. J. Cell Biol 128:5–14

Greber, UF, and Kasamatsu, H (1996) Nuclear targeting of adenovirus and simian virus SV40. Trends Cell Biol 6:189–195

Greber, UF, Singh, I, and Helenius, A (1994) Mechanisms of virus uncoating. Trends Microbiol 2:52–56

Greber, UF, Suomalainen, M, Stidwill, RP, Boucke, K, Ebersold, M, and Helenius, A (1997) The role of the nuclear pore complex in adenovirus DNA entry. EMBO J. 16:5998–6007

Greber, UF, Webster, P, Weber, J, and Helenius, A (1996) The role of the adenovirus protease in virus entry into cells. EMBO J. 15:1766–1777

Greber, UF, Willetts, M, Webster, P, and Helenius, A (1993) Stepwise dismantling of adenovirus 2 during entry into cells. Cell 75:477–486

Haller, O, and Kochs, G (2002) Interferon-induced mx proteins: dynamin-like GTPases with antiviral activity. Traffic 3:710–7

Hang, J, and Dasso, M (2002) Association of the human SUMO-1 protease SENP2 with the nuclear pore. J Biol Chem 277:19961–6

Hansen, J, Qing, K, and Srivastava, A (2001) Infection of purified nuclei by adeno-associated virus 2. Mol Ther:4:289–296

Henderson, BR, and Fagotto, F (2002) The ins and outs of APC and beta-catenin nuclear transport. EMBO Rep 3:834–9

Howley, PM (1996) Fundamental Virology. In: BN Fields, DM Knipe and PM Howley (eds). Lippincott-Raven, New York, pp. 947–978

Howley, PM, and Lowy, DR (2001) Papillomaviruses and their replication. In: DM Knipe and PM Howley (eds) Fundamental Virology. Lippincott-Raven, Philadelphia, pp. 1019–51

Huber, J, Dickmanns, A, and Luhrmann, R (2002) The importin-beta binding domain of snurportin1 is responsible for the Ran- and energy-independent nuclear import of spliceosomal U snRNPs in vitro. J Cell Biol 156:467–79

Hurwitz, ME, and Blobel, G (1995) Nup82 is an essential yeast nucleoporin required for poly(A)$^+$ RNA export. J Cell Biol 130:1275–1281

Jaggi, RD, Franco-Obregon, A, Muhlhausser, P, Thomas, F, Kutay, U, and Ensslin, K (2003) Modulation of nuclear pore topology by transport modifiers. Biophys J 84:665–70

Jarnik, M, and Aebi, U (1991) Toward a more complete 3-D structure of the nuclear pore complex. J Struct Biol 107:291–308

Jenkins, Y, McEntee, M, Weis, K, and Greene, WC (1998) Characterization of HIV-1 vpr nuclear import: analysis of signals and pathways. J Cell Biol 143:875–85

Kann, M, Sodeik, B, Vlachou, A, Gerlich, WH, and Helenius, A (1999) Phosphorylation-dependent binding of hepatitis B virus core particles to the nuclear pore complex. J Cell Biol 145:45–55

Kasamatsu, H, and Nakanishi, A (1998) How do animal DNA viruses get to the nucleus? Annu Rev Microbiol 52:627–86

Katahira, J, Strasser, K, Podtelejnikov, A, Mann, M, Jung, JU, and Hurt, E (1999) The Mex67p-mediated nuclear mRNA export pathway is conserved from yeast to human. EMBO J 18:2593–2609

Kehlenbach, RH, Dickmanns, A, Kehlenbach, A, Guan, T, and Gerace, L (1999) A role for RanBP1 in the release of CRM1 from the nuclear pore complex in a terminal step of nuclear export. J Cell Biol 145:645–57

Kehlenbach, RH, and Gerace, L (2000) Phosphorylation of the nuclear transport machinery down-regulates nuclear protein import in vitro. J Biol Chem 275:17848–56

Khokhlatchev, AV, Canagarajah, B, Wilsbacher, J, Robinson, M, Atkinson, M, Goldsmith, E, and Cobb, MH (1998) Phosphorylation of the MAP kinase ERK2 promotes its homodimerization and nuclear translocation. Cell 93:605–15

Knipe, DM, and Smith, JL (1986) A mutant herpesvirus protein leads to a block in nuclear localization of other viral proteins. Mol Cell Biol 6:2371–2381

Kochs G, and Haller O (1999) Interferon-induced human MxA GTPase blocks nuclear import of Thogoto virus nucleocapsids. Proc Natl Acad Sci U S A 96:2082–6

Kochs G, and Haller O. (1999) GTP-bound human MxA protein interacts with the nucleocapsids of Thogoto virus (Orthomyxoviridae). J Biol Chem 274:4370–6

Kochs, G, Janzen, C, Hohenberg, H, and Haller, O (2002) Antivirally active MxA protein sequesters La Crosse virus nucleocapsid protein into perinuclear complexes. Proc Natl Acad Sci U S A 99:3153–8

Kose, S, Imamoto, N, Tachibana, T, Shimamoto, T, and Yoneda, Y (1997) Ran-unassisted nuclear migration of a 97-kD component of nuclear pore-targeting complex. J Cell Biol 139:841–9

Kose, S, Imamoto, N, Tachibana, T, Yoshida, M, and Yoneda, Y (1999) beta-Subunit of nuclear pore-targeting complex (importin-beta) can be exported from the nucleus in a Ran-independent manner. J Biol Chem 274:3946–52

Kuersten, S, Ohno, M, and Mattaj, IW (2001) Nucleocytoplasmic transport:Ran, beta and beyond. Trends Cell Biol 11:497–503

Lamb, RA, Holsinger, LJ, and Pinto, LH (1994) The influenza virus M2 ion channel protein and its role in the Influenza virus life cycle. In: E Wimmer (ed) Cellular receptors for animal viruses. Cold Spring Harbor Laboratory Press, Cold Spring Harbor, NY, pp. 303–321

Lamb, RA, and Krug, RM (2001) Orthomyxoviridae: the viruses and their replication. In: DM Knipe and PM Howley eds) Fundamental virology. Lippincott-Raven, Philadelphia, pp. 725–769

Le Rouzic, E, Mousnier, A, Rustum, C, Stutz, F, Hallberg, E, Dargemont, C, and Benichou, S (2002) Docking of HIV-1 Vpr to the nuclear envelope is mediated by the interaction with the nucleoporin hCG1. J Biol Chem 277:45091–8

Levin, HL, Weaver, DC, and Boeke, JD (1990) Two related families of retrotransposons from *Schizosaccharomyces pombe*. Mol Cell Biol 10:6791–8

Levin, HL, Weaver, DC, and Boeke, JD (1993) Novel gene expression mechanism in a fission yeast retroelement:Tf1 proteins are derived from a single primary translation product. EMBO J 12:4885–95

Lin, SS, Nymark-McMahon, MH, Yieh, L, and Sandmeyer, SB (2001) Integrase mediates nuclear localization of Ty3. Mol Cell Biol 21:7826–38

Lombardo, E, Ramirez, JC, Garcia, J, and Almendral, JM (2002) Complementary roles of multiple nuclear targeting signals in the capsid proteins of the parvovirus minute virus of mice during assembly and onset of infection. J Virol 76:7049–59

Lycke, E, Hamark, B, Johansson, M, Krotochwil, A, Lycke, J, and Svennerholm, B (1988) Herpes simplex virus infection of the human sensory neuron. An electron microscopy study. Arch Virol 101:87–104

Lyman, SK, Guan, T, Bednenko, J, Wodrich, H, and Gerace, L (2002) Influence of cargo size on Ran and energy requirements for nuclear protein import. J Cell Biol 159:55–67

Mabit, H, Breiner, KM, Knaust, A, Zachmann-Brand, B, and Schaller, H (2001) Signals for bidirectional nucleocytoplasmic transport in the duck hepatitis B virus capsid protein. J Virol 75:1968–1977

Mabit, H, Nakano, MY, Prank, U, Saam, B, Döhner, K, Sodeik, B, and Greber, UF (2002) Intact microtubules support Adenovirus and Herpes simplex virus infections. J Virol 76:9962–9971

Matsubayashi, Y, Fukuda, M, and Nishida, E (2001) Evidence for existence of a nuclear pore complex-mediated, cytosol-independent pathway of nuclear translocation of ERK MAP kinase in permeabilized cells. J Biol Chem 276:41755–60

McDonald, D, Vodicka, MA, Lucero, G, Svitkina, TM, Borisy, GG, Emerman, M, and Hope, TJ (2002) Visualization of the intracellular behavior of HIV in living cells. J Cell Biol 159:441–52

Meier, O, and Greber, UF (2003) Adenovirus endocytosis. J. Gene Med. 5:451–462

Merle, E, Rose, RC, LeRoux, L, and Moroianu, J (1999) Nuclear import of HPV11 L1 capsid protein is mediated by karyopherin alpha2beta1 heterodimers. J Cell Biochem 74:628–37

Mettenleiter, TC (2002) Herpesvirus assembly and egress. J Virol 76:1537–47

Miller, MD, Farnet, CM, and Bushman, FD (1997) Human immunodeficiency virus type 1 preintegration complexes: studies of organization and composition. J Virol 71:5382–90

Miyamoto, Y, Hieda, M, Harreman, MT, Fukumoto, M, Saiwaki, T, Hodel, AE, Corbett, AH, and Yoneda, Y (2002) Importin alpha can migrate into the nucleus in an importin beta- and Ran-independent manner. EMBO J 21:5833–42

Morgan, C, Rosenkranz, HS, and Mednis, B (1969) Structure and development of viruses as observed in the electron microscope: X. Entry and uncoating of Adenovirus. J Virol 4:777–796

Nachury, MV, and Weis, K (1999) The direction of transport through the nuclear pore can be inverted. Proc Natl Acad Sci U S A 96:9622–7

Nakanishi, A, Shum, D, Morioka, H, Otsuka, E, and Kasamatsu, H (2002) Interaction of the Vp3 nuclear localization signal with the importin alpha 2/beta heterodimer directs nuclear entry of infecting simian virus 40. J Virol 76:9368–77

Nakano, MY, Boucke, K, Suomalainen, M, Stidwill, RP, and Greber, UF (2000) The first step of adenovirus type 2 disassembly occurs at the cell surface, independently of endocytosis and escape to the cytosol. J Virol 74:7085–95

Nakielny, S, and Dreyfuss, G (1998) Import and export of the nuclear protein import receptor transportin by a mechanism independent of GTP hydrolysis. Curr Biol 8:89–95

Nakielny, S, and Dreyfuss, G (1999) Transport of proteins and RNAs in and out of the nucleus. Cell 99:677–90

Nakielny, S, Shaikh, S, Burke, B, and Dreyfuss, G (1999) Nup153 is an M9-containing mobile nucleoporin with a novel ran-binding domain. EMBO J 18:1982–95

Nelson, LM, Rose, RC, and Moroianu, J (2002) Nuclear import strategies of high risk HPV16 L1 major capsid protein. J Biol Chem 277:23958–64

Nelson, LM, Rose, RC, and Moroianu, J (2003) The L1 major capsid protein of human papillomavirus type 11 interacts with Kap beta2 and Kap beta3 nuclear import receptors. Virology 306:162–9

Nermut, MV, and Fassati, A (2003) Structural analyses of purified human immunodeficiency virus type 1 intracellular reverse transcription complexes. J Virol 77:8196–8206

Newcomb, WW, Juhas, RM, Thomsen, DR, Homa, FL, Burch, AD, Weller, SK, and Brown, JC (2001) The UL6 gene product forms the portal for entry of DNA into the herpes simplex virus capsid. J Virol 75:10923–32

O'Neill, RE, Jaskunas, R, Blobel, G, Palese, P, and Moroianu, J (1995) Nuclear import of influenza virus RNA can be mediated by viral nucleoprotein and transport factors required for protein import. J Biol Chem 270:22701–4

Ojala, PM, Sodeik, B, Ebersold, MW, Kutay, U, and Helenius, A (2000) Herpes simplex virus type 1 entry into host cells: reconstitution of capsid binding and uncoating at the nuclear pore complex in vitro. Mol Cell Biol 20:4922–4931

Pante, N, and Kann, M (2002) Nuclear pore complex is able to transport macromolecules with diameters of about 39 nm. Mol Biol Cell 13:425–434

Pelkmans, L, and Helenius, A (2002) Endocytosis via caveolae. Traffic 3:311–20

Pelkmans, L, Kartenbeck, J, and Helenius, A (2001) Caveolar endocytosis of simian virus 40 reveals a new two-step vesicular-transport pathway to the ER. Nat Cell Biol 3:473–483

Petit, C, Schwartz, O, and Mammano, F (2000) The karyophilic properties of human immunodeficiency virus type 1 integrase are not required for nuclear import of proviral DNA. J Virol 74:7119–26

Pho, MT, Ashok, A, and Atwood, WJ (2000) JC virus enters human glial cells by clathrin-dependent receptor-mediated endocytosis. J Virol 74:2288–2292

Pichler, A, Gast, A, Seeler, JS, Dejean, A, and Melchior, F (2002) The nucleoporin RanBP2 has SUMO1 E3 ligase activity. Cell 108:109–20

Piller, SC, Caly, L, and Jans, DA (2003) Nuclear import of the pre-integration complex (PIC):the Achilles heel of HIV? Curr Drug Targets 4:409–29

Ploubidou, A, and Way, M (2001) Viral transport and the cytoskeleton. Curr Opin Cell Biol 13:97–105

Popov, S, Rexach, M, Ratner, L, Blobel, G, and Bukrinsky, M (1998) Viral protein R regulates docking of the HIV-1 preintegration complex to the nuclear pore complex. J Biol Chem 273:13347–52

Poranen, MM, Daugelavicius, R, and Bamford, DH (2002) Common principles in viral entry. Annu Rev Microbiol 56:521–38
Rabe, B, Vlachou, A, Panté, N, Helenius, A, and Kann, M (2003) Nuclear import of Hepatitis B virus capsids and release of the viral genome. Proc Natl Acad Sci USA in press
Reichelt, R, Holzenburg, A, Buhle, EJ, Jarnik, M, Engel, A, and Aebi, U (1990) Correlation between structure and mass distribution of the nuclear pore complex and of distinct pore complex components. J Cell Biol 110:883–894
Ribbeck, K, and Gorlich, D (2001) Kinetic analysis of translocation through nuclear pore complexes. EMBO J 20:1320–30
Ribbeck, K, and Gorlich, D (2002) The permeability barrier of nuclear pore complexes appears to operate via hydrophobic exclusion. EMBO J 21:2664–71
Ribbeck, K, Kutay, U, Paraskeva, E, and Gorlich, D (1999) The translocation of transportin-cargo complexes through nuclear pores is independent of both Ran and energy. Curr Biol 9:47–50
Rodriguez, MS, Dargemont, C, and Hay, RT (2001) SUMO-1 conjugation in vivo requires both a consensus modification motif and nuclear targeting. J Biol Chem 276:12654–9
Roizman, B, and Knipe, DM (2001) Herpes simplex viruses and their replication. In: DM Knipe and PM Howley eds) Fundamental virology. Lippincott-Raven, Philadelphia, pp. 1123–1183
Rollenhagen, C, Muhlhausser, P, Kutay, U, and Pante, N (2003) Importin beta-depending nuclear import pathways: role of the adapter proteins in the docking and releasing steps. Mol Biol Cell 14:2104–15
Rout, MP, Aitchison, JD, Suprapto, A, Hjertaas, K, Zhao, Y, and Chait, BT (2000) The yeast nuclear pore complex: composition, architecture, and transport mechanism. J. Cell Biol. 148:635–51
Rux, JJ, and Burnett, RM (2000) Type-specific epitope locations revealed by X-ray crystallographic study of adenovirus type 5 hexon. Mol Ther 1:18–30
Ryan, KJ, and Wente, SR (2000) The nuclear pore complex: a protein machine bridging the nucleus and cytoplasm. Curr Opin Cell Biol 12:361–71
Saavedra, CA, Hammell, CM, Heath, CV, and Cole, CN (1997) Yeast heat shock mRNAs are exported through a distinct pathway defined by Rip1p. Genes Dev 11:2845–56
Saphire, ACS, Guan, TL, Schirmer, EC, Nemerow, GR, and Gerace, L (2000) Nuclear import of adenovirus DNA in vitro involves the nuclear protein import pathway and hsc70. J Biol Chem 275:4298–4304
Schwoebel, ED, Talcott, B, Cushman, I, and Moore, MS (1998) Ran-dependent signal-mediated nuclear import does not require GTP hydrolysis by Ran. J Biol Chem 273:35170–5
Selinka, HC, Giroglou, T, and Sapp, M (2002) Analysis of the infectious entry pathway of human papillomavirus type 33 pseudovirions. Virology 299:279–287
Shah, S, Tugendreich, S, and Forbes, D (1998) Major binding sites for the nuclear import receptor are the internal nucleoporin Nup153 and the adjacent nuclear filament protein Tpr. J Cell Biol 141:31–49

Sherman, MP, de Noronha, CM, Heusch, MI, Greene, S, and Greene, WC (2001) Nucleocytoplasmic shuttling by human immunodeficiency virus type 1 Vpr. J Virol 75:1522–32

Shibayama, S, Shibata-Seita, R, Miura, K, Kirino, Y, and Takishima, K (2002) Identification of a C-terminal region that is required for the nuclear translocation of ERK2 by passive diffusion. J Biol Chem 277:37777–82

Sieczkarski, SB, and Whittaker, GR (2002) Dissecting virus entry via endocytosis. J Gen Virol 83:1535–45

Smith, AE, Slepchenko, BM, Schaff, JC, Loew, LM, Macara, IG, Smith, A, and Brownawell, A (2002) Systems analysis of Ran transport. Science 295:488–91

Sodeik, B (2000) Mechanisms of viral transport in the cytoplasm. Trends Microbiol 8:465–472

Sodeik, B, Ebersold, MW, and Helenius, A (1997) Microtubule-mediated transport of incoming Herpes Simplex Virus 1 capsids to the nucleus. J Cell Biol 136:1007–1021

Spear, PG (2002) Viral interactions with receptors in cell junctions and effects on junctional stability. Dev Cell 3:462–464

Stehle, T, and Dermody, TS (2003) Structural evidence for common functions and ancestry of the reovirus and adenovirus attachment proteins. Rev Med Virol 13:123–32

Stoffler, D, Fahrenkrog, B, and Aebi, U (1999) The nuclear pore complex: from molecular architecture to functional dynamics. Cur Opin Cell Biol 11:391–401

Strahm, Y, Fahrenkrog, B, Zenklusen, D, Rychner, E, Kantor, J, Rosbach, M, and Stutz, F (1999) The RNA export factor Gle1p is located on the cytoplasmic fibrils of the NPC and physically interacts with the FG-nucleoporin Rip1p, the DEAD-box protein Rat8p/Dbp5p and a new protein Ymr 255p. EMBO J 18:5761–77

Strom, AC, and Weis, K (2001) Importin-beta-like nuclear transport receptors. Genome Biol 2:REVIEWS3008

Suntharalingam, M, and Wente, SR (2003) Peering through the pore. Nuclear pore complex structure, assembly, and function. Dev Cell 4:775–89

Trono, D (2003) Virology. Picking the right spot. Science 300:1670–1

Trotman, LC, Achermann, DP, Keller, S, Straub, M, and Greber, UF (2003) Non-classical export of an Adenovirus structural protein. Traffic 4:390–402

Trotman, LC, Mosberger, N, Fornerod, M, Stidwill, RP, and Greber, UF (2001) Import of adenovirus DNA involves the nuclear pore complex receptor CAN/Nup214 and histone H1. Nat Cell Biol 3:1092–1100

Turelli, P, Doucas, V, Craig, E, Mangeat, B, Klages, N, Evans, R, Kalpana, G, and Trono, D (2001) Cytoplasmic recruitment of INI1 and PML on incoming HIV preintegration complexes: interference with early steps of viral replication. Mol Cell 7:1245–54

Vasu, SK, and Forbes, DJ (2001) Nuclear pores and nuclear assembly. Curr Opin Cell Biol 13:363–75

Vihinen-Ranta, M, Wang, D, Weichert, WS, and Parrish, CR (2002) The VP1N-terminal sequence of canine parvovirus affects nuclear transport of capsids and efficient cell infection. J Virol 76:1884–1891

Vodicka, MA, Koepp, DM, Silver, PA, and Emerman, M (1998) Hiv-1 Vpr interacts with the nuclear transport pathway to promote macrophage infection. Genes Dev 12:175–185

Walters, RW, Freimuth, P, Moninger, TO, Ganske, I, Zabner, J, and Welsh, MJ (2002) Adenovirus fiber disrupts CAR-mediated intercellular adhesion allowing virus escape. Cell 110:789–799

Walther, TC, Fornerod, M, Pickersgill, H, Goldberg, M, Allen, TD, and Mattaj, IW (2001) The nucleoporin Nup153 is required for nuclear pore basket formation, nuclear pore complex anchoring and import of a subset of nuclear proteins. EMBO J 20:5703–5714

Walther, TC, Pickersgill, HS, Cordes, VC, Goldberg, MW, Allen, TD, Mattaj, IW, and Fornerod, M (2002) The cytoplasmic filaments of the nuclear pore complex are dispensable for selective nuclear protein import. J Cell Biol 158:63–77

Wang, P, Palese, P, and Oneill, RE (1997) The Npi-1/Npi-3 (karyopherin alpha) binding site on the influenza a virus nucleoprotein np is a nonconventional nuclear localization signal. J Virol 71:1850–1856

Weichert, WS, Parker, JSL, Wahid, ATM, Chang, SF, Meier, E, and Parrish, CR (1998) Assaying for structural variation in the parvovirus capsid and its role in infection. Virology 250:106–117

Wente, SR (2000) Gatekeepers of the nucleus. Science 288:1374–7

Whittaker, GR (2003) Virus nuclear import. Adv Drug Deliv Rev 55:733–47

Whittaker, GR, Kann, M, and Helenius, A (2000) Viral entry into the nucleus. Annu Rev Cell Dev Biol 16:627–51

Wiegand, HL, Coburn, GA, Zeng, Y, Kang, Y, Bogerd, HP, and Cullen, BR (2002) Formation of Tap/NXT1 heterodimers activates Tap-dependent nuclear mRNA export by enhancing recruitment to nuclear pore complexes. Mol Cell Biol 22:245–56

Wimmer, E (1994) Cellular receptors for animal viruses. E Wimmer, ed., Cold Spring Harbor Laboratory Press, Cold Spring Harbor, NY,

Wisnivesky, JP, Leopold, PL, and Crystal, RG (1999) Specific binding of the adenovirus capsid to the nuclear envelope. Hum Gene Ther 10:2187–95

Xiao, W, Warrington, KH, Hearing, P, Hughes, J, and Muzyczka, N (2002) Adenovirus-facilitated nuclear translocation of adeno-associated virus type 2. J Virol 76:11505–11517

Yokoya, F, Imamoto, N, Tachibana, T, and Yoneda, Y (1999) beta-Catenin can be transported into the nucleus in a Ran-unassisted manner. Mol Biol Cell 10:1119–31

Zennou, V, Petit, C, Guetard, D, Nerhbass, U, Montagnier, L, and Charneau, P (2000) HIV-1 genome nuclear import is mediated by a central DNA flap. Cell 101:173–85

Zhang, H, Saitoh, H, and Matunis, MJ (2002) Enzymes of the SUMO modification pathway localize to filaments of the nuclear pore complex. Mol Cell Biol 22:6498–508

zur Hausen, H (2002) Papillomaviruses and cancer: from basic studies to clinical application. Nat Rev Cancer 2:342–50

Viral RNA Replication in Association with Cellular Membranes

A. Salonen · T. Ahola · L. Kääriäinen (✉)

Program in Cellular Biotechnology, Institute of Biotechnology, Viikki Biocenter, University of Helsinki, P.O. Box 56, 00014 Helsinki, Finland
leevi.kaariainen@helsinki.fi

1	Introduction	140
2	Alphaviruses as Models	141
2.1	RNA Replication in Cytoplasmic Vacuoles Derived from Endosomes	141
2.2	nsP1 as the Membrane Anchor of the Replication Complex	143
2.3	Membrane Binding Mechanism of nsP1	144
2.4	Polyprotein Conducts the Assembly and Targeting of the Replication Complex	147
3	Alphavirus-Like Superfamily	150
4	Picornavirus-Like Superfamily	154
4.1	Poliovirus as a Model	154
4.2	Other Members of the Picornavirus Superfamily	158
5	Flavivirus-Like Superfamily	159
6	Nidoviruses	162
7	Concluding Remarks	163
7.1	Targeting of the Replicase Complex	163
7.2	Mechanisms of Membrane Binding	164
7.3	Membrane Modification by Replicase Proteins	164
7.4	Functional Implications of Membrane Attachment	167
	References	167

Abstract All plus-strand RNA viruses replicate in association with cytoplasmic membranes of infected cells. The RNA replication complex of many virus families is associated with the endoplasmic reticulum membranes, for example, picorna-, flavi-, arteri-, and bromoviruses. However, endosomes and lysosomes (togaviruses), peroxisomes and chloroplasts (tombusviruses), and mitochondria (nodaviruses) are also used as sites for RNA replication. Studies of individual nonstructural proteins, the virus-specific components of the RNA replicase, have revealed that the replication complexes are associated with the membranes and targeted to the respective organelle by the ns proteins rather than RNA. Many ns proteins have hydrophobic sequences and may transverse the membrane like polytopic integral membrane proteins, whereas others interact with membranes monotopically. Hepatitis C virus ns proteins offer examples of polytopic transmembrane proteins (NS2, NS4B), a "tip-anchored" protein attached to the membrane by an amphipathic α-helix (NS5A) and a "tail-anchored" posttranslationally inserted protein (NS5B). Semliki Forest virus nsP1 is attached to the plasma membrane by a specific binding peptide in the middle of the protein, which forms an amphipathic α-helix. Interaction of nsP1 with membrane lipids is essential for its capping enzyme activities. The other soluble replicase proteins are directed to the endo-lysosomal membranes only as part of the initial polyprotein. Poliovirus ns proteins utilize endoplasmic reticulum membranes from which vesicles are released in COPII coats. However, these vesicles are not directed to the normal secretory pathway, but accumulate in the cytoplasm. In many cases the replicase proteins induce membrane invaginations or vesicles, which function as protective environments for RNA replication.

1
Introduction

The genome replication of all plus-strand RNA viruses infecting eukaryotic cells is associated with cellular membranes. The membranes can be derived from the endoplasmic reticulum (ER), or other organelles of the secretory pathway, mitochondria, chloroplasts, or from the endo-lysosomal compartment. The membrane association provides a structural framework for replication, it fixes the RNA replication process to a spatially confined place, increasing the local concentration of necessary components, and it offers protection for the alien RNA molecules against host defense mechanisms. The theme of immobilized polymerase and moving template may in fact be common also to most cellular DNA replication and transcription systems (Cook 1999) and might therefore reflect a primordial pathway in nucleic acid replication. The modes of membrane binding and targeting to specific intracellular organelles of the replication complexes of different viruses are so far poorly understood. However, this field at the interface of virology, cell biology, and biochemistry is attracting increased interest, as it represents an ancient feature shared by many virus groups. Some aspects have been nicely treated in earlier reviews (de Graaff and Jaspars 1994; Buck 1996). We will present in some detail relevant studies on alphaviruses, especially Semliki Forest virus (SFV), which has been the object of our own interest. We will then review recent work on other viruses based on their classification in three superfamilies (Koonin and Dolja 1993), except that we have treated nidoviruses as a separate group.

2
Alphaviruses as Models

2.1
RNA Replication in Cytoplasmic Vacuoles Derived from Endosomes

Association of SFV-specific RNA synthesis with membranes was demonstrated in several studies starting in the late 1960s (for reviews see Kääriäinen and Söderlund 1978; Kääriäinen and Ahola 2002). Simple fractionation of membranes derived from the postnuclear supernatant fraction of alphavirus-infected cells showed that essentially all RNA polymerase activity was associated with a "mitochondrial" pellet fraction sedimenting at 15,000 x g. On the other hand, early electron microscopic (EM) studies had revealed cytoplasmic structures typical for alphavirus-infected cells. These were designated as "cytopathic vacuoles type I" (CPV-I), hereafter referred as CPVs. Their size varied from 600 nm to 2,000 nm, and their surface consisted of small vesicular invaginations or spherules, of homogenous size, with a diameter of about 50 nm. EM autoradiography of SFV-infected cells pulse-labeled with tritiated uridine already suggested that CPVs, and possibly the spherules, were involved in virus-specific, actinomycin D-resistant RNA synthesis (Grimley et al. 1968).

However, the origin, nature, and function of CPVs remained unclear for about two decades until Froshauer et al. (1988) demonstrated that they were modified endosomes and lysosomes. Immunofluorescence and immuno-EM techniques showed that Sindbis virus-specific nonstructural proteins nsP3 and nsP4 were associated with CPVs. The authors proposed that CPVs were derived from endosomes, which were participating in the internalization of virus particles. This would have nicely explained the endosomal origin of CPVs as a direct consequence of fusion of the virus envelope with the endo/lysosomal membrane, which would bring the virus nucleocapsid and genome directly to the cytoplasmic surface of the organelle. Thus genome uncoating and subsequent synthesis of replicase components would result in the modification of endosomes to virus-specific CPVs. However, this hypothesis cannot explain why the amount of CPVs was not dependent on the amount of infecting virions. Moreover, typical CPVs were seen also in cells transfected with the genomic RNA of SFV, demonstrating that the endosomal targeting of the replication complexes must be a posttranslational event (Peränen and Kääriäinen 1991).

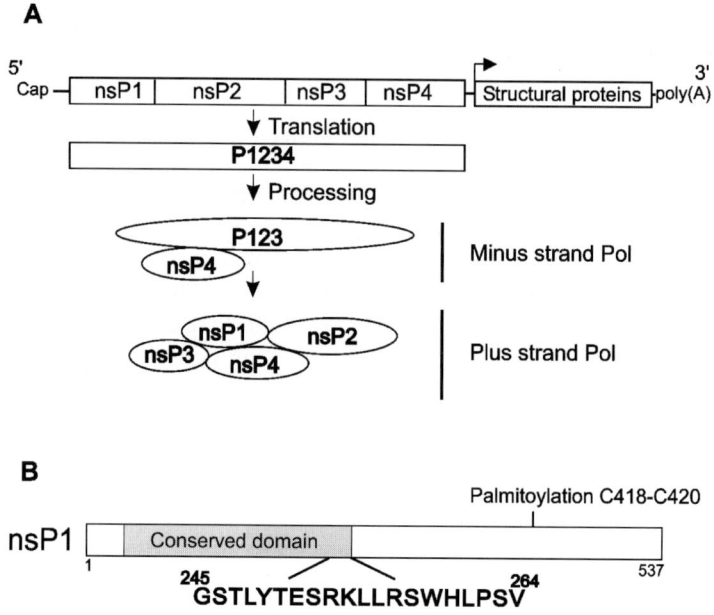

Fig. 1. A Genome organization of SFV. The translation and processing products relevant for SFV replication are shown. Physical interactions have been identified between nsP1 and nsP4, as well as nsP1 and nsP3 (Salonen et al. 2003). B Scheme of SFV nsP1 showing the two regions responsible for membrane binding. The amino acid sequence of the lipid binding peptide is given in *single-letter code*, and the position of palmitoylated cysteines is marked

Alphavirus nonstructural (=replicase) proteins are synthesized as a polyprotein precursor P1234, which is processed in a highly regulated manner into the individual components nsP1–nsP4 (Fig. 1A). Genetic and biochemical experiments have revealed many of the functions of the nsPs (reviewed in Strauss and Strauss 1994; Kääriäinen and Ahola 2002). Thus nsP4 is the catalytic RNA-dependent RNA polymerase subunit, nsP2 is involved in the regulation of the synthesis of the subgenomic 26S mRNA coding for structural proteins of the virion, whereas nsP1 is needed in the synthesis of the complementary (minus strand) RNA early in infection. nsP3 is essential for infection, but no specific function has been assigned for it as yet. Expression of the individual nsPs in *E. coli* and in insect cells revealed further functions of nsPs. nsP1 is an RNA capping enzyme with unique methyltransferase and guanylyltransferase activities (Mi and Stollar 1991; Ahola and Kääriäinen 1995), whereas nsP2 turned out to be a NTPase and RNA helicase (Gomez de Cedrón at al. 1999), RNA triphosphatase (Vasiljeva et al. 2000), as well as the protease responsible for the processing of the nonstructural polyprotein precursor (Vasiljeva et al. 2001).

Creation of potent monospecific antibodies allowed the identification and localization of the ns proteins in SFV-infected cells (Peränen et al. 1988; 1995). After crude cell fractionation most of nsP1, nsP3, and nsP4 were associated with the P15 membrane fraction, whereas about 50% of nsP2 was found in the nucleus (Peränen et al. 1990). Pairwise double staining with different anti-nsP antibodies revealed costaining of CPV-like structures in immunofluorescence microscopy, suggesting that all four nsPs were associated with CPVs. This was confirmed by double-labeling in cryo-immuno EM (Kujala et al. 2001). Moreover, bromouridine given in short pulses also localized to CPVs and spherules together with the nsPs, indicating that these structures were the sites of RNA replication. The CPVs costained with late endosomal (lamp-1, lamp-2, and rab7) and lysosomal markers (LBPA and LysoTracker). Interestingly, all nsPs had localization sites also outside of the CPVs. nsP2 was found in the nucleus, nsP1 at the plasma membrane, nsP3 in cytoplasmic spotlike structures, and nsP4 diffusely in the cytoplasm (Kujala et al. 2001). Therefore, only a fraction of nsPs are present in the actual replication complexes.

2.2
nsP1 as the Membrane Anchor of the Replication Complex

As none of the alphavirus nsPs has sequences typical for transmembrane proteins, we have studied their membrane binding by expressing them individually in BHK, HeLa, and insect cells. These studies revealed that only nsP1 had a specific association with membranes (Peränen et al. 1995), whereas nsP2 on its own was transported almost quantitatively to the nucleus and nsP3 was in cytoplasmic aggregates (Salonen et al. 2003), which in light microscopy gives an impression of vesicles of variable size (Vihinen et al. 2001). Finally, nsP4 was distributed diffusely in the cytoplasm.

Thus nsP1 was a promising candidate as the membrane anchor of the SFV replication complex. nsP1 turned out to be very tightly membrane bound, as the association was not sensitive to high salt, EDTA, or alkaline sodium carbonate treatments, which release peripheral membrane proteins (Peränen et al. 1995). The tight binding was due to palmitoylation of cysteine residues 418–420 (Fig. 1B). When these residues were mutated to alanines, nsP1 was still membrane associated, but less tightly, as it could now be released by high-salt treatment. Thus elimination of palmitoylation altered nsP1 from an "integral" to a "peripheral" membrane protein (Laakkonen et al. 1996).

To study the significance of the palmitoylation of nsP1, the C418–420A mutation was introduced to the infectious cDNA of SFV, followed by transcription of genomic RNA, which was used for transfection of BHK cells. Infectious virus was released to the medium, indicating that palmitoylation of nsP1 was not essential for virus replication. However, there was some retardation in the kinetics of virus growth. Analysis of the membrane association of wild-type and palmitoylation-negative mutant (1pa−) replicase proteins of SFV showed that 1pa− nsP1 was bound less tightly to the membranes than the wild-type protein. Typical CPVs with spherules, indistinguishable from those in wild-type SFV-infected cells, were seen in EM. The same results were obtained when the single palmitoylated cysteine residue (C420) of Sindbis virus nsP1 was mutated to alanine. However, the SFV 1pa− mutant was apathogenic for mouse. After intraperitoneal infection blood viremia was detected, but no infectious virus was found in the brain (Ahola et al. 2000).

2.3
Membrane Binding Mechanism of nsP1

Because palmitoylation was not the decisive mechanism for membrane binding of nsP1, we studied the peripheral binding by producing the wild-type protein in *E. coli*, which cannot palmitoylate proteins. Enzymatically active nsP1 was associated with bacterial membranes as judged by flotation in sucrose gradients. In vitro translated nsP1 also associated with liposomes containing 20%–50% phosphatidylserine (PS) or other anionic phospholipids, but not with liposomes containing only phosphatidylcholine (PC). Solubilization of membranes containing nsP1 by detergents, such as Triton X-100 or octylglucoside, resulted in loss of the protein's methyltransferase and guanylyltransferase activities, which could be reactivated by reconstitution of nsP1 into vesicle membranes or into mixed detergent-lipid micelles containing anionic phospholipids (Fig. 2A). Detergents also inhibited the binding of the methyl donor S-adenosylmethionine to nsP1, and the binding was resumed under the same conditions as enzymatic activity. Thus binding to anionic phospholipids causes a conformational change, which activates the protein (Ahola et al. 1999).

The membrane-binding site in nsP1 was identified by site-directed mutagenesis and deletion mapping in both bacterial and in vitro expres-

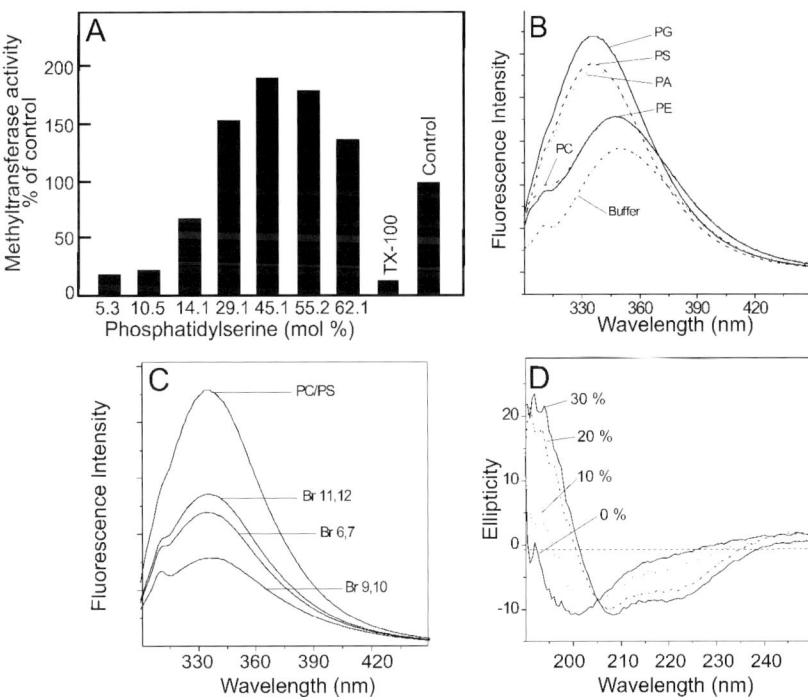

Fig. 2. Biochemical characterization of the lipid binding of SFV nsP1. **A** Inactivation and activation of nsP1 methyltransferase activity. Triton X-100 inactivates nsP1, compared with the control reaction in the absence of detergent (on the *right*). When TX-100 micelles containing increasing amounts of phosphatidylserine are added, nsP1 regains activity, even exceeding control levels at optimal concentration of the lipid. (Reproduced from Ahola et al. 1999, with permission). **B** Negatively charged phospholipids increase the intensity of tryptophan fluorescence of the lipid binding peptide. Tryptophan emission spectrum was recorded in the buffer or in the presence of small unilamellar vesicles consisting of phosphatidylcholine (*PC*) or PC with 30 mol % of phosphatidylglycerol (*PG*), phosphatidylserine (*PS*), phosphatidic acid (*PA*), or phosphatidylethanolamine (*PE*). **C** Depth of tryptophan W259 penetration to membrane measured by quenching of tryptophan fluorescence by brominated PCs. The position of bromine in the acyl chains is indicated. **D** The peptide adopts an α-helical conformation in the presence of PC-containing vesicles, as measured by circular dichroism spectroscopy. The numbers indicate the mol % of PS in the vesicles. (**B–D** reproduced from Lampio et al. 2000 with permission by the American Society of Biochemistry and Molecular Biology)

sion systems. Flotation of nsP1 with membranes or liposomes in discontinuous sucrose gradients was used as a criterion for membrane association. By this means a putative binding region of about 20 amino acid residues, starting from Gly245, was identified (Fig. 1B). The corresponding synthetic peptide consisting of Gly245–Val264 (GSTLYTESRKLLR-SWHLPSV) was able to compete with the binding of in vitro synthesized nsP1 to liposomes containing PS, strongly suggesting that this region of nsP1 is responsible for its membrane association (Ahola et al. 1999).

The interaction of the synthetic membrane binding peptide with liposomes was assayed by utilizing the fluorescence of tryptophan residue W259 (Lampio et al. 2000). Tryptophan emission spectrum changes when it is embedded into an apolar environment. There was a marked increase in the fluorescence intensity and a blue shift of the emission in the presence of monolamellar liposomes, which consisted of PC and negatively charged phospholipids (PS, phosphatidylglycerol, or phosphatidic acid) (Fig. 2B). By using phospholipids with bromide substitution in different carbon atoms of the acyl chain, for quenching of the tryptophan fluorescence, it was possible to estimate that W259 penetrated to the level of carbon atoms 9 and 10 of the PC acyl chains in the outer leaflet of the liposomes (Figs. 2C and 3A). The circular dichroism spectrum of the binding peptide was dependent on the content of the apolar constituents (liposomes or trifluoroethanol). In a buffer solution the binding peptide was mostly in a random coil, whereas in the presence of liposomes with 20%–30% PS or in 30%–50% trifluoroethanol the peptide attained an α-helical conformation (Fig. 2D). The solution structure of the binding peptide, determined by NMR spectroscopy in 30% trifluoroethanol, revealed an amphipathic α-helix (Fig. 3A). One face consisted of hydrophobic residues, leucines 248, 255, 256, and 261, valine 264, and residues S252 and W259 interacting with the apolar fatty acid chains on the cytoplasmic leaflet of the membrane (Fig. 3). The other face contained positively charged residues R253, K254, and R257 lying parallel to the polar head groups of the bilayer surface. The hydrophobic surface of the peptide is rather stable, whereas the polar residues show considerable mobility in trifluoroethanol.

Point mutations R253E and W259A in the nsP1 protein, when expressed in *E. coli*, resulted in the loss of enzymatic activity and the lack of ability to float with membranes in sucrose gradients (Ahola et al. 1999). Thus these two residues were considered to be essential in the interaction of nsP1 with membranes. This view was supported by competition experiments done with synthetic mutant peptides. Both were unable to inhibit the binding of in vitro synthesized nsP1 to liposomes, in

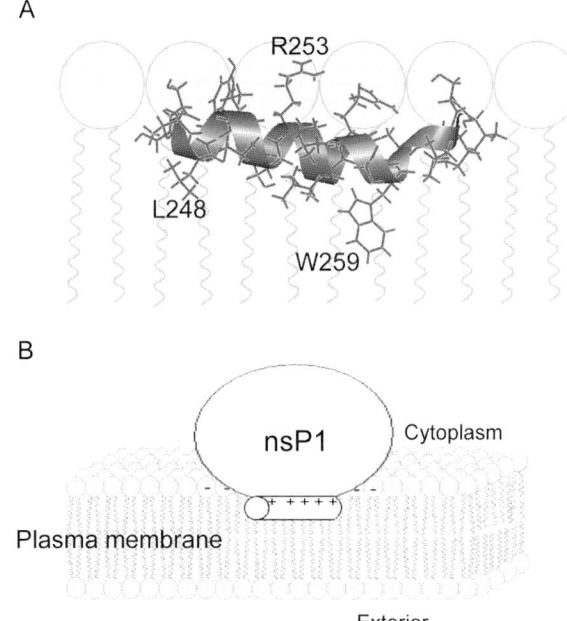

Fig. 3A, B. Monotopic binding of nsP1 to membrane via amphipathic α-helical peptide. **A** The NMR structure of the peptide is shown in interaction with the cytoplasmic leaflet of the lipid bilayer. **B** Highly schematic overview of interaction of nsP1 with a lipid bilayer

contrast to the synthetic wild-type binding peptide (Lampio et al. 2000). When the corresponding mutations were introduced to the SFV genome, neither W259A nor R253E was able to produce infectious virus after transfection (Salonen et al., unpublished data). Altogether, these results indicate that the interaction of nsP1 binding peptide with membranes is an essential and structurally finely tuned process, dependent on interaction with anionic phospholipids.

2.4
Polyprotein Conducts the Assembly and Targeting of the Replication Complex

The nsPs are derived from a common precursor P1234, the initial cleavage products of which (P123 plus nsP4) are necessary for the first step in the RNA replication, the synthesis of complementary RNA (Lemm et al. 1994, 1998) (Fig. 1A). To understand the role of this and other cleavage intermediates, we have produced them both in wild-type form and as noncleavable polyprotein variants, in which the autoprotease of nsP2 was inactivated by a mutation of the active site cysteine to alanine (superscript "CA"). The constructs were expressed in insect and mammalian cells, and the localization of individual proteins was followed by confocal microscopy and the complex formation by immunoprecipitation (Salonen et al. 2003).

The cleavable polyproteins (P12, P23, P123, P1234) containing an active nsP2 protease were processed to their constituents, most of which were distributed in the cell as though they were expressed alone. For instance, P12 yielded nsP1 and nsP2, which were targeted to the plasma membrane and nucleus, respectively. However, flotation analysis and immunoprecipitation recapture experiments showed that expression of P123 and P1234 resulted in membrane-bound complexes, containing all the individual proteins. This was different from the coexpression of all four nsPs individually, allowing us to conclude that the membrane association of the complex is guided by the polyprotein intermediate.

The uncleavable polyproteins were palmitoylated and had enzymatic activities typical for nsP1 and nsP2, and those containing nsP3 were phosphorylated, suggesting that the individual domains had folded properly in the context of the polyprotein. When $P12^{CA}$ was expressed in HeLa cells, it was localized exclusively at the cytoplasmic side of the plasma membrane and in long filopodia-like extensions (Fig. 4A), indistinguishable from those previously described for cells expressing nsP1 alone (Laakkonen et al. 1998). This indicated that the affinity to plasma membrane of nsP1 in the polyprotein overruled the attraction of nsP2 for nuclear transport. However, an interesting change in the localization was seen when $P12^{CA}3$ was expressed (Fig. 4B). No filopodia-like extensions were seen, and instead intracellular vesicular staining was observed, which in immuno-EM resembled CPVs (Fig. 4C and D). Double immunofluorescence with antisera against nsPs and lamp-2 suggested that at least a fraction of the vesicular structures were late endosomes or lysosomes (Salonen et al. 2003). Thus it seems that endosomal targeting is a joint action of nsP1 and nsP3 domains in the nonstructural polyprotein. The polyprotein is attached to the membrane first by the nsP1 binding peptide, which adopts α-helical structure. Concomitantly the nsP1 domain undergoes a conformational change, which activates the methyltransferase and guanylyltransferase. Palmitoylation of cysteine residues 418–420 thereafter anchors the protein irreversibly to the membrane. We propose that the targeting of nsP1 and polyproteins with this

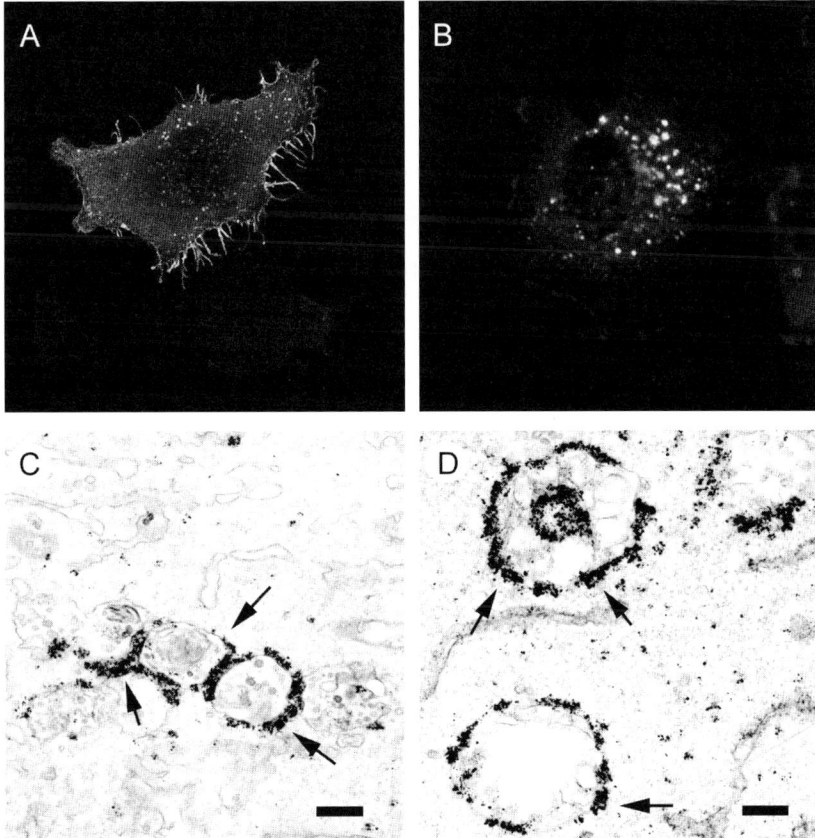

Fig. 4. Immunolocalization of SFV nonstructural polyproteins expressed by the aid of recombinant adenovirus vectors (**A–C**) and nsP3 during SFV infection (**D**) in HeLa cells. Cleavage-deficient P12^{CA} (**A**) localizes to the plasma membrane and filopodia, whereas P12^{CA}3 (**B**) displays vesicular staining. At the ultrastructural level P12^{CA}3 (**C**) localizes to the outer membrane of cytoplasmic vesicles (*arrows*), which resemble the characteristic CPV structures carrying the viral replication complex in SFV-infected cells (**D**). *Bars* 200 nm

domain to the plasma membrane may simply be dictated by the optimal PS concentration in its cytoplasmic leaflet.

Early in alphavirus infection the minus-strand RNA synthesis is regulated by the processing of the nonstructural polyprotein. The first cleavage releases nsP4 from P1234 giving rise to the minus-strand polymerase (Fig. 1A). The further processing of P123 is regulated by the slow *in cis* cleavage of the nsP1/2 site, which is essential for the next cleavage at the P2/3 site (Vasilieva et al. 2003). Thus the polyprotein has time to fold properly and bind to membranes by the aid of the nsP1 domain. The proper folding of the complex enables protein-protein interactions, which cannot be achieved when the components are expressed individually (Salonen et al. 2003). The polyprotein has a half-life of about 15 min before it is processed into the final components. During this time a replication complex synthesizes possibly only one minus-strand RNA before it is transformed into a stable plus-strand polymerase, which operates as the unit of replication within the spherule (Kujala et al. 2001).

3
Alphavirus-Like Superfamily

Rubella virus belongs to the *Togaviridae* family together with alphaviruses, and rubella virus replication complexes resemble in many ways those of SFV. Spherule-lined endo-lysosomal vacuoles are also found in rubella virus-infected cells (Magliano et al. 1998). Rubella virus replicase protein and newly synthesized RNA are located on the vacuoles, and specifically in spherule structures (Kujala et al. 1999). The role of spherules as sites of RNA replication is supported by localization of double-stranded RNA to them by antibodies against dsRNA (Lee et al. 1994).

Plant viruses belonging to the alphavirus-like superfamily replicate on various intracellular membranes, for instance, brome mosaic virus (BMV) and tobacco mosaic virus (TMV) on the ER, alfalfa mosaic virus on the vacuolar (tonoplast) membrane, and turnip yellow mosaic virus on the chloroplast envelope (Restrepo-Hartwig and Ahlquist 1996; Más and Beachy 1999; Prod'homme et al. 2001; van der Heijden et al. 2001). For BMV (Fig. 5A; Table 1), the targeting determinant of the replication complex has been mapped to the 1a protein, and more precisely to its N-terminal domain, part of which is distantly related to alphavirus nsP1. 1a is peripherally but tightly bound to membranes and exposed to the cytoplasm. Relatively large regions of 1a are needed for membrane association and ER targeting, but the exact molecular basis for membrane binding is not yet known (den Boon et al. 2001). Further comparative studies are needed to determine whether replicase proteins in the alphavirus-like superfamily share similar mechanisms of membrane association and targeting, but in the case of several viruses the capping enzyme domain binds to membranes (Magden et al. 2001). Interestingly,

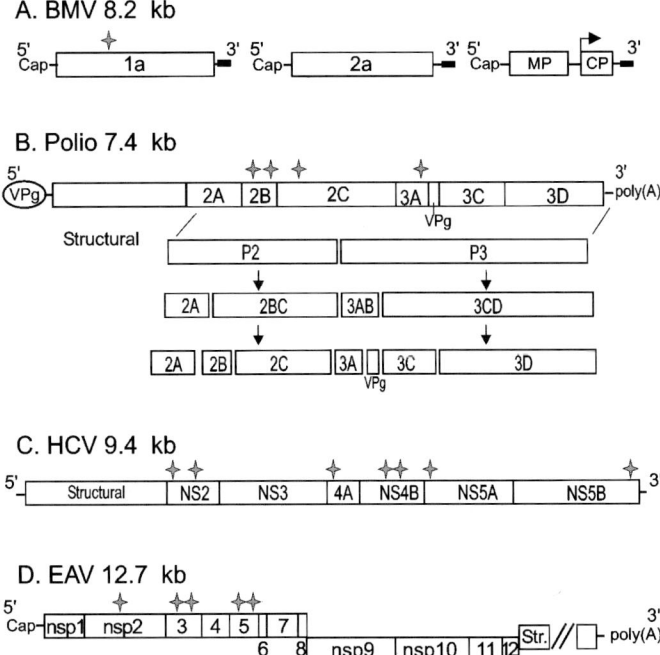

Fig. 5A–D. Genome organization of model viruses described in the text, representing the major groups of plus-strand RNA viruses. The different genomes are not in the same scale, and the structural region of the EAV genome is not represented. Regions attaching proteins to membranes are marked with *star symbols* (see also Table 1)

TMV replication in *A. thaliana* specifically and absolutely requires host genes TOM1 or TOM3. They encode related multipass transmembrane proteins, which seem to interact with the TMV replicase and are speculated to participate in its membrane anchoring (Yamanaka et al. 2002).

In cells infected with these plant viruses structures closely resembling the spherules, described above for alphaviruses, have been detected by EM-techniques (Prod'homme et al. 2001; Schwartz et al. 2002 and references therein). The spherules produced by BMV in yeast cells have recently been characterized in exquisite detail (Schwartz et al. 2002). BMV 1a protein alone, in the absence of other viral components, can induce spherule formation. When viral RNA is coexpressed with 1a, it is apparently protected inside the spherule in a membrane-associated, nuclease-resistant state. When the viral polymerase protein 2a is coexpressed, it associates with the spherules through interaction with 1a, and viral minus- and plus-sense RNA synthesis takes place in close association with

Table 1. Properties of replicase proteins

Virus	Replicase Proteins	Size aas	Function	Mode of membrane binding	Localization alone	Remarks	References
Togaviridae							
SFV	nsP1	537	MT, GT	α-Helix; monotopic[1,2]	PM[3]	RC is in spherules	[1]Ahola et al. 1999
	nsP2	799	NTPase, TP, Hel, Pro	–	Nucleus	In CPVs and at PM[5]	[2]Lampio et al. 2000
	nsP3	482	Unknown	–	CP; aggregates[4]		[3]Laakkonen et al. 1996
	nsP4	614	RdRp	–	CP	P123 is required for targeting the RC[4]	[4]Salonen et al. 2003
	P123	1818	Early RC	Via nsP1	PM, endosomes[4]		[5]Kujala et al. 2001
Bromoviridae							
BMV	1a	961	MT,GT,(Hel?)	Monotopic?[1]	ER	RC is in spherules, which can be induced by 1a alone[2]	[1]den Boon et al. 2001
	2a	822	RdRp	–	CP		[2]Schwartz et al. 2002
Picornaviridae							
Poliovirus	2A	149	Protease	–	CP	RC is in DMVs[4], which can be induced by the coexpression of 2BC and 3A[5]	[1]de Jong et al. 2003
	2B	97	Unknown	α-Helix[1] + TM[2]	ER, Golgi[1]		[2]Agirre et al. 2002
	2C	329	NTPase	α-Helix; monotopic[3,4]	ER		[3]Paul et al. 1994
	3A	87	Unknown	HP C-term.[5]	ER		[4]Echeverri&Dasgupta 1995
	3B	22	VPg	–	CP		[5]Towner et al. 1996
	3C	182	Protease	–	CP		[6]Cho et al. 1996
	3D	461	RdRp	–	CP		[7]Lyle et al. 2002
	2BC	426	Unknown	Via 2B and 2C[6]	ER, Golgi, PM?		
	3AB	109	RNA priming	Via 3A[5,7]	ER, Golgi, PM?		
	3CD	643	Protease (main)	–	CP		
Flaviridae							
HCV	NS2	149	Protease	TM; polytopic[1]	ER	Membranous web is the potential site of RC	[1]Yamaga et al. 2002
	NS3	631	Pro, NTPase,Hel	–	CP, Nucleus	NS4B alone is able to induce the web[6]	[2]Wölk et al. 2000
	NS4A	54	Pro cofactor	HP N-term.[2]	ER		[3]Hügle et al. 2001
	NS4B	261	Unknown	TM; polytopic[3]	ER		[4]Brass et al. 2002
	NS5A	447	Unknown	α-Helix; monotopic[4]	ER		[5]Ivashkina et al. 2002
	NS5B	591	RdRp	TM, tail-anchored[5]	ER		[6]Egger et al. 2002

Table 1. (continued)

Virus	Replicase Proteins	Size aas	Function	Mode of membrane binding	Localization alone	Remarks	References
Arteriviridae							
EAV	nsp1	260	Protease	–	Nucleus, CP	RC is in perinuclear DMVs[2], which can also be formed by expression of self-cleaving nsp2-nsp3[3]	[1]van der Meer et al. 1998 [2]Pedersen et al. 1999 [3]Snijder et al. 2001
	nsp2	570	Protease	HP[1]	ER		
	nsp3	232	Unknown	HP[1]	ER		
	nsp4	204	Protease (main)	–	CP		
	nsp5	163	Unknown	HP[1]	ER?		
	nsp6–8	302	Unknown	–	CP		
	nsp9	643	RdRp	–	CP		
	nsp10	467	Hel, NTPase	–	CP		
	nsp11–12	338	Endonuclease?	–	CP		

Abbreviations: MT, methyltransferase; GT, guanylyltransferase; TP, triphosphatase; Hel, helicase; Pro, protease; RdRp, RNA-dependent RNA polymerase; RC, replication complex; PM, plasma membrane; CP, cytoplasm; CPV, cytopathic vacuole; ER, endoplasmic reticulum; HP, hydrophobic; DMV, double membrane vesicle; TM, transmembrane.

the spherules, possibly in their interior, from where plus-sense RNA is released to the cytoplasm. Calculations based on immunolabeling suggest that many (maximally a few hundred) 1a proteins may be present in a spherule; leading to a hypothesis that 1a may form a shell-like structure coating the inside of the spherule (Schwartz et al. 2002). BMV replication in yeast requires a certain concentration of unsaturated fatty acids, as demonstrated through a mutation in the host fatty acid desaturase gene and its complementation by addition of unsaturated fatty acids. Under restrictive conditions, 1a can still normally recruit viral RNA and 2a to membranes, but minus-strand synthesis is strongly inhibited. Unsaturated fatty acids, present in membrane lipids, generally increase membrane fluidity and plasticity. Therefore, proper assembly or function of the BMV replication complex seems to require a relatively fluid membrane (Lee et al. 2001).

Uniquely among positive-sense RNA viruses, a highly purified detergent-solubilized replication complex of cucumber mosaic virus can catalyze a complete cycle of minus-strand and plus-strand synthesis on an exogenously provided specific template (Hayes and Buck 1990). However, this preparation seems to be relatively unstable and it has not been characterized further. In contrast, partially purified template-dependent preparations, such as that isolated from TMV-infected cells (Osman and Buck 1996), will be useful in further analyzing the role of membranes in RNA replication.

4
Picornavirus-Like Superfamily

4.1
Poliovirus as a Model

Poliovirus is one of the best-characterized viruses. Its structure and replication have been described in recent reviews (Pfister et al. 1999; Racaniello 2001; Semler and Wimmer 2002). Even though poliovirus does not have an envelope, the synthesis of the structural and nonstructural proteins takes place in association with cytoplasmic membranes in close vicinity to the RNA replication site (Caliguiri and Tamm 1970). The entire positive-strand RNA genome is translated to a single polyprotein, which is nascently cleaved into three polyproteins, P1, P2, and P3. P1 is the precursor of virion structural proteins, whereas P2 and P3 represent nonstructural proteins participating in the replication of viral

RNA (Fig. 5B; Table 1). Protease 2A cleaves P1 from the nascent polyprotein, while cleavages between nonstructural and structural proteins are carried out by protease $3C^{pro}$ (or rather $3CD^{pro}$). Nonstructural protein P2 yields protease $2A^{pro}$ and 2BC, which in turn is cleaved to 2B and the NTPase 2C. P3 yields 3AB and $3CD^{pro}$, which are processed to 3A and 3B (=VPg) and to 3C and $3D^{pol}$, respectively (Fig. 5B). VPg, a terminal protein of 22 aa, is linked to the 5' end of the genome. The 5' nontranslated region consists of a cloverleaf structure and an internal ribosome entry site.

The incoming poliovirus genomes seem to migrate to specific perinuclear sites, where replication starts. RNA recombination, which occurs during the synthesis of the complementary (minus-strand) RNA, takes place in these perinuclear sites (Egger and Bienz 2002). Throughout infection plus- and minus-strand RNAs are synthesized in the same replication complexes approximately in a ratio of 100 to 1 (Bolten et al. 1998). Poliovirus replication complexes consist of clusters of vesicles of 70–400 nm in diameter, which after isolation are associated as large "rosettelike structures" of numerous vesicles interconnected with tubular extensions. The rosettes can dissociate reversibly into tubular vesicles, which carry poliovirus nonstructural proteins on their surface and synthesize poliovirus RNA in vitro (Semler and Wimmer 2002). Immunoisolated poliovirus-specific vesicles contain cellular markers for the ER, lysosomes, and trans-Golgi network, suggesting complex biogenesis of the RNA replication complexes (Schlegel et al. 1996).

Involvement of the secretory route in the biogenesis of poliovirus replication complexes was suggested by the finding that brefeldin A, which inhibits the transport of secretory proteins from the ER to the Golgi complex, also inhibits poliovirus replication both in vivo and in vitro (Racaniello 2001). The importance of ER as the primary source for poliovirus replication complexes was confirmed recently by experiments in which COPII coat components were shown to colocalize with poliovirus nonstructural proteins on budding vesicles upon their exit from ER. Resident ER proteins were excluded from the released vesicles, which were not destined to the Golgi complex, but accumulated in the cytoplasm (Rust et al. 2001). These results are in conformity with previous findings, which showed that poliovirus infection inhibits the transport of secretory and plasma membrane proteins (Doedens and Kirkegaard 1995). Thus it seems that in poliovirus-infected cells a continuous proliferation and loss of ER membranes takes place. This process does not supply the Golgi complex with its normal lipids. The sensitivity of poliovirus replication to lipid synthesis inhibitors such as cerulenin could be explained by this scenario (Racaniello 2001; Pfister et al. 1999). Overexpression of viral or cellular ER-associated proteins also inhibits poliovirus replication, possibly by competing for the capacity of ER to generate new membrane material (Egger et al. 2000).

Expression of P2 and P3 without structural proteins results in membrane alterations similar to those seen in infected cells (Teterina et al. 2001). However, vesicles formed from nonreplicating poliovirus RNA could not be recruited to support the replication of superinfecting poliovirus RNA, suggesting that functional replication complexes are formed only *in cis* by the direction of the incoming RNA. This must be first translated to yield the replicase proteins for its own replication in situ (Egger et al. 2000). Assuming that the newly synthesized plus-strands create in turn new replication complexes by a similar *in cis* process, the "rosette structures" might well consist of closely packed assemblies of a parent replication complex and its numerous daughters, which are loosely bound to each other (Semler and Wimmer 2002).

Numerous studies, in which poliovirus nonstructural proteins were expressed individually or in combinations, in the absence of RNA synthesis, have helped to understand the biochemical functions of nonstructural proteins and their role in membrane association during the biogenesis of the replication complexes (Racaniello 2001; Semler and Wimmer 2002). The 2B protein is targeted to ER membranes and to the Golgi complex. It interferes with the secretory pathway in mammalian and yeast cells (Barco and Carrasco 1995; Doedens and Kirkegaard 1995; de Jong et al. 2003). It has been reported to disassemble the Golgi complex (Sandoval and Carrasco 1997). 2B has a predicted cationic amphipathic α-helix within the N-terminal 34–49 aa and a potential transmembrane domain (aa 61–81), which according to modeling form tetrameric aqueous pores, which could be responsible, for example, for the observed hygromycin sensitivity and increased permeability of poliovirus-infected cells. To cause these effects 2B has to be transported to the plasma membrane, evidently on the cytoplasmic surface of the transport elements (Agirre et al. 2002; de Jong et al. 2003).

When 2C is expressed alone in mammalian cells it is localized to the ER, causing its expansion into tubular structures. As opposed to 2B, it does not prevent the transport of VSV G-protein to the plasma membrane (Suhy et al. 2000). The fragment responsible for the membrane binding of 2C has been mapped to the N-terminal part of 2C within aa 18–54 (Pfister et al. 1999). It has been predicted that this region has an amphipathic α-helix, which starts either from residue 10 (Paul et al. 1994) or 21 (Echeverri and Dasgupta 1995). 2C has NTPase activity, which can be inhibited by guanidine, the well-known specific inhibitor of poliovirus replication (Pfister and Wimmer 1999). The ATPase activity of 2C is needed specifically in the initiation of minus-strand RNA synthesis, the only step inhibited by guanidine (Barton and Flanegan 1997). Because 2C binds specifically to the 3′ end of the minus-strand RNA, it may also have a function in the synthesis of the plus-strand RNAs, which takes place even in the presence of guanidine, that is, without ATPase activity. Anyhow, the tight membrane association of 2C and its intimate participation in minus-strand RNA synthesis mean that this process must also take place in association with membranes, although it has been difficult to prove (Egger and Bienz 2002). 2BC, like 2B, is a membrane protein, which interferes with the vesicular transport in both animal and yeast cells. Thus the 2B moiety in 2BC is responsible for the transport inhibition (Doedens and Kirkegaard 1995). 2BC induces vesicles similar to those seen in poliovirus-infected cells and causes permeability increase of the plasma membrane, like 2B (Teterina et al. 1997).

3A expressed alone efficiently inhibits the vesicular transport of secretory proteins from the ER to the Golgi. It remains associated with ER membranes but can be mobilized into secretory vesicles, similar to those in poliovirus-infected cells, by coexpression with 2BC (Dodd et al. 2001). In poliovirus-infected cells 3AB delivers the 22-aa-long VPg peptide to the 5′ end of both minus- and plus-strand RNA molecules. Only membrane-associated 3AB can be cleaved by the viral proteases ($3C^{pro}$ and $3CD^{pro}$), and thus serve as the source of VPg (Pfister et al. 1999). 3AB associates tightly with cellular membranes, resembling the binding of integral membrane proteins. The binding region has been mapped to the C-terminal amino acids 59–80 of 3A, specifically to a hydrophobic region consisting of aa 73–80. However, the exact binding mechanism is not known (Towner et al. 1996). The 3B (=VPg) portion of 3AB has affinity to the catalytic subunit $3D^{pol}$, and its precursor 3CD, which in turn recruits the template RNA into the membrane-associated replication complex by interaction with 3C and 3D (Egger et al. 2000; Pfister et al. 1999).

In summary, the assembly of the poliovirus replication apparatus is a complex process of specific membrane recognition, followed by protein-protein and RNA-protein interactions. At the same time the vesicles develop by a poorly understood autophagocytosis-like process to double-membrane vesicles (DMVs) and large rosettes containing proteins from ER, Golgi, and lysosomes (Schlegel et al. 1996; Suhy et al. 2000). Because of the extreme proliferation of the ER, the secretory apparatus becomes exhausted. The Golgi complex disappears, probably through retrograde transport that is not compensated by normal lipid flow from the ER. Development of poliovirus-induced vesicles must be associated with multiple fusion events directed either by viral or cellular proteins. Another possibility would be that the membrane proteins from the Golgi complex, and perhaps beyond it, would be enclosed to the poliovirus-specific vesicles through retrograde transport via ER. In any case, the result is that the membranes of the secretory and endocytotic apparatuses become mixed.

4.2
Other Members of the Picornavirus Superfamily

Many plant viruses in the picornavirus superfamily appear to replicate in association with membranes derived from the ER. In comovirus- and nepovirus-infected cells the ER is proliferated and vesiculized, but in contrast to poliovirus-infected cells, the Golgi complex remains normal. Replicase proteins and newly synthesized RNA are associated with the ER-derived structures. Sensitivity to cerulenin, as an inhibitor of RNA synthesis, seems to be a common property in the picornavirus superfamily (Carette et al. 2000; Ritzenthaler et al. 2002). Cowpea mosaic comovirus 32-kDa and 60-kDa replicase proteins are both targeted to subregions of the ER when individually expressed, and they also cause morphological alterations of the membrane system. The 32-kDa protein is a hydrophobic component specific for comoviruses, whereas the 60-kDa protein may contain membrane binding regions analogous to poliovirus 2C and 3A (Carette et al. 2002). For tobacco etch potyvirus, the 6-kDa protein (analogous to poliovirus 3A) appears to be decisive in directing the replicase to the ER. The 6-kDa protein associates tightly with ER membranes by a single central hydrophobic domain (Schaad et al. 1997). The protein may be inserted to the membrane posttranslationally, but its exact binding mechanism and topology are not known.

Although the polymerase of the insect nodaviruses appears to be distantly related to the picornavirus-like superfamily (Koonin and Dolja 1993), these viruses have capped mRNAs and the ultrastructure of the replication complex resembles that of the alphaviruses. The outer mitochondrial membranes of flock house virus (FHV)-infected *Drosophila* cells contain numerous spherulelike invaginations, connected to the cytoplasm by narrow necks. The number of spherules increases during infection, leading finally to disruption of mitochondrial structure. The single virus-encoded replicase component, 112-kDa protein A, localizes to the outer mitochondrial membrane, which is also the site of viral

RNA synthesis (Miller et al. 2001). The N-terminal 46 aa of protein A contain a mitochondrial targeting signal and a transmembrane helix, such that the N-terminus is embedded in the mitochondrial matrix while most of protein remains on the cytoplasmic side (Miller and Ahlquist 2002). This transmembrane topology distinguishes nodavirus replicase from the replicase proteins of alphaviruses.

An interesting result has been obtained with the crude membrane-bound replication complex isolated from FHV-infected cells. When treated with micrococcal nuclease and supplied with an exogenous template, the FHV replicase synthesizes a complementary minus-strand resulting in a dsRNA product. However, when glycerophospholipids are added to the mixture, relatively large quantities of plus-strand RNAs are also produced, that is, a complete RNA replication cycle takes place. Several phospholipid species can stimulate this reaction, for instance, phosphatidylcholine bearing acyl chains of 14–18 carbons. It has been speculated that glycerophospholipid might directly interact with a component of the crude membrane preparation, perhaps activating an enzymatic function, or alternatively, that the lipid might facilitate a membrane modification or assembly process required specifically for plus-strand synthesis (Wu et al. 1992). Because of these advances and the simplicity of the nodavirus replicase, this virus group is promising for further analysis of membrane-associated replication.

5
Flavivirus-Like Superfamily

Of the members of the *Flaviviridae*, hepatitis C virus (HCV) and Kunjin virus are the best studied in the context of membrane-associated replication. Here they will be discussed only briefly, because these aspects of HCV and Kunjin virus have been reviewed recently (Dubuisson et al. 2002; Westaway et al. 2002).

Analogous to picornaviruses, the whole positive-strand RNA genome of flaviviruses is translated to a large polyprotein. The structural proteins consist of a capsid protein and envelope glycoproteins followed by nonstructural proteins (Fig. 5C; Table 1). Because the envelope proteins are translocated to and glycosylated in the ER, it would be expected that the nonstructural proteins would associate directly to the ER membrane. Nevertheless, in heterologous expression systems HCV nonstructural proteins NS2, NS4A, NS4B, NS5A, and NS5B each alone bind to the ER, whereas the soluble protease/helicase (NS3) associates with the membrane by interaction with NS4A, a cofactor for the protease domain of NS3 (Dubuisson et al. 2002; Wölk et al. 2000).

NS2 is a polytopic integral membrane protein introduced to the ER membrane by internal signal sequences. It is a protease responsible for the cleavage between NS2 and NS3, but it is not essential for RNA replication (Yamaga et al. 2002). NS4B is also a polytopic membrane protein cotranslationally inserted into the ER membrane with its own internal signal sequences (Hügle et al. 2001). Although the exact function of NS4B is not known, its interaction with NS3 and NS5B modulates the RNA polymerase activity (Piccininni et al. 2002). NS5A phosphoprotein is tightly associated with membranes through an N-terminal amphipathic helix of about 30 residues. When these residues are joined to GFP, the fusion protein is associated with ER membranes, suggesting that the N-terminus of NS5A has also an address for the ER, in addition to membrane binding (Brass et al. 2002; Dubuisson et al. 2002). The monotopic binding of NS5A to membranes resembles the situation in alphaviruses, except that the binding peptide of NS5A is "tip-anchored," rather than residing in the middle of the protein like in nsP1 (Ahola et al. 1999). NS5B, the catalytic subunit of the HCV RNA polymerase, has a C-terminal membrane insertion sequence of 21 aa, which is targeted to the ER membrane posttranslationally, like typical tail-anchored membrane proteins (Ivashkina et al. 2002).

Expression of the entire HCV polyprotein induced a special ER-derived membranous web, where a cluster of tiny vesicles was embedded in a membranous matrix, often accompanied by tightly associated vesicles surrounding the web. According to immuno-EM analysis, all HCV proteins were associated with the web, but not with the vesicles. When the ns proteins were expressed individually or in combinations, NS4B alone induced the web, whereas NS3-NS4A complex induced a multitude of single vesicles having no direct analog in polyprotein-expressing cells (Egger et al. 2002). NS5A and NS5B did not modify the ER. In cells expressing functional HCV subgenomic replicons all ns proteins associated with ER membranes according to light and electron microscopic analysis (Mottola et al. 2002).

Even though the genome organization of HCV and flaviviruses is similar, there are some differences as well. Flaviviruses have a nonstructural glycoprotein NS1 (46 kDa), which is translocated into the lumen of ER and transported through the secretory route to the exterior of the cell. NS1 also plays an essential role in RNA replication, evidently by recognizing portions of the other replicase proteins penetrating the ER membrane (Westaway et al. 2002). Another difference is that there are two small membrane-associated proteins, NS2A (25 kDa) and NS2B (14 kDa), preceding the soluble NS3 protein (60 kDa), which has protease-helicase-RNA triphosphatase activities. NS2B acts as a cofactor for the NS3 protease. NS4A (16 kDa) and NS4B (27 kDa) are poorly conserved membrane proteins. Finally, NS5 (104 kDa) protein is the catalytic subunit of the RNA polymerase complex. It is a soluble protein, unlike its homologous counterpart NS5B of HCV.

Extensive immunofluorescence microscopy studies of Kunjin virus-infected cells have established that NS1, NS2A, NS3, NS4A, and NS5 are regularly associated with dsRNA, which has served as marker for genuine replication complexes. These markers also colocalize with cellular markers of trans-Golgi membranes (Mackenzie et al. 1999), even in cells that do not express the viral glycoproteins (Mackenzie et al. 2001).

Electron microscopy of Kunjin virus-infected cells has revealed dramatic changes in the organization of the ER membrane. Proliferation of the ER leads to convoluted membranes (CM) and paracrystalline structures (PC) and to vesicle packets of smooth membranes (VP) (Westaway et al. 1997, 2002). The majority of NS proteins, dsRNA, as well as nascent labeled viral RNA have been immunolocalized to VPs, which are derived from trans-Golgi membranes late in infection. VPs are "vesicle sacs" consisting of a cluster of individual vesicles (diameter about 50–100 nm) surrounded by a membrane. Interestingly, VPs were not detected on expression of Kunjin replicons, whereas CMs and PCs were formed. In these cells the dsRNA was scattered throughout the cytoplasm in small isolated foci, suggesting that all membrane structures induced by the replicase proteins are not necessarily sites of RNA replication (Mackenzie et al. 2001; Westaway et al. 1999). Comparison between replicon cell lines producing RNA and NS proteins with different efficiencies suggests that the induction of virus-specific membranes is dose dependent and requires a certain level or concentration of viral products to manifest (Mackenzie et al. 2001).

Tombusviruses, plant viruses classified in the same supergroup with flaviviruses, replicate on peroxisomal or chloroplast membranes, or on the mitochondrial outer membrane depending on the virus species. Tombusvirus infection induces multivesicular bodies, where the limiting membrane of the organelle is transformed into numerous spherules (Rochon 1999). A putative polymerase of carnation Italian ringspot virus is a 92-kDa protein translated by read-through of an amber termination codon at the end of a 36-kDa protein. Both the 36-kDa and 92-kDa proteins are targeted to the mitochondrial membranes and anchored there via two hydrophobic domains located close to the N-terminus

(Weber-Lotfi et al. 2002). However, the expression of the 36-kDa protein alone was not sufficient to induce the vesiculation of mitochondria and hence the formation of spherules (Rubino et al. 2000).

6
Nidoviruses

Coronaviruses and arteriviruses, which are grouped in the order *Nidovirales,* express their replicase genes from two large open reading frames through complex proteolytic processing, leading to 12 or more end products, depending on the virus (Snijder and Meulenberg 2001; Lai and Holmes 2001). The replicase proteins of equine arteritis virus (EAV) (Fig. 5D; Table 1), as well as newly synthesized RNA, accumulate in perinuclear granules and vesicles, which are of ER origin (van der Meer et al. 1998). An electron microscopic study of EAV-infected cells revealed DMVs of approximately 80-nm diameter, carrying the replication complex (Pedersen et al. 1999). Usually the inner and outer membranes of the DMVs were tightly apposed but clearly separate. The mechanism for DMV formation appears to be a protrusion of paired ER-membranes, because DMVs having the outer membrane continuous with ER could be seen. These profiles sometimes contained a neck between the paired ER-membrane and a forming DMV, which had not yet pinched off. The formation of DMVs is not dependent on RNA synthesis, because DMVs strikingly resembling those seen in EAV-infected cells can be induced by heterologous expression of the nsp2-nsp3 region of the polyprotein (Snijder et al. 2001). On individual expression of these proteins, DMVs were not observed. The large nsp2 has a long central hydrophobic sequence, which may represent its membrane anchor, whereas nsp3 and nsp5 have several hydrophobic sequences, suggesting that they are polytopic membrane proteins. They all, and their precursors, also behave biochemically as integral membrane proteins (van der Meer et al. 1998).

DMVs are also the sites of coronavirus RNA replication (Gosert et al. 2002). Coronavirus-induced DMVs are larger than those induced by EAV, over 200 nm in diameter, and they are surrounded by tightly apposed membranes that have fused into a lipid trilayer. Viral RNA and replicase proteins are found on the surface of DMVs, where RNA synthesis also takes place. The coronavirus replicase proteins are membrane associated and include the integrally bound components p210 and p44, the coronavirus counterparts of EAV nsp2 and nsp3 (Gosert et al. 2002). At the moment, the origin of the coronavirus replication complexes (DMVs) is not clear; involvement of membranes from secretory and endosomal compartments has been suggested. It should be noted that the majority of the viral replicase proteins may be located elsewhere than in the active replication complexes, as in SFV-infected cells (Kujala et al. 2001).

7
Concluding Remarks

7.1
Targeting of the Replicase Complex

Many viruses replicate on the cytoplasmic side of the ER membrane. For instance, picornavirus nonstructural proteins are targeted to the ER through 2B, 2C, 3A, and their precursors, bromovirus replication complex by the 1a protein, and arterivirus replicase by the nsp2-nsp3 complex (Table 1). However, the targeting mechanisms remain to be solved, as no specific receptors for viral components on the ER have been identified. Because all HCV and flavivirus proteins are translated from the same polyprotein as their envelope glycoproteins, it might be expected that their ns proteins also remain at the ER membrane. However, their location is guaranteed by their own independent affinity for ER membranes. In the case of Kunjin virus the NS1 glycoprotein, which is translocated to the cisternal side of the ER, may be responsible for the transport of the replication complex to the trans-Golgi region by transmembrane contact.

For some viral replicase proteins, classic targeting sequences directing them to a specific compartment have been identified. FHV virus protein A is directed to the mitochondrial outer membrane by a specific targeting sequence. A similar mechanism seems to operate in the targeting of the tombusvirus replication complex to either mitochondria or chloroplasts, depending on the virus species. The RNA replication of alphaviruses on the membranes of the endosomal apparatus (plasma membrane, endosomes, and lysosomes) might be explained by direct interaction of the amphipathic α-helix of nsP1 with these PS-rich membranes. It is an interesting question whether specific lipid components, or the lipid composition of the target membranes, might attract replicase components of other viruses as well.

The genomes of animal viruses discussed in this review are expressed as polyproteins. Picornaviruses and flaviviruses express both structural and nonstructural proteins in the same polyprotein, whereas togaviruses and nidoviruses express their nonstructural proteins as a separate polyprotein. The polyprotein strategy evidently guarantees the proper targeting and assembly of the membrane-associated RNA replication complex. For instance, in SFV-infected cells the delayed proteolysis of the ns polyprotein enables the folding and assembly of "soluble" components (nsP2-nsP4) with the nsP1 membrane anchor. In the case of picornaviruses and flaviviruses several membrane anchors are present.

7.2
Mechanisms of Membrane Binding

The modes of membrane attachment are also variable. Monotopic binding by amphipathic α-helix has been suggested for several replicase proteins, which lack continuous hydrophobic anchor sequences (Table 1). Except for SFV nsP1 (Lampio et al. 2000) and HCV 5A (Brass et al. 2002), these conclusions have been based only on sequence-based predictions and mutagenesis studies. The amphipathic α-helix strategy has been proposed for components of picornavirus and for several plant virus replication complexes. Monotopic hydrophobic anchors have been proposed for poliovirus 3A and related plant virus proteins, and polytopic membrane anchors have been demonstrated for HCV proteins NS2 and NS4B. Recent results have shown that the catalytic subunit of HCV RNA polymerase is a typical tail-anchored ER protein, whereas the polymerases of other viruses are soluble proteins (Table 1).

7.3
Membrane Modification by Replicase Proteins

Alphaviruses and rubella virus give rise to specific cytoplasmic vesicles with regular membrane invaginations, spherules (Fig. 6A and B), that

Fig. 6A–E. Ultrastructure of membranes involved in virus replication. A Typical cytopathic vesicles (CPVs) in SFV-infected BHK cell (4 h p.i.) with characteristic spherules inside. B A single spherule showing a neck opening to the cytoplasm with filled electron-dense material (courtesy of Dr. Ari Helenius and Dr. Jürgen Kartenbeck). C Poliovirus type 1-infected COS-1 cell (4 h p.i.) showing typical cytoplasmic vesicles ranging from 70 to 400 nm (courtesy of Dr. Karla Kirkegaard and Dr. Thomas Giddings). D EAV-infected BHK cell (4 h p.i) showing double membrane vesicles (DMVs) in close vicinity to RER (courtesy of Dr. Eric Snijder and Dr. Ketil Pedersen). E Membraneous web in UHCV-57.3 cell expressing the entire HCV open reading frame, 48 h after tetracycline removal (courtesy of Dr. Kurt Bienz). *Bars* 200 nm (**A**), 60 nm (**B**), 100 nm (**D**), 500 nm (**E**)

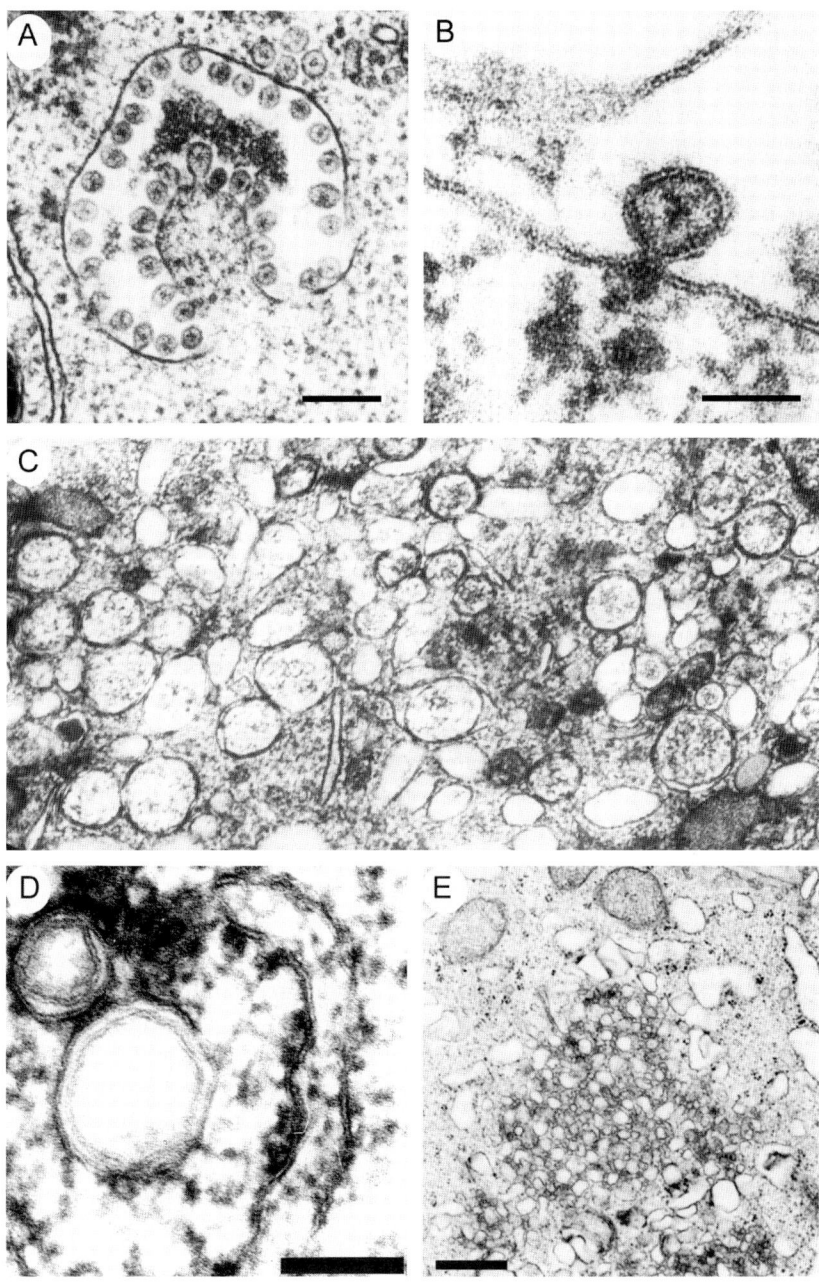

seem to be the actual units of RNA replication (Kujala et al. 2001). So far, it is not known how these structures arise. Similar spherules have been described in BMV-infected plant and in yeast cells expressing nonstructural proteins of the virus. It has been proposed that the internal surface of the spherules is covered by the replicase protein 1a of BMV (Schwartz et al. 2002). As the spherules seem to be a general feature for the members of the alphavirus superfamily, it will be interesting to see whether the suggested protein coating of the inner surface of spherules is a general mechanism within the whole superfamily. Involvement of host cell proteins cannot be excluded, as membrane bending and vesiculation in cells is a complex process requiring several factors (Hurley and Wendland 2002).

Membrane vesicles and multivesicular bodies induced by poliovirus (Fig. 6C) and other picornaviruses contain proteins from the Golgi complex, endosomes, and lysosomes, suggesting multiple fusion events during their development. Isolated membrane structures, "rosettes," consist of clusters of vesicles, which can be separated from each other at low ionic strength and low temperature. The individual vesicles represent units of replication (Egger et al. 1996). They have a tubular extension, which resembles the neck of a spherule through which the nascent RNA is proposed to be "secreted" from the site of synthesis (Froshauer et al. 1988). The DMVs (Fig. 6D) seem to the sites of arterivirus and coronavirus RNA synthesis as well. Evidently the vesicles wrap the template RNA, which in the case of poliovirus, alphaviruses, and flaviviruses is probably double-stranded, shielding it from host nucleases. During virus infection, translation and assembly of virus particles take place in close association with the replication complexes, utilizing nascent RNAs immediately after their synthesis.

A common feature for the viruses discussed in this review is that the nonstructural proteins, in the absence of RNA template, seem to be sufficient to induce the membrane modifications seen in the infected cells. Such is also the case for the membranous web seen during HCV polyprotein expression (Fig. 6E). Future studies at the molecular level should reveal how the different viruses and proteins can cause these fundamental structural changes in the membranes. A suitable lipid composition of the membrane may also be required for these dynamic membrane assembly and modification processes (Lee et al. 2001).

7.4
Functional Implications of Membrane Attachment

As the polymerase complex itself is firmly attached to the membrane, the template apparently has to move through the complex, which often also contains helicase, capping enzyme, and other subunits. In most instances, this would also mean that the same template would be repeatedly utilized by circling through the same replication complex. The dimensions of the membrane vesicles seen in EM images are such that the RNA would be relatively tightly packed within them, analogous to dsRNA virus cores. For picornavirus-like viruses there are special challenges, as for each round of RNA synthesis a protein component is consumed as a primer. It should also be emphasized that membrane lipids provide active components for the replication complex. They may directly bind to replicase proteins, thereby changing their conformation and activating them (Ahola et al. 1999).

Acknowledgements We thank Drs. Eija Jokitalo for the EM figures and Ilkka Kilpeläinen for help in presenting the peptide structure and Marja Makarow for critical reading of the manuscript. The work has been supported by the Academy of Finland (grants 8397 and 201687), Biocentrum Helsinki, and Helsinki University Research Funds. Drs. Kurt Bienz (University of Basel), Ari Helenius, (ETH, Zürich), Jürgen Kartenbeck (DKFZ, Heidelberg), Karla Kirkegaard (Stanford University), Thomas Giddings (University of Colorado), Eric Snijder (University of Leiden), and Ketil Pedersen (University of Oslo) are gratefully acknowledged for the electron micrographs.

References

Agirre A, Barco A, Carrasco L, Nieva JL (2002) Viroporin-mediated membrane permeabilization. Pore formation by nonstructural poliovirus 2B protein. J Biol Chem 277:40434–40441

Ahola T, Kääriäinen L (1995) Reaction in alphavirus mRNA capping: formation of a covalent complex of nonstructural protein nsP1 with 7-methyl-GMP. Proc Natl Acad Sci USA 92:507–511

Ahola T, Lampio A, Auvinen P, Kääriäinen L (1999) Semliki Forest virus mRNA capping enzyme requires association with anionic membrane phospholipids for activity. EMBO J 18:3164–3172

Ahola T, Kujala P, Tuittila M, Blom T, Laakkonen P, Hinkkanen A, Auvinen P (2000) Effects of palmitoylation of replicase protein nsP1 on alphavirus infection. J Virol 74:6725–6733

Barco A, Carrasco L (1995) A human virus protein, poliovirus protein 2BC, induces membrane proliferation and blocks the exocytic pathway in the yeast *Saccharomyces cerevisiae*. EMBO J 14:3349–3364

Barton DJ, Flanegan JB (1997) Synchronous replication of poliovirus RNA: initiation of negative-strand RNA synthesis requires the guanidine-inhibited activity of protein 2C. J Virol 71:8482–8489

Bolten R, Egger D, Gosert R, Schaub G, Landmann L, Bienz K (1998) Intracellular localization of polio-virus plus- and minus-strand RNA visualized by strand-specific fluorescent in situ hybridization. J Virol 72:8578–8585

Brass V, Bieck E, Montserret R, Wolk B, Hellings JA, Blum HE, Penin F, Moradpour D (2002) An amino-terminal amphipathic alpha-helix mediates membrane association of the hepatitis C virus nonstructural protein 5A. J Biol Chem 277:8130–8139

Buck KW (1996) Comparison of the replication of positive-stranded RNA viruses of plants and animals. Adv Virus Res 47:159–251

Caliguiri LA, Tamm I (1970) The role of cytoplasmic membranes in poliovirus biosynthesis. Virology 42:100–111

Carette JE, Stuiver M, van Lent J, Wellink J, van Kammen A (2000) Cowpea mosaic virus infection induces a massive proliferation of endoplasmic reticulum but not Golgi membranes and is dependent on de novo membrane synthesis. J Virol 74:6556–6563

Carette JE, van Lent J, MacFarlane SA, Wellink J, van Kammen A (2002) Cowpea mosaic virus 32- and 60-kilodalton replication proteins target and change the morphology of endoplasmic reticulum membranes. J Virol 76:6293–6301

Cho MW, Teterina N, Egger D, Bienz K, Ehrenfeld E (1994) Membrane rearrangement and vesicle induction by recombinant poliovirus 2C and 2BC in human cells. Virology 202:129–145

Cook PR (1999) The organization of replication and transcription. Science 284:1790–1795

de Graaff M, Jaspars EMJ (1994) Plant viral RNA synthesis in cell-free systems. Annu Rev Phytopathol 32:311–335

de Jong AS, Wessels E, Dijkman HB, Galama JM, Melchers WJ, Willems PH, van Kuppeveld FJ (2003) Determinants for membrane association and permeabilization of the coxsackievirus 2B protein and the identification of the Golgi complex as the target organelle. J Biol Chem 278:1012–1021

den Boon JA, Chen J, Ahlquist P (2001) Identification of sequences in brome mosaic virus replicase protein 1a that mediate association with endoplasmic reticulum membranes. J Virol 75:12370–12381

Dodd DA, Giddings THJ, Kirkegaard K (2001) Poliovirus 3A protein limits interleukin-6 (IL-6), IL-8, and beta interferon secretion during viral infection. J Virol 75:8158–8165

Doedens JR, Kirkegaard K (1995) Inhibition of cellular protein secretion by poliovirus proteins 2B and 3A. EMBO J 14:894–907

Dubuisson J, Penin F, Moradpour D (2002) Interaction of hepatitis C virus proteins with host cell membranes and lipids. Trends Cell Biol 12:517–523

Echeverri AC, Dasgupta A (1995) Amino terminal regions of poliovirus 2C protein mediate membrane binding. Virology 208:540–553

Egger D, Bienz K (2002) Recombination of poliovirus RNA proceeds in mixed replication complexes originating from distinct replication start sites. J Virol 76:10960–10971

Egger D, Teterina N, Ehrenfeld E, Bienz K (2000) Formation of the poliovirus replication complex requires coupled viral translation, vesicle production, and viral RNA synthesis. J Virol 74:6570–6580

Egger D, Wolk B, Gosert R, Bianchi L, Blum HE, Moradpour D, Bienz K (2002) Expression of hepatitis C virus proteins induces distinct membrane alterations including a candidate viral replication complex. J Virol 76:5974–5984

Froshauer S, Kartenbeck J, Helenius A (1988) Alphavirus RNA replicase is located on the cytoplasmic surface of endosomes and lysosomes. J Cell Biol 107:2075–2086

Gomez de Cedrón M, Ehsani N, Mikkola ML, García JA, Kääriäinen L (1999) RNA helicase activity of Semliki Forest virus replicase protein NSP2. FEBS Lett 448:19–22

Gosert R, Kanjanahaluethai A, Egger D, Bienz K, Baker SC (2002) RNA replication of mouse hepatitis virus takes place at double-membrane vesicles. J Virol 76:3697–3708

Grimley PM, Berezesky IK, Friedman RM (1968) Cytoplasmic structures associated with an arbovirus infection: loci of viral ribonucleic acid synthesis. J Virol 2:1326–1338

Hayes RJ, Buck KW (1990) Complete replication of a eukaryotic virus RNA in vitro by a purified RNA-dependent RNA polymerase. Cell 63:363–368

Hügle T, Fehrmann F, Bieck E, Kohara M, Krausslich HG, Rice CM, Blum HE, Moradpour D (2001) The hepatitis C virus nonstructural protein 4B is an integral endoplasmic reticulum membrane protein. Virology 284:70–81

Hurley JH, Wendland B (2002) Endocytosis: driving membranes around the bend. Cell 111:143–146

Ivashkina N, Wolk B, Lohmann V, Bartenschlager R, Blum HE, Penin F, Moradpour D (2002) The hepatitis C virus RNA-dependent RNA polymerase membrane insertion sequence is a transmembrane segment. J Virol 76:13088–13093

Kääriäinen L, Ahola T (2002) Functions of alphavirus nonstructural proteins in RNA replication. Prog Nucleic Acid Res Mol Biol 71:187–222

Kääriäinen L, Söderlund H (1978) Structure and replication of alphaviruses. Curr Top Microbiol Immunol 82:15–69

Koonin EV, Dolja VV (1993) Evolution and taxonomy of positive-strand RNA viruses: Implications of comparative analysis of amino acid sequences. Crit Rev Biochem Mol Biol 28:375–430

Kujala P, Ahola T, Ehsani N, Auvinen P, Vihinen H, Kääriäinen L (1999) Intracellular distribution of rubella virus nonstructural protein P150. J Virol 73:7805–7811

Kujala P, Ikäheimonen A, Ehsani N, Vihinen H, Auvinen P, Kääriäinen L (2001) Biogenesis of the Semliki Forest virus RNA replication complex. J Virol 75:3873–3884

Laakkonen P, Ahola T, Kääriäinen L (1996) The effects of palmitoylation on membrane association of Semliki Forest virus RNA capping enzyme. J Biol Chem 271:28567–28571

Laakkonen P, Auvinen P, Kujala P, Kääriäinen L (1998) Alphavirus replicase protein Nsp1 induces filopodia and rearrangement of actin filaments. J Virol 72:10265–10269

Lai MC, Holmes KV (2001) *Coronaviridae*: the viruses and their replication. In: Knipe DM, Howley PM (eds) Fields virology. Lippincott Williams & Wilkins, Philadelphia, pp 1163–1185

Lampio A, Kilpeläinen I, Pesonen S, Karhi K, Auvinen P, Somerharju P, Kääriäinen L (2000) Membrane binding mechanism of an RNA virus-capping enzyme. J Biol Chem 275:37853–37859

Lee J-Y, Marshall JA, Bowden DS (1994) Characterization of rubella virus replication complexes using antibodies to double-stranded RNA. Virology 200:307–312

Lee WM, Ishikawa M, Ahlquist P (2001) Mutation of host delta9 fatty acid desaturase inhibits brome mosaic virus RNA replication between template recognition and RNA synthesis. J Virol 75:2097–2106

Lemm JA, Rümenapf T, Strauss EG, Strauss JH, Rice CM (1994) Polypeptide requirements for assembly of functional Sindbis virus replication complexes: a model for the temporal regulation of minus- and plus-strand RNA synthesis. EMBO J 13:2925–2934

Lemm JA, Bergqvist A, Read CM, Rice CM (1998) Template-dependent initiation of Sindbis virus RNA replication in vitro. J Virol 72:6546–6553

Lyle JM, Clewell A, Richmond K, Richards OC, Hope DA, Schultz SC, Kirkegaard K (2002) Similar structural basis for membrane localization and protein priming by an RNA-dependent RNA polymerase. J Biol Chem 277:16324–16331

Mackenzie JM, Jones MK, Westaway EG (1999) Markers for trans-Golgi membranes and the intermediate compartment localize to induced membranes with distinct replication functions in flavivirus-infected cells. J Virol 73:9555–9567

Mackenzie JM, Khromykh AA, Westaway EG (2001) Stable expression of noncytopathic Kunjin replicons simulates both ultrastructural and biochemical characteristics observed during replication of Kunjin virus. Virology 279:161–172

Magden J, Takeda N, Li T, Auvinen P, Ahola T, Miyamura T, Merits A, Kääriäinen L (2001) Virus-specific mRNA capping enzyme encoded by hepatitis E virus. J Virol 75:6249–6255

Magliano D, Marshall JA, Bowden DS, Vardaxis N, Meanger J, Lee J-Y (1998) Rubella virus replication complexes are virus-modified lysosomes. Virology 240:57–63

Más P, Beachy RN (1999) Replication of tobacco mosaic virus on endoplasmic reticulum and role of the cytoskeleton and virus movement protein in intracellular distribution of viral RNA. J Cell Biol 147:945–958

Mi S, Stollar V (1991) Expression of Sindbis virus nsP1 and methyltransferase activity in *Escherichia coli*. Virology 184:423–427

Miller DJ, Ahlquist P (2002) Flock house virus RNA polymerase is a transmembrane protein with amino-terminal sequences sufficient for mitochondrial localization and membrane insertion. J Virol 76:9856–9867

Miller DJ, Schwartz MD, Ahlquist P (2001) Flock house virus RNA replicates on outer mitochondrial membranes in *Drosophila* cells. J Virol 75:11664–11676

Mottola G, Cardinali G, Ceccacci A, Trozzi C, Bartholomew L, Torrisi MR, Pedrazzini E, Bonatti S, Migliaccio G (2002) Hepatitis C virus nonstructural proteins are localized in a modified endoplasmic reticulum of cells expressing viral subgenomic replicons. Virology 293:31–43

Osman TAM, Buck KW (1996) Complete replication in vitro of tobacco mosaic virus RNA by a template-dependent, membrane-bound RNA polymerase. J Virol 70:6227–6234

Paul AV, Molla A, Wimmer E (1994) Studies of a putative amphipathic helix in the N-terminus of poliovirus protein 2C. Virology 199:188–199.

Pedersen KW, van der Meer Y, Roos N, Snijder EJ (1999) Open reading frame 1a-encoded subunits of the arterivirus replicase induce endoplasmic reticulum-derived double-membrane vesicles which carry the viral replication complex. J Virol 73:2016–2026

Peränen J, Kääriäinen L (1991) Biogenesis of type I cytopathic vacuoles in Semliki Forest virus-infected BHK cells. J Virol 65:1623–1627

Peränen J, Takkinen K, Kalkkinen N, Kääriäinen L (1988) Semliki Forest virus-specific non-structural protein nsP3 is a phosphoprotein. J Gen Virol 69:2165–2178

Peränen J, Rikkonen M, Liljeström P, Kääriäinen L (1990) Nuclear localization of Semliki Forest virus-specific nonstructural protein nsP2. J Virol 64:1888–1896

Peränen J, Laakkonen P, Hyvönen M, Kääriäinen L (1995) The alphavirus replicase protein nsP1 is membrane-associated and has affinity to endocytic organelles. Virology 208:610–620

Pfister T, Wimmer E (1999) Characterization of the nucleoside triphosphatase activity of poliovirus protein 2C reveals a mechanism by which guanidine inhibits poliovirus replication. J Biol Chem 274:6992–7001

Pfister T, Mirzayan C, Wimmer E (1999) Polioviruses (*Picornaviridae*): molecular biology. In: Granoff A, Webster RG (eds) Encyclopedia of virology, 2nd edition. Academic Press, San Diego, pp 1330–1348

Piccininni S, Varaklioti A, Nardelli M, Dave B, Raney KD, McCarthy JE (2002) Modulation of the hepatitis C virus RNA-dependent RNA polymerase activity by the non-structural (NS) 3 helicase and the NS4B membrane protein. J Biol Chem 277:45670–45679

Prod'homme D, Le Panse S, Drugeon G, Jupin I (2001) Detection and subcellular localization of the turnip yellow mosaic virus 66 K replication protein in infected cells. Virology 281:88–101

Racaniello VR (2001) *Picornaviridae*: the viruses and their replication. In: Knipe DM, Howley PM (eds) Fields virology, 4th edition. Lippincott Williams & Wilkins, Philadelphia, pp 685–722

Restrepo-Hartwig MA, Ahlquist P (1996) Brome mosaic virus helicase- and polymerase-like proteins colocalize on the endoplasmic reticulum at sites of viral RNA synthesis. J Virol 70:8908–8916

Ritzenthaler C, Laporte C, Gaire F, Dunoyer P, Schmitt C, Duval S, Piequet A, Loudes AM, Rohfritsch O, Stussi-Garaud C, Pfeiffer P (2002) Grapevine fanleaf virus replication occurs on endoplasmic reticulum-derived membranes. J Virol 76:8808–8819

Rochon DM (1999) Tombusviruses. In: Granoff A, Webster RG (eds) Encyclopedia of virology, 2nd edition. Academic Press, San Diego, pp 1789–1798

Rubino L, Di Franco A, Russo M (2000) Expression of a plant virus non-structural protein in *Saccharomyces cerevisiae* causes membrane proliferation and altered mitochondrial morphology. J Gen Virol 81:279–286

Rust RC, Landmann L, Gosert R, Tang BL, Hong W, Hauri HP, Egger D, Bienz K (2001) Cellular COPII proteins are involved in production of the vesicles that form the poliovirus replication complex. J Virol 75:9808–9818

Salonen A, Vasiljeva L, Merits A, Magden J, Jokitalo E, Kääriäinen L (2003) Properly folded nonstructural polyprotein directs the Semliki Forest virus replication complex to endosomal compartment. J Virol 77:1691–1702

Sandoval IV, Carrasco L (1997) Poliovirus infection and expression of the poliovirus protein 2B provoke the disassembly of the Golgi complex, the organelle target for the antipoliovirus drug Ro-090179. J Virol 71:4679–4693

Schaad MC, Jensen PE, Carrington JC (1997) Formation of plant RNA virus replication complexes on membranes: role of an endoplasmic reticulum-targeted viral protein. EMBO J 16:4049–4059

Schlegel A, Giddings JTH, Ladinsky MS, Kirkegaard K (1996) Cellular origin and ultrastructure of membranes induced during poliovirus infection. J Virol 70:6576–6588

Schwartz M, Chen J, Janda M, Sullivan M, den Boon J, Ahlquist P (2002) A positive-strand RNA virus replication complex parallels form and function of retrovirus capsids. Mol Cell 9:505–514

Semler BL, Wimmer E (eds) (2002) Molecular biology of picornaviruses. ASM Press, Washington, DC

Snijder EJ, Meulenberg JJM (2001) Arteriviruses. In: Knipe DM, Howley PM (eds) Fields virology. Lippincott Williams & Wilkins, Philadelphia, pp 1205–1220

Snijder EJ, van Tol H, Roos N, Pedersen KW (2001) Non-structural proteins 2 and 3 interact to modify host cell membranes during the formation of the arterivirus replication complex. J Gen Virol 82:985–994

Strauss JH, Strauss EG (1994) The alphaviruses: gene expression, replication, and evolution. Microbiol Rev 58:491–562

Suhy DA, Giddings THJ, Kirkegaard K (2000) Remodeling the endoplasmic reticulum by poliovirus infection and by individual viral proteins: an autophagy-like origin for virus-induced vesicles. J Virol 74:8953–8965

Teterina N, Gorbalenya AE, Egger D, Bienz K, Ehrenfeld E (1997) Poliovirus 2C protein determinants of membrane binding and rearrangements in mammalian cells. J Virol 71:8962–8972

Teterina NL, Egger D, Bienz K, Brown DM, Semler BL, Ehrenfeld E (2001) Requirements for assembly of poliovirus replication complexes and negative-strand RNA synthesis. J Virol 75:3841–3850

Towner JS, Ho TV, Semler BL (1996) Determinants of membrane association for poliovirus protein 3AB. J Biol Chem 271:26810–26818

van der Heijden MW, Carette JE, Reinhoud PJ, Haegi A, Bol JF (2001) Alfalfa mosaic virus replicase proteins P1 and P2 interact and colocalize at the vacuolar membrane. J Virol 75:1879–1887

van der Meer Y, van Tol H, Krijnse Locker J, Snijder EJ (1998) ORF1a-encoded replicase subunits are involved in the membrane association of the arterivirus replication complex. J Virol 72:6689–6698

Vasiljeva L, Merits A, Auvinen P, Kääriäinen L (2000) Identification of a novel function of the *Alphavirus* capping apparatus—RNA 5' triphosphatase activity of Nsp2. J Biol Chem 275:17281–17287

Vasiljeva L, Valmu L, Kääriäinen L, Merits A (2001) Site-specific protease activity of the carboxyl-terminal domain of Semliki Forest virus replicase protein nsP2. J Biol Chem 276:30786–30793

Vasiljeva L, Merits A, Golubtsov A, Sizemskaja V, Kääriäinen L, Ahola T (2003) Regulation of the sequential processing of Semliki Forest virus replicase polyprotein. J Biol Chem 278:41636–41645

Vihinen H, Ahola T, Tuittila M, Merits A, Kääriäinen L (2001) Elimination of phosphorylation sites of Semliki Forest virus replicase protein nsP3. J Biol Chem 276:5745–5752

Weber-Lotfi F, Dietrich A, Russo M, Rubino L (2002) Mitochondrial targeting and membrane anchoring of a viral replicase in plant and yeast cells. J Virol 76:10485–10496

Westaway EG, Mackenzie JM, Kenney MT, Jones MK, Khromykh AA (1997) Ultrastructure of Kunjin virus-infected cells: colocalization of NS1 and NS3 with double-stranded RNA, and NS2B with NS3, in virus-induced membrane structures. J Virol 71:6650–6661

Westaway EG, Khromykh AA, Mackenzie JM (1999) Nascent flavivirus RNA colocalized in situ with double-stranded RNA in stable replication complexes. Virology 258:108–117

Westaway EG, Mackenzie JM, Khromykh AA (2002) Replication and gene function in Kunjin virus. Curr Top Microbiol Immunol 267:323–351

Wölk B, Sansonno D, Krausslich HG, Dammacco F, Rice CM, Blum HE, Moradpour D (2000) Subcellular localization, stability, and trans-cleavage competence of the hepatitis C virus NS3-NS4A complex expressed in tetracycline-regulated cell lines. J Virol 74:2293–2304

Wu S-X, Ahlquist P, Kaesberg P (1992) Active complete in vitro replication of nodavirus RNA requires glycerophospholipid. Proc Natl Acad Sci USA 89:11136–11140

Yamaga AK, Ou J (2002) Membrane topology of the hepatitis C virus NS2 protein. J Biol Chem 277:33228–33234

Yamanaka T, Imai T, Satoh R, Kawashima A, Takahashi M, Tomita K, Kubota K, Meshi T, Naito S, Ishikawa M (2002) Complete inhibition of tobamovirus multiplication by simultaneous mutations in two homologous host genes. J Virol 76:2491–2497

Synthesis and Quality Control of Viral Membrane Proteins

C. Maggioni · I. Braakman (✉)

University of Utrecht, Padualaan 8, 3584 CH, Utrecht, The Netherlands
I.Braakman@chem.uu.nl

1	Introduction	176
2	**Protein Biosynthesis**	176
2.1	Entering the Endoplasmic Reticulum: Targeting and Translocation	176
2.2	Protein Folding in the ER	178
2.2.1	Chaperones and Folding Enzymes	178
2.2.2	Glycosylation and the CNX/CRT Cycle	182
2.2.3	Disulfide Bond Formation	184
3	**Quality Control**	185
3.1	Retention in the ER	185
3.2	Degradation and the Unfolded Protein Response	186
4	**How Viruses Evolve to Fool the Host**	188
5	**Conclusions**	191
	References	191

Abstract Viruses use the host cellular machinery to translate viral proteins. Similar to cellular proteins directed to the secretory pathway, viral (glyco)proteins are synthesized on polyribosomes and targeted to the endoplasmic reticulum (ER). For viruses that encode polyproteins, folding of the individual proteins of the precursor often is coordinated. Translocation and the start of folding coincide and are assisted by cellular folding factors present in the lumen of the ER. The protein concentration a newborn protein finds in this compartment is enormous (hundreds of mg/ml) and the action of molecular chaperones is essential to prevent aggregation. Viral envelope proteins also undergo the cellular quality control mechanisms, which ensure, with variable stringency, that only proteins with the correct structure will proceed through the secretory pathway. Proteins that are misfolded, or not yet folded, are retained in the ER until they reach the native conformation or until their retrotranslocation into the cytosol for degradation. Peculiar characteristic of viruses is their ability to interfere with the cellular machinery to ensure virus production and, moreover, to pass through the body unobserved by the host immune system. This section describes some mechanisms of genetic variation and viral immune evasion that involve the secretory pathway.

1
Introduction

The membranes of enveloped virus particles contain one or more types of virally encoded integral membrane protein. The most abundant of these proteins contain a single transmembrane domain, a large ectodomain, that is localized at the outside of the viral envelope and is frequently glycosylated, and a small cytoplasmic domain. Almost all viral membrane proteins are oligomeric, either homo-oligomeric or hetero-oligomeric. Each subunit consists of one or two polypeptide chains that can be held together by noncovalent interactions or by covalent disulfide bridges. In the case of two polypeptide chains, they are often derived from a single-chain precursor that is proteolytically cleaved during transport of the protein to the plasma membrane. The cleavage is accompanied by conformational changes that prime the biological activity. For most of the viral proteins the cleavage is performed in trans by cellular or viral proteases. Viral envelope proteins have different functions: They mediate binding to the receptor(s) on the host cell's plasma membrane, membrane fusion, and penetration. For some viruses all these functions are combined in a single glycoprotein. In addition, some small viral envelope proteins have been shown to possess ion channel activity (reviewed by Fischer and Sansom 2002). The M2 of Influenza A for example, possesses proton conductance activity and is responsible for a pH change in the interior of the virion that is necessary for disassembly of the virus particle. This class of viral membrane proteins contains one or more transmembrane domains and either lacks or has a lower number of glycans and disulfide bonds. To be functional, as is true for all proteins, viral envelope glycoproteins must reach the correct three-dimensional structure, a process that is referred to as protein folding. To translate and fold viral envelope proteins, viruses use the host cellular machinery and they are subjected to the same quality control systems as endogenous proteins.

2
Protein Biosynthesis

2.1
Entering the Endoplasmic Reticulum: Targeting and Translocation

The biogenesis of most secretory and membrane proteins involves targeting of the nascent protein to the endoplasmic reticulum (ER), translocation across or integration into the ER membrane, and maturation into a functional product. Leader peptides of nascent chains are recognized in the cytosol by signal recognition particle (SRP) as soon as they emerge from the ribosome. The SRP-nascent chain-ribosome complex interacts with the SRP receptor, which targets it to the ER. Once targeted, the nascent chain is transferred to the translocation channel in a GTP-dependent step. The ribosome-nascent chain complex is tightly bound to the translocon without the ER luminal and cytosolic contents coming into communication. The translocation channel is the heterotrimeric Sec61p complex, composed of α-, β-, and γ-subunits (reviewed by Matlack et al. 1998). The channel is not a passive hole but is a dynamic structure that accomplishes different functions including targeting, cotranslational translocation, and cotranslational integration (Johnson and van Waes 1999).

Other proteins that are part of the translocation machinery are the translocon-associated protein (TRAP) complex with unknown function (Hartmann et al. 1993; Wang and Dobberstein 1999), the translocating chain associating membrane protein (TRAM), the small ribosome-associated membrane protein 4 (RAMP4) (Gorlich and Rapoport 1993), the signal peptidase (SP), and oligosaccharyl transferase (OST). TRAM is necessary for translocation of most substrates and regulates which domains of the nascent chain are "accessible" from the cytosol during a translocational pause (Hegde et al. 1998). Signal peptidase is necessary for cleavage of leader peptides and OST for transfer of oligosaccharide chains to nascent proteins, two processes that as a rule occur cotransl(oc)ationally. Exceptions to this rule have been found for some viral proteins. The leader peptide of the HIV-1 Envelope protein (Berman et al. 1988; Li et al. 1994; Land et al. 2003) and the signal peptide of a portion of the newly synthesized HCMV US11 (Rehm et al. 2001) are removed only after synthesis of these proteins is completed. Glycosylation of the hepatitis C virus (HCV) envelope protein E1 can occur also posttranslationally, albeit in a mannosylphosphoryldolichol-deficient CHO mutant cell (Duvet et al. 2002). Because molecular details of biosynthetic processes are more frequently studied in viral proteins than in mammalian proteins, we can only speculate on the specificity of these exceptions for viral proteins, and on their abundance in both the viral and the mammalian world.

Not only the translocon and the associated proteins but also ER luminal proteins have been shown to be necessary for proper translocation. For example BiP, the Hsp70 homolog in the ER, was found to be responsible for sealing the luminal end of not only the ribosome-free but also the ribosome-docked translocon (Haigh and Johnson 2002).

2.2
Protein Folding in the ER

The ER is the entry point for newly synthesized proteins that are directed into the secretory pathway. Folding can occur spontaneously because all information needed to reach the proper three-dimensional structure is present in the primary sequence (Anfinsen 1973). Still, within the cell the process needs assistance, which is provided by helper proteins known as molecular chaperones and folding enzymes. The ER is specialized for protein folding, providing an optimized environment: of oxidizing conditions and a high concentration of chaperones and folding enzymes. Folding begins cotranslationally and continues for minutes to hours after termination of polypeptide synthesis.

Some viral glycoproteins need not only cotranslational folding but also cotranslational assembly to reach their proper, native structure. For example, correct folding of the envelope glycoprotein E1 of HCV requires the presence of E2 (Michalak et al. 1997) through an interaction between the transmembrane domains of the two proteins (Cocquerel et al. 2001; Patel et al. 2001). In addition, another flanking protein of E1, the core protein, seems to play a role for correct folding of E1 (Merola et al. 2001). The folding process of p62 and E1, the envelope glycoproteins of Semliki Forest virus, is coordinated as well. The p62 protein can efficiently fold without E1, but E1 is found in aggregates in the absence of p62 (Andersson et al. 1997). Sindbis, another alphavirus, has an envelope composed of 80 trimers of E1-E2 dimers. The precursor of E2, pE2, dimerizes with E1 but, in this case, p62 interaction is needed for assembly and exit from the ER rather than for proper folding of E1 (Carleton et al. 1997). A recent study on the flavivirus tick-borne encephalitis (TBE) virus (Lorenz et al. 2002) clearly shows perhaps the first example of a mutual need for proper folding of the two envelope glycoproteins E and the precursor of M (prM), suggesting a chaperone-like function for prM. At the same time, E is also necessary for rapid signal sequence cleavage of prM and the two proteins can be each other's helpers even when expressed in trans, from different constructs.

2.2.1
Chaperones and Folding Enzymes

The main challenge a newly synthesized protein encounters within a cell is the intracellular environment, in particular the extremely high protein concentration of more than 200 mg/ml (Ellis 2001). A nascent chain is prone to aggregation and misfolding because of this high concentration (molecular crowding) and because of the close proximity of other nascent polypeptides emerging from a polyribosome. Considering that ribosomes on the same mRNA were found at a distance of only 80 nucleotides (Hesketh and Pryme 1991), or even 27 nucleotides during ribosome pausing (Wolin and Walter 1988), nascent chains in the lumen of the ER should emerge at a distance of about 15–40 nm, corresponding to a chain length of only a hundred amino acid residues or less. In the living cell, nonproductive protein folding generally is prevented by the action of molecular chaperones. Two types of chaperones can be recognized: the general kind, with a highly promiscuous interaction pattern, and the private kind, which caters for one particular protein (family) only.

In the ER, these general chaperones, present also at high concentration (Fig. 1A), associate with the growing nascent chain during translocation and continue to assist folding until a protein has acquired its native structure. Molecular chaperones act by facilitating rate-limiting steps, by stabilizing unfolded proteins, and by preventing undesired inter- and intrachain interactions that could lead to aggregation. They may recognize hydrophobic surface patches, mobile loops, and lack of compactness, or they may recognize a specific amino acid sequence in the case of the private chaperones. These features are exposed transiently during folding, resulting in only transient associations with ER folding factors if the newly synthesized protein folds correctly.

The general chaperones identified so far in the ER are among the most abundant proteins in a cell, and even more so in the ER. Well-known and well-studied are the ER Hsp70 homolog BiP, the Hsp90 homolog Grp94, and the lectin chaperones calnexin (CNX) and calreticulin (CRT). Long-known folding enzymes include protein disulfide isomerase (PDI) and two other oxidoreductases, ERp72 and ERp57. More recently, ERp44 was identified in mammalian ER (Anelli et al. 2002), and the *Saccharomyces cerevisiae* genome revealed the existence of a total of five PDI family members in the yeast ER (Norgaard et al. 2001). Peptidyl-prolyl cis-trans isomerases (PPIases) constitute another class of enzymes present in the ER. They catalyze the isomerisation of peptide bonds between any amino acid and a proline in a polypeptide chain; their role in folding was mainly demonstrated in vitro. A very high number of PPIases were discovered in mammals, grouped in three families (cyclophilins, FK506 binding proteins, and parvulins). Their functions reach beyond the assistance of newly synthesized proteins, being at the intersection between protein folding, signal transduction, trafficking, as-

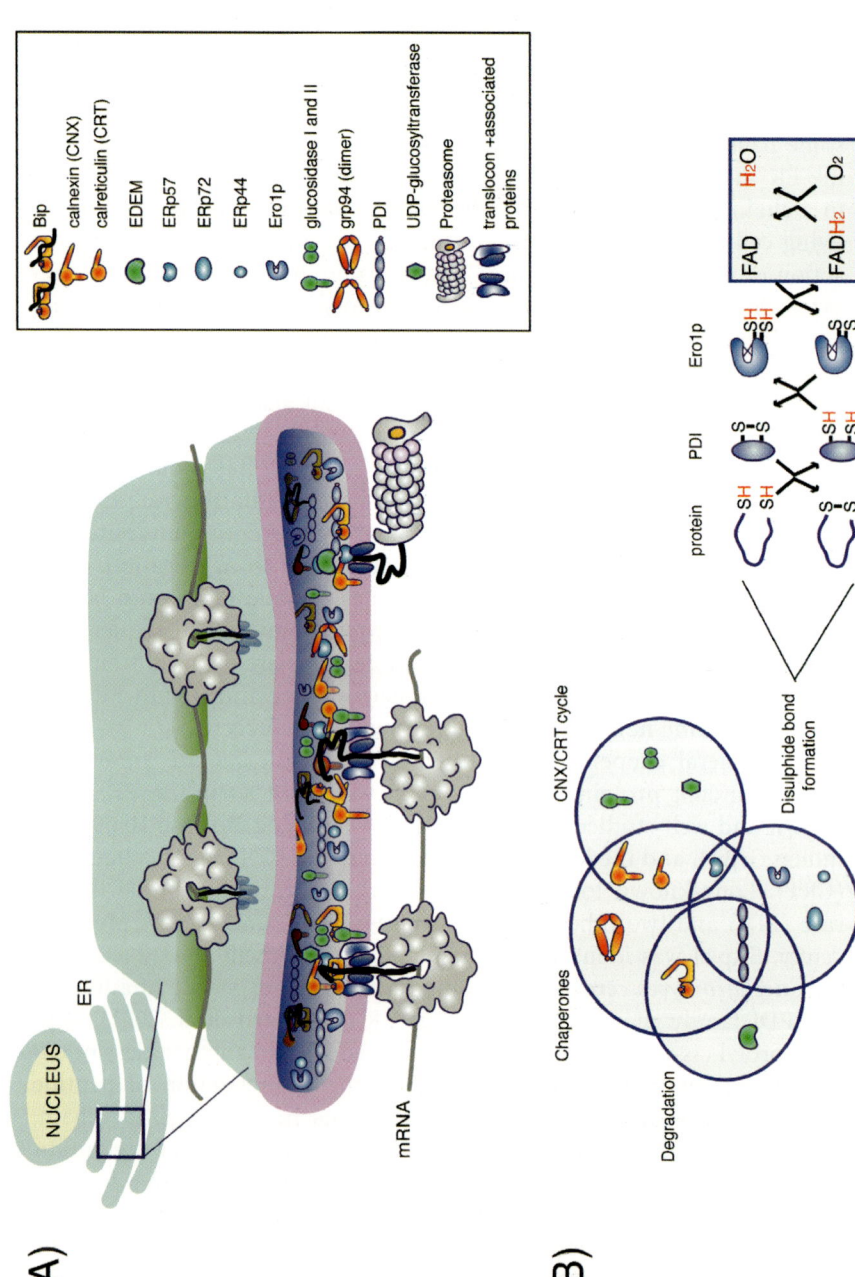

sembly, and cell cycle regulation (Gothel and Marahiel 1999). Mechanistic details of some of the ER folding factors are described below.

The question of why particular newly synthesized proteins have a preference for particular chaperones and folding enzymes has not been answered yet. Helenius and colleagues postulated that a hierarchy exists that depends on the position of the first glycan (Molinari and Helenius 2000). Important is the concept of redundancy. It exists everywhere in biology and is prominent among chaperones and folding enzymes. If the favored chaperone or folding enzyme is not available (because of a defect or because of competition) another ER folding factor takes over. This second folding factor may act through a different mechanism, but the end result, a high yield of properly folded protein, may be the same. A consequence of redundancy is the lower chance for misfolding, and at the same time an increased likelihood of competition. The ER protein folding capacity therefore is likely to depend on the relative concentration of the various ER folding factors, as well as the nature of the proteins synthesized by that cell.

Viruses have evolved to exploit the cell's machineries, including those specialized in protein folding and quality control. Some viruses may have strategies to stop host protein synthesis altogether, but any virus will cause a shift in the balance in the ER. Viral proteins as a rule are much more abundant than endogenous newly synthesized proteins, such that they compete out other substrates for the ER folding factors. This is clearly illustrated when comparing the number of proteins coimmunoprecipitated with CNX from a noninfected cell lysate with the almost exclusive coimmunoprecipitation of X31 influenza virus hemagglutinin (HA) from an influenza-infected cell lysate (Peterson et al. 1995), even though host protein synthesis is barely inhibited by this influenza strain.

The private chaperones are protein-specific factors that have evolved to assist the folding of protein (families) with atypical physical structures, or of proteins that exist in unusual situations or conditions. Colla-

---

Fig. 1. Inside the crowded ER. **A** Model for ER crowding: in the lumen of the ER, newly synthesized proteins are received into a very crowded environment. Proteins are depicted according to their relative size and shape. **B** Venn diagram showing the overlapping functions of some folding factors. The *inset* shows one of the known processes in protein folding: the formation of disulfide bonds in a nascent protein. The final electron acceptors (*gray box*) have been shown for yeast but are still unclear for mammalian cells

gen, for example, needs HSP47 to prevent premature fiber formation (Nagata 1996; Tasab et al. 2000), and RAP is considered to be specific for the LDL receptor family (Bu et al. 1995). Another example is the ER-resident molecule tapasin (Tpn) that is uniquely dedicated to tethering MHC class I molecules jointly with the chaperone calreticulin and the oxidoreductase ERp57 to MHC-encoded peptide transporter TAP (Momburg and Tan 2002).

Whereas viral proteins are found abundantly associated with general folding factors, they are of course less likely to use such private chaperones, except when they would be virally encoded. On the other hand, we cannot exclude that viral proteins coevolving with cells acquire(d) or exploit private chaperones. As yet, the number of proteins (both viral and mammalian) for which biosynthesis is extensively studied is too low to allow general conclusions.

2.2.2
Glycosylation and the CNX/CRT Cycle

Glycosylation has been shown to be important for protein folding, for protein stability, for immune evasion, and for receptor usage. Most viral envelope proteins are glycoproteins, and they bind during folding to the lectin chaperones CNX and CRT (Parodi 2000). The process of N-linked glycosylation of secretory proteins is characterized by enzymatic reactions occurring on both sides of the ER membrane. Monosaccharides are added to the lipid intermediate dolichol pyrophosphate to make Man5GlcNac2-PP-Dol. Further elongation of the glycan chain depends on specific translocators called flippases (Helenius et al. 2002). Once "flipped" over to the ER luminal side, the precursor is elongated to Glc3Man9GlcNac2-PP-Dol, for which the OST complex has optimal affinity. The enzyme transfers the oligosaccharide 14-mer en bloc to asparagine residues on the nascent chain within the consensus sequence Asn-X-Ser or Asn-X-Thr, with X being any amino acid except proline or aspartic acid.

General dogma says that glycans are crucial for proper glycoprotein folding. The reasons are understood to a large extent. Addition of a glycan adds a very large, strongly hydrophilic moiety to an otherwise perhaps more hydrophobic stretch of polypeptide. This will force this region of the protein away from the hydrophobic core of the protein. Another, direct, effect of glycosylation was found in peptides. A glycosylated peptide was shown to adopt a different conformation than the peptide without glycan (Imperiali and Rickert 1995), indicating a direct change of protein structure. Studies with the glucosidase inhibitor *n*-butyl deoxynojirimycin (nBuDNJ) showed that the assembly of viral envelopes of viruses, such as for example HIV, was not affected. However, the virus could not enter the host cell because the rearrangement of the V1/V2 loops necessary to release gp120 from gp160 could not take place (Fischer et al. 1996) with gp41 unable to mediate fusion. This demonstrates the importance of correct glycosylation for viral proteins.

A more indirect role for glycans is their recognition by the lectin chaperones CNX and CRT. These proteins bind the carbohydrate chain of newly synthesized glycoproteins only when in the monoglucosylated form, reached after trimming of the N-linked glycan by action of ER glucosidases I and II. A prolonged interaction with CNX and CRT is established via reversible glucosylation of the N-linked glycan after glucosidase II has removed the important glucose. In this way, CNX and CRT bind and release substrate in cycles, giving a prolonged chance to the substrate to fold correctly. The enzyme responsible for reglucosylation is the UDP-glucose:glycoprotein glucosyltransferase (GT), which behaves as a sensor for glycoprotein conformation (Parodi 2000; Ellgaard and Helenius 2001). The interaction with CNX and CRT exposes the folding protein to the associated cochaperone ERp57, a thiol oxidoreductase of the PDI family. During folding of viral glycoproteins, ERp57 has been shown to form transient intermolecular disulfide bonds with glycoprotein substrates bound to CNX and CRT (Molinari and Helenius 1999). The interaction between the substrate and one of the lectins seems to be required for the interaction with ERp57. Recently the three-dimensional structures of the CRT P-domain and the CNX ectodomain have been solved (Ellgaard et al. 2001, 2002; Schrag et al. 2001). A structural characterization of the binding site of ERp57 on CRT has been defined by NMR studies (Frickel et al. 2002). Whether ERp57 functions as a thiol oxidase and/or a disulfide isomerase still needs to be determined.

The lectin specificity of CNX and CRT are identical, and they can bind to the same substrate, sometimes even simultaneously. VSV G protein, which has two N-linked glycans, binds to CNX but not to CRT (Hammond et al. 1994). In contrast, Influenza HA associates with both lectins, albeit through different glycans: CRT binds preferentially with earlier folding intermediates and CNX associates also with a native monomeric form of the HA (Hebert et al. 1997). Such differences between CNX and CRT may reflect differential accessibility of the glycans in relation to the ER membrane, because CNX has a transmembrane anchor and CRT is a soluble protein (Danilczyk et al. 2000).

2.2.3
Disulfide Bond Formation

Many proteins that fold in the ER contain disulfide bonds, which are required for the protein's folding, stability, and function. Disulfide bond formation starts during synthesis; during the folding process, both native and nonnative disulfide bonds may form. One of the special features of the ER is its oxidizing environment, similar to the extracellular space. The formation and isomerization of disulfide bonds is catalyzed by protein thiol-disulfide oxidoreductases in the ER. The activity of this class of proteins depends on a pair of cysteines arranged in a Cys-X-X-Cys motif that has become a hallmark of all proteins involved in the formation or breakage of disulfide bonds. PDI was one of the first identified proteins belonging to this class of enzymes, and it can catalyze the formation, reduction, or isomerization of disulfide bonds, depending on the redox environment (Noiva 1999; Freedman et al. 2002).

The ER needs an oxidizing milieu to be able to support disulfide bond formation. Oxidized glutathione (GSSG) originally was thought to be responsible for generation and maintenance of the redox conditions in the ER, because of the higher ratio of oxidized to reduced glutathione in the ER (1:3) compared with the cytosol (1:60) (Hwang et al. 1992). Glutathione's assumed role, however, was completely changed by the identification of the Ero1p protein in yeast (Frand and Kaiser 1998; Pollard et al. 1998) and Ero1-Lα and -β in mammalian cells (Cabibbo et al. 2000; Pagani et al. 2000). This protein binds to the main oxidoreductase PDI (Frand and Kaiser 1999; Benham et al. 2000;) and drives disulfide bond formation by maintaining PDI, and possibly other PDI family members such as Erp44 (Anelli et al. 2002), in the oxidized state necessary for disulfide transfer (Tu et al. 2000).

In yeast, electrons flow from the oxidized substrate via PDI to Ero1p to FAD, with molecular oxygen as final electron acceptor (Fig. 1B, inset) (Tu and Weissman 2002). Mammalian Ero1 has not been found to bind FAD (yet), and downstream electron acceptors still need to be identified. A second, Ero1-independent, pathway for disulfide formation is proposed to involve a small ER oxidase known as Erv2, a 22-kDa protein that is noncovalently bound to FAD (Gerber et al. 2001; Sevier et al. 2001). The exact role of Erv2 in disulfide bond formation still remains unclear (Tu and Weissman 2002). In vaccinia virus, three cytoplasmic thiol oxidoreductases were identified that comprise a complete pathway for disulfide bond formation for viral proteins in the relatively reducing cytosol. Interestingly, the upstream component, E10R, is an Erv-like protein. These redox proteins are conserved in all poxviruses (Senkevich et al. 2002).

Disulfide bonds are usually highly conserved and critical for folding. They contribute to the stability of folding intermediates perhaps even more than to the stability of folded native proteins. Folding and disulfide bond formation coincide, meaning that folding can be followed by assaying formation and isomerization of disulfide bonds, for instance, through pulse-chase experiments (Braakman and Herbert 1996). Almost everything we know about the role of cysteines and disulfide bonds in protein folding comes from studies on viral proteins. The redox conditions in the ER can be easily manipulated by addition of DTT in the culture medium (Alberini et al. 1990; Braakman et al. 1992): Disulfide bond formation is inhibited, but the effect is reversed when DTT is removed. This allowed postponing of disulfide bond formation until after synthesis: completely posttranslationally. Cotranslational folding did increase efficiency of folding but turned out to be not essential, at least not in an intact cell where all ER factors are available.

Mutagenesis of cysteines involved in disulfide bonds has often been applied to study their influence on folding, maturation, and intracellular transport. Often, deletion of one cysteine residue from a pair is more deleterious than removal of both because a single free cysteine is very reactive and can interfere with other cysteines normally involved in other disulfide bonds. The fact that highly variable viral envelope proteins contain completely conserved sets of cysteine residues, and therefore most likely highly conserved disulfide bonds, demonstrates the importance of these covalent links for these proteins' functions.

3
Quality Control

3.1
Retention in the ER

During folding, proteins are subjected to a quality control machinery that allows exit from the ER only for proteins that have reached the correct three-dimensional structure. Misfolded and incompletely assembled proteins are retained in the ER and eventually degraded. The association of newly synthesized proteins with resident ER folding factors not only provides assistance during folding but also represents the main mechanism for quality control, both at the co- and posttranslational level.

Folding factors are indeed localized in the ER, and association with them results in retention of the folding substrate in the ER until proper folding is reached (Ellgaard et al. 1999). Some incompletely assembled substrates can also be retained in the ER through exposure of a more specific feature. An example of this is thiol-mediated retention, which was observed first for the immunoglobulin IgM (Sitia et al. 1990). During oligomerization, the exposure of a cysteine residue in the tailpiece (cys 575), involved in an interchain disulfide bond in the polymer, is a signal for ER retention as well as for degradation (Fra et al. 1993).

For glycoproteins, the CNX/CRT cycle is a common quality control system, which allows the protein multiple chances to fold correctly. The exit of certain glycoproteins from the ER to the Golgi complex is assisted by another membrane-bound lectin, ERGIC-53, that recognizes mannose residues (Hauri et al. 2000). In addition to cellular factors, proper folding of viral proteins may require complex interactions with neighboring viral proteins (see Sect. 2.2; Braakman and van Anken 2000). Many viral envelopes consist of highly ordered scaffolds of strongly interacting structural proteins. These regular interactions may require more stringent conformations for the structural proteins, resulting in perhaps a more stringent quality control compared with the quality control of cellular proteins. It is not easy to conclude this from available data, because of the limited set of proteins for which folding and quality control have been studied. Actually, viral glycoproteins frequently serve as model proteins for studies on protein folding in the ER (Doms et al. 1993). Not only for viral but also for cellular proteins, large variations exist concerning retention in the ER: For example, 50% of newly synthesized wt cystic fibrosis transmembrane conductance regulator (CFTR) protein is degraded (Ward and Kopito 1994), whereas mutants of the low-density lipoprotein (LDL) receptor often leave the ER to reach the plasma membrane (Hobbs et al. 1992).

3.2
Degradation and the Unfolded Protein Response

Prolonged retention of misfolded and incompletely folded proteins in the ER leads to their degradation (ER-associated degradation, or ERAD) (Brodsky and McCracken 1999). This process involves the translocon, cytosolic proteasomes, and ER chaperones such as CNX, BiP, and PDI (Zhang et al. 2001; Molinari et al. 2002). The substrate to be degraded is retrotranslocated through the Sec61 channel into the cytosol, deglycosylated by a cytosolic N-glycanase, and, often, polyubiquitinated before proteasomal degradation (Tsai et al. 2002). The criteria that determine how proteins are identified and sent to the degradation machinery are not completely understood. In eukaryotes, misfolded proteins are recognized by the enzyme mannosidase I, which removes the terminal mannose residue, producing the sugar moiety Man8GlcNAc2. The Man_8 species (including Glc_1Man_8) are recognized by a lectin, which is related to α-mannosidase but lacks enzymatic activity (called Htm1or Mnl1 in yeast and EDEM in mammals) (Jakob et al. 2001; Nakatsukasa et al. 2001). The lectin is thought to target the protein to the retrotranslocation pathway. Inhibition of mannosidase I, or deletion of the lectin, stabilizes some misfolded glycoproteins. Which mechanism targets nonglycosylated proteins is still not known.

Together the CNX/CRT cycle and the ER mannosidase allow incorrectly folded proteins multiple chances over prolonged periods of time to acquire the correct conformation. To handle the accumulation of incorrectly folded proteins in the ER, however, a signaling machinery has evolved that communicates to the nucleus that protein expression needs to be modulated to alleviate cellular stress, referred to as the unfolded protein response (UPR) (Ma and Hendershot 2001; Patil and Walter 2001; Harding et al. 2002). The result is an upregulation of genes encoding ER folding factors. Studies in *S. cerevisiae* have shown that the processes of UPR and ERAD are functionally linked to each other (Friedlander et al. 2000; Travers et al. 2000). In this way, cells can at the same time increase ER protein folding capacity and reduce ER protein folding load.

It is not surprising that viral infection induces cellular stress and the heavy load of viral protein that needs to be handled leads to an upregulation of ER. However, it remains unclear whether it is the high viral activity within the cell or the accumulation of viral proteins that induces stress. Two examples of viruses inducing UPR are HCV and Japanese encephalitis virus (JEV). HCV replication induces ER stress with the activation of UPR through transcription factor ATF6 and increased transcript levels of BiP (Tardif et al. 2002). During JEV infection, the lumen of the ER rapidly accumulates substantial amounts of viral proteins that trigger the UPR with the activation of CHOP/GADD153, a distinctive transcription factor often induced by UPR (Su et al. 2002).

4
How Viruses Evolve to Fool the Host

For a productive infection, viruses need to reproduce as many particles as possible. They can subvert the host cell translation apparatus, affecting several steps in the process of protein synthesis: degradation of host mRNA, competition for the host translation apparatus, changing the specificity of the host translation apparatus. In some cases, cellular protein synthesis is almost completely shut off, but this becomes rapidly toxic to the cell. A more limited inhibition of cellular functions usually is preferred to allow a long and productive infection.

When virus production is ensured in an infected cell, the next problem is to escape the organism's bodyguards. The immune system evolved to counterattack viral infections, but viruses have the possibility to evolve more rapidly and they have developed many tricks to evade immune surveillance. The immune system is mainly based on humoral and cellular responses, and both are targets for viral evasion.

The humoral response is based on recognition of viruses and viral proteins in solution. Envelope glycoproteins, being on the viral surface, are very immunogenic. Mammalian viruses therefore are subjected to tremendous selective pressures to continually change their molecular profiles. In many cases, natural selection produces viral strains that vary considerably in the antigenic regions of their spike proteins. The main strategy to mask the envelope glycoproteins from the immune system is by antigenic variation. RNA viruses are far more susceptible to genetic variation than DNA viruses. Three main mechanisms have been described: (1) point mutations (antigenic drift), (2) recombination, and (3) reassortment (antigenic shift).

1. RNA viruses lack a proofreader for replication, allowing the virus to mutate rapidly and frequently. RNA polymerases are at least 1,000–10,000 times more prone to error than DNA polymerases, resulting in higher mutation rates for RNA viruses. Hot spots for mutation often occur in the viral genome, coinciding in particular with antigenic sites recognized by virus-neutralizing antibodies. Influenza and HIV virus are two well-characterized examples of viruses that use this strategy.
Another attempt to mask the viral envelope proteins to the humoral immune system is by using host glycosylation as a means to cover or change potential antigenic epitopes. Most viral envelope glycoproteins are heavily glycosylated, and glycosylation sites are easily added or deleted during evolution through antigenic drift. HIV is an example of a virus with a highly glycosylated envelope protein (~30 glycans). In this

case it is not the position of the glycans in the protein but rather the gross number of glycans that is conserved. For folding of a glycoprotein, the exact position is less important than the presence per se of a glycan in that region of the polypeptide chain. Moving a glycan over the surface of the protein often does not affect folding and maturation but does effect shielding of antigenic sites, or disappearance of an antigenic epitope if the glycan was part of that epitope. It is possible that the virus maintains the number of glycans because they indeed help the virus to pass through cells and body unobserved.

2. Viruses can undergo recombination, during which genetic material is exchanged with related viral or cellular sequences through cutting and splicing of nucleic acids. This leads to a sudden change of the expressed proteins. This feature is prominent in virus families that contain positively stranded RNA, such as the *Picornaviridae* and the *Coronaviridae*.

3. Finally, reassortment of the individual pieces of RNA may occur during a mixed infection. For influenza virus, which does not show recombination as defined in (2), this frequently occurring antigenic shift is sometimes called recombination. Reassortment, however, leads to the production of a virus encoding proteins of different origin. This is evident in viruses in which the mRNA genome is present in multiple different segments that can be exchanged without any need to cut and splice nucleic acids. Other virus families with multiple genome segments include the *Rotaviridae* and the *Bunyaviridae*. This mechanism is postulated to account for the major antigenic shift of influenza A and is responsible for epidemic outbreaks of this virus.

Using a cellular response, the immune system not only recognizes virions in solution but also infected cells. Like cellular proteins, intracellular viral proteins are degraded and the resulting peptides are presented by specific molecules on the cell surface. The T cell receptor, present on T cells, recognizes the antigen-presenting molecules MHC class I and II associated with peptide. Recognition of a foreign antigen will result, through different mechanims, in cell death. Almost every step in the antigen presentation pathway (by both MHC I and II) has been taken as a target for viral subversion (for complete reviews, see Tortorella et al. 2000; Vossen et al. 2002).

We will describe here only the viral products that interfere with mechanisms related to the secretory pathway. Most examples involve the MHC class I presentation route. First, peptides are produced through proteasomal degradation of viral proteins. Some viral products (EBNA 1 of Epstein-Barr virus, for example) avoid their own degradation by insertion of Gly-Ala repeats in their sequence. These repeats work as cis-acting inhibitors of proteasomal proteolytic enzymes (Levitskaya et al. 1997; Leonchiks et al. 2002). Once peptides are formed, they need to be transferred into the ER through a channel, formed by the MHC-encoded peptide transporter TAP. The ER-resident protein tapasin is associated with TAP. Some viruses produce a protein that interacts with the cytosolic part of TAP (such as herpes simplex virus ICP47; York et al. 1994); others, such as human cytomegalovirus US6, interact with TAP on the ER luminal face (Hengel et al. 1997; see also the chapter by Döhner and Sodeik, this volume). Binding of US6 to TAP stabilizes a conformation of TAP that is unable to bind ATP and hence cannot translocate peptide (Hewitt et al. 2001). US6 itself does not contain a retention signal to stably reside in the ER, but it was suggested to bind to calnexin, through which it would be retained in the ER (Hengel et al. 1997).

After assembly of MHC class I with peptide the complex is normally transported to the cell surface, but many virus products can cause its intracellular retention. MHC class I can be retained in the ER, for example, by binding with the adenoviral product E3-19k (Cox et al. 1991), which in addition binds to TAP and acts as a competitive inhibitor of tapasin (Bennett et al. 1999). Retention can also occur in the ERGIC/cis-Golgi compartment by binding with the murine cytomegalovirus (MCMV) protein gp40 (m152), which has a retention signal (Ziegler et al. 1997). Two other proteins, gp48 (m06) and gp34 (m04), made by the same MCMV, interfere with MHC-I molecules. gp48 directs class I molecules to the lysosomes, whereas gp34 interacts with MHC-I in the ER; in addition, the gp34-class I complex is found on the cell surface. The exact function of gp34 is not completely understood. It may function to silence the NK response and/or alter the interaction with CD8+ T cells (Holtappels et al. 2000).

HCMV uses other tricks to prevent MHC-I from reaching the plasma membrane: The US3 protein retains MHC-I in the ER by a still-unknown mechanism. The same US3 can also bind to MHC II reducing their association with invariant chain (Ii) (Hegde et al. 2002). Another HCMV gene product, US2, is implicated in degradation of two important proteins of MHC class II (Tomazin et al. 1999) and, together with US11, efficiently directs class I molecules for proteasomal degradation with a halftime of less than 1 min. These viral studies also showed the first example of what turned out to be a more general process, where misfolded proteins in the ER need dislocation into the cytosol to be degraded by the proteasome (Wiertz et al. 1996; van der Wal et al. 2002).

HIV also interferes with MHC-I through two gene products: Nef and Vpu. Nef accelerates endocytosis of MHC-I (Schwartz et al. 1996) and

CD4 (Rhee and Marsh 1994). The MHC-I-Nef complex accumulates in the TGN and is then delivered to lysosomes. The mechanism of endocytosis for MHC-I and CD4 is different. Endocytosis of CD4 involves clathrin-coated pits (Piguet et al. 1998; Williams et al. 2002), whereas MHC-I was recently connected to the ARF6 endocytic pathway (Blagoveshchenskaya et al. 2002; see also the chapter by Döhner and Sodeik, this volume). Another HIV protein, Vpu, induces destabilization of newly synthesized MHC-I and induces proteasomal degradation of newly synthesized CD4 by phosphorylation of the protein on two specific residues (Fujita et al. 1997; Kerkau et al. 1997; Paul and Jabbar 1997).

5
Conclusions

Viruses have evolved into enormously efficient infectious agents: they exploit the cellular machinery to translate and fold their proteins and they keep their host alive for as long as possible to allow production of the maximum amount of progeny virus. Studies of every strategy that viruses have developed to survive increase our knowledge of the cellular and immunological processes involved in these strategies and will improve the chances of developing immunotherapies and other therapies for the treatment of viral infections.

Acknowledgements We thank Marije Liscalijet and Eelco van Anken for critical reading of the manuscript and the audio visual department, especially Aloys Lurvink, for assistance with the figure.

References

Alberini CM, P Bet, C Milstein and R Sitia (1990) Secretion of immunoglobulin M assembly intermediates in the presence of reducing agents. Nature 347(6292): 485–7

Andersson H, BU Barth, M Ekstrom and H Garoff (1997) Oligomerization-dependent folding of the membrane fusion protein of Semliki Forest virus. J Virol 71(12): 9654–63

Anelli T, M Alessio, A Mezghrani, T Simmen, F Talamo, A Bachi and R Sitia (2002) ERp44, a novel endoplasmic reticulum folding assistant of the thioredoxin family. EMBO J 21(4): 835–44

Anfinsen CB (1973) Principles that govern the folding of protein chains. Science 181(96): 223–30

Benham AM, A Cabibbo, A Fassio, N Bulleid, R Sitia and I Braakman (2000) The CXXCXXC motif determines the folding, structure and stability of human Ero1-Lα. EMBO J 19(17): 4493–502

Bennett EM, JR Bennink, JW Yewdell and FM Brodsky (1999) Cutting edge: adenovirus E19 has two mechanisms for affecting class I MHC expression. J Immunol 162(9): 5049–52

Berman PW, WM Nunes and OK Haffar (1988) Expression of membrane-associated and secreted variants of gp160 of human immunodeficiency virus type 1 in vitro and in continuous cell lines. J Virol 62(9): 3135–42

Blagoveshchenskaya AD, L Thomas, SF Feliciangeli, CH Hung and G Thomas (2002) HIV-1 Nef downregulates MHC-I by a PACS-1- and PI3K-regulated ARF6 endocytic pathway. Cell 111(6): 853–66

Braakman I, J Helenius and A Helenius (1992) Manipulating disulfide bond formation and protein folding in the endoplasmic reticulum. EMBO J 11(5): 1717–22

Braakman, I., D. Hebert. Disulfide (–SS–) bond formation overview. In: Current Protocols in Protein Science, Chapter 14.1 (eds. J. Coligan, B. Dunn, H. Ploegh, D. Speicher, P. Wingfield; John Wiley and Sons, Inc, New York, 1996): 14.1.1–14.1.15

Braakman I and E van Anken (2000) Folding of viral envelope glycoproteins in the endoplasmic reticulum. Traffic 1(7): 533–9

Brodsky JL and AA McCracken (1999) ER protein quality control and proteasome-mediated protein degradation. Semin Cell Dev Biol 10(5): 507–13

Bu G, HJ Geuze, GJ Strous and AL Schwartz (1995) 39 kDa receptor-associated protein is an ER resident protein and molecular chaperone for LDL receptor-related protein. EMBO J 14(10): 2269–80

Cabibbo A, M Pagani, M Fabbri, M Rocchi, MR Farmery, NJ Bulleid and R Sitia (2000) ERO1-L, a human protein that favors disulfide bond formation in the endoplasmic reticulum. J Biol Chem 275(7): 4827–33

Carleton M, H Lee, M Mulvey and DT Brown (1997) Role of glycoprotein PE2 in formation and maturation of the Sindbis virus spike. J Virol 71(2): 1558–66

Cocquerel L, JC Meunier, A Op de Beeck, D Bonte, C Wychowski and J Dubuisson (2001) Coexpression of hepatitis C virus envelope proteins E1 and E2 in cis improves the stability of membrane insertion of E2. J Gen Virol 82(7): 1629–35

Cox JH, JR Bennink and JW Yewdell (1991) Retention of adenovirus E19 glycoprotein in the endoplasmic reticulum is essential to its ability to block antigen presentation. J Exp Med 174(6): 1629–37

Danilczyk UG, MF Cohen-Doyle and DB Williams (2000) Functional relationship between calreticulin, calnexin, and the endoplasmic reticulum luminal domain of calnexin. J Biol Chem 275(17): 13089–97

Doms RW, RA Lamb, JK Rose and A Helenius (1993) Folding and assembly of viral membrane proteins. Virology 193(2): 545–62

Duvet S, A Op De Beeck, L Cocquerel, C Wychowski, R Cacan and J Dubuisson (2002) Glycosylation of the hepatitis C virus envelope protein E1 occurs post-translationally in a mannosylphosphoryldolichol-deficient CHO mutant cell line. Glycobiology 12(2): 95–101

Ellgaard L, P Bettendorff, D Braun, T Herrmann, F Fiorito, I Jelesarov, P Guntert, A Helenius and K Wuthrich (2002) NMR structures of 36 and 73-residue fragments of the calreticulin P-domain. J Mol Biol 322(4): 773–84

Ellgaard L and A Helenius (2001) ER quality control: towards an understanding at the molecular level. Curr Opin Cell Biol 13(4): 431–7

Ellgaard L, M Molinari and A Helenius (1999) Setting the standards: quality control in the secretory pathway. Science 286(5446): 1882–8

Ellgaard L, R Riek, D Braun, T Herrmann, A Helenius and K Wuthrich (2001) Three-dimensional structure topology of the calreticulin P-domain based on NMR assignment. FEBS Lett 488(1–2): 69–73

Ellis RJ (2001) Macromolecular crowding: an important but neglected aspect of the intracellular environment. Curr Opin Struct Biol 11(1): 114–9

Fischer PB, GB Karlsson, TD Butters, RA Dwek and FM Platt (1996) *n*-Butyldeoxynojirimycin-mediated inhibition of human immunodeficiency virus entry correlates with changes in antibody recognition of the V1/V2 region of gp120. J Virol 70(10): 7143–52

Fischer WB and MS Sansom (2002) Viral ion channels: structure and function. Biochim Biophys Acta 1561(1): 27–45

Fra AM, C Fagioli, D Finazzi, R Sitia and CM Alberini (1993) Quality control of ER synthesized proteins: an exposed thiol group as a three-way switch mediating assembly, retention and degradation. EMBO J 12(12): 4755–61

Frand AR and CA Kaiser (1998) The ERO1 gene of yeast is required for oxidation of protein dithiols in the endoplasmic reticulum. Mol Cell 1(2): 161–70

Frand AR and CA Kaiser (1999) Ero1p oxidizes protein disulfide isomerase in a pathway for disulfide bond formation in the endoplasmic reticulum. Mol Cell 4(4): 469–77

Freedman RB, P Klappa and LW Ruddock (2002) Protein disulfide isomerases exploit synergy between catalytic and specific binding domains. EMBO Rep 3(2): 136–40

Frickel EM, R Riek, I Jelesarov, A Helenius, K Wuthrich and L Ellgaard (2002) TROSY-NMR reveals interaction between ERp57 and the tip of the calreticulin P-domain. Proc Natl Acad Sci U S A 99(4): 1954–9

Friedlander R, E Jarosch, J Urban, C Volkwein and T Sommer (2000) A regulatory link between ER-associated protein degradation and the unfolded-protein response. Nat Cell Biol 2(7): 379–84

Fujita K, S Omura and J Silver (1997) Rapid degradation of CD4 in cells expressing human immunodeficiency virus type 1 Env and Vpu is blocked by proteasome inhibitors. J Gen Virol 78(3): 619–25

Gerber J, U Muhlenhoff, G Hofhaus, R Lill and T Lisowsky (2001) Yeast ERV2p is the first microsomal FAD-linked sulfhydryl oxidase of the Erv1p/Alrp protein family. J Biol Chem 276(26): 23486–91

Gorlich D and TA Rapoport (1993) Protein translocation into proteoliposomes reconstituted from purified components of the endoplasmic reticulum membrane. Cell 75(4): 615–30

Gothel SF and MA Marahiel (1999) Peptidyl-prolyl cis-trans isomerases, a superfamily of ubiquitous folding catalysts. Cell Mol Life Sci 55(3): 423–36

Haigh NG and AE Johnson (2002) A new role for BiP: closing the aqueous translocon pore during protein integration into the ER membrane. J Cell Biol 156(2): 261–70

Hammond C, I Braakman and A Helenius (1994) Role of N-linked oligosaccharide recognition, glucose trimming, and calnexin in glycoprotein folding and quality control. Proc Natl Acad Sci U S A 91(3): 913-7

Harding HP, M Calfon, F Urano, I Novoa and D Ron (2002) Transcriptional and translational control in the mammalian unfolded protein response. Annu Rev Cell Dev Biol 18:575-99

Hartmann E, D Gorlich, S Kostka, A Otto, R Kraft, S Knespel, E Burger, TA Rapoport and S Prehn (1993) A tetrameric complex of membrane proteins in the endoplasmic reticulum. Eur J Biochem 214(2): 375-81

Hauri H, C Appenzeller, F Kuhn and O Nufer (2000) Lectins and traffic in the secretory pathway. FEBS Lett 476(1-2): 32-7

Hebert DN, JX Zhang, W Chen, B Foellmer and A Helenius (1997) The number and location of glycans on influenza hemagglutinin determine folding and association with calnexin and calreticulin. J Cell Biol 139(3): 613-23

Hegde NR, RA Tomazin, TW Wisner, C Dunn, JM Boname, DM Lewinsohn and DC Johnson (2002) Inhibition of HLA-DR assembly, transport, and loading by human cytomegalovirus glycoprotein US3: a novel mechanism for evading major histocompatibility complex class II antigen presentation. J Virol 76(21): 10929-41

Hegde RS, S Voigt, TA Rapoport and VR Lingappa (1998) TRAM regulates the exposure of nascent secretory proteins to the cytosol during translocation into the endoplasmic reticulum. Cell 92(5): 621-31

Helenius J, DT Ng, CL Marolda, P Walter, MA Valvano and M Aebi (2002) Translocation of lipid-linked oligosaccharides across the ER membrane requires Rft1 protein. Nature 415(6870): 447-50

Hengel H, JO Koopmann, T Flohr, W Muranyi, E Goulmy, GJ Hammerling, UH Koszinowski and F Momburg (1997) A viral ER-resident glycoprotein inactivates the MHC-encoded peptide transporter. Immunity 6(5): 623-32

Hesketh JE and IF Pryme (1991) Interaction between mRNA, ribosomes and the cytoskeleton. Biochem J 277(1): 1-10

Hewitt EW, SS Gupta and PJ Lehner (2001) The human cytomegalovirus gene product US6 inhibits ATP binding by TAP. EMBO J 20(3): 387-96

Hobbs HH, MS Brown and JL Goldstein (1992) Molecular genetics of the LDL receptor gene in familial hypercholesterolemia. Hum Mutat 1(6): 445-66

Holtappels R, D Thomas, J Podlech, G Geginat, HP Steffens and MJ Reddehase (2000) The putative natural killer decoy early gene m04 (gp34) of murine cytomegalovirus encodes an antigenic peptide recognized by protective antiviral CD8 T cells. J Virol 74(4): 1871-84

Hwang C, AJ Sinskey and HF Lodish (1992) Oxidized redox state of glutathione in the endoplasmic reticulum. Science 257(5076): 1496-502

Imperiali B and KW Rickert (1995) Conformational implications of asparagine-linked glycosylation. Proc Natl Acad Sci U S A 92(1): 97-101

Jakob CA, D Bodmer, U Spirig, P Battig, A Marcil, D Dignard, JJ Bergeron, DY Thomas and M Aebi (2001) Htm1p, a mannosidase-like protein, is involved in glycoprotein degradation in yeast. EMBO Rep 2(5): 423-30

Johnson AE and MA van Waes (1999) The translocon: a dynamic gateway at the ER membrane. Annu Rev Cell Dev Biol 15:799-842

Kerkau T, I Bacik, JR Bennink, JW Yewdell, T Hunig, A Schimpl and U Schubert (1997) The human immunodeficiency virus type 1 (HIV-1) Vpu protein interferes with an early step in the biosynthesis of major histocompatibility complex (MHC) class I molecules. J Exp Med 185(7): 1295–305

Land A, D Zonneveld and I Braakman (2003) Folding of HIV-1 envelope glycoprotein involves extensive isomerization of disulfide bonds and conformation-dependent signal peptide cleavage. FASEB J in press

Leonchiks A, V Stavropoulou, A Sharipo and MG Masucci (2002) Inhibition of ubiquitin-dependent proteolysis by a synthetic glycine-alanine repeat peptide that mimics an inhibitory viral sequence. FEBS Lett 522(1–3): 93–8

Levitskaya J, A Sharipo, A Leonchiks, A Ciechanover and MG Masucci (1997) Inhibition of ubiquitin/proteasome-dependent protein degradation by the Gly-Ala repeat domain of the Epstein-Barr virus nuclear antigen 1. Proc Natl Acad Sci USA 94(23): 12616–21

Li Y, L Luo, DY Thomas and CY Kang (1994) Control of expression, glycosylation, and secretion of HIV-1 gp120 by homologous and heterologous signal sequences. Virology 204(1): 266–78

Lorenz IC, SL Allison, FX Heinz and A Helenius (2002) Folding and dimerization of tick-borne encephalitis virus envelope proteins prM and E in the endoplasmic reticulum. J Virol 76(11): 5480–91

Ma Y and LM Hendershot (2001) The unfolding tale of the unfolded protein response. Cell 107(7): 827–30

Matlack KE, W Mothes and TA Rapoport (1998) Protein translocation: tunnel vision. Cell 92(3): 381–90

Merola M, M Brazzoli, F Cocchiarella, JM Heile, A Helenius, AJ Weiner, M Houghton and S Abrignani (2001) Folding of hepatitis C virus E1 glycoprotein in a cell-free system. J Virol 75(22): 11205–17

Mezghrani A, A Fassio, A Benham, T Simmen, I Braakman and R Sitia (2001) Manipulation of oxidative protein folding and PDI redox state in mammalian cells. EMBO J 20(22): 6288–96

Michalak JP, C Wychowski, A Choukhi, JC Meunier, S Ung, CM Rice and J Dubuisson (1997) Characterization of truncated forms of hepatitis C virus glycoproteins. J Gen Virol 78(9): 2299–306

Molinari M, C Galli, V Piccaluga, M Pieren and P Paganetti (2002) Sequential assistance of molecular chaperones and transient formation of covalent complexes during protein degradation from the ER. J Cell Biol 158(2): 247–57

Molinari M and A Helenius (1999) Glycoproteins form mixed disulphides with oxidoreductases during folding in living cells. Nature 402(6757): 90–3

Molinari M and A Helenius (2000) Chaperone selection during glycoprotein translocation into the endoplasmic reticulum. Science 288(5464): 331–3

Momburg F and P Tan (2002) Tapasin-the keystone of the loading complex optimizing peptide binding by MHC class I molecules in the endoplasmic reticulum. Mol Immunol 39(3–4): 217–33

Nagata K (1996) Hsp47: a collagen-specific molecular chaperone. Trends Biochem Sci 21(1): 22–6

Nakatsukasa K, S Nishikawa, N Hosokawa, K Nagata and T Endo (2001) Mnl1p, an alpha -mannosidase-like protein in yeast *Saccharomyces cerevisiae*, is required for endoplasmic reticulum-associated degradation of glycoproteins. J Biol Chem 276(12): 8635-8

Noiva R (1999) Protein disulfide isomerase: the multifunctional redox chaperone of the endoplasmic reticulum. Semin Cell Dev Biol 10(5): 481-93

Norgaard P, V Westphal, C Tachibana, L Alsoe, B Holst and JR Winther (2001) Functional differences in yeast protein disulfide isomerases. J Cell Biol 152(3): 553-62

Pagani M, M Fabbri, C Benedetti, A Fassio, S Pilati, NJ Bulleid, A Cabibbo and R Sitia (2000) Endoplasmic reticulum oxidoreductin 1-lbeta (ERO1-Lbeta), a human gene induced in the course of the unfolded protein response. J Biol Chem 275(31): 23685-92

Parodi AJ (2000) Protein glucosylation and its role in protein folding. Annu Rev Biochem 69:69-93

Patel J, AH Patel and J McLauchlan (2001) The transmembrane domain of the hepatitis C virus E2 glycoprotein is required for correct folding of the E1 glycoprotein and native complex formation. Virology 279(1): 58-68

Patil C and P Walter (2001) Intracellular signaling from the endoplasmic reticulum to the nucleus: the unfolded protein response in yeast and mammals. Curr Opin Cell Biol 13(3): 349-55

Paul M and MA Jabbar (1997) Phosphorylation of both phosphoacceptor sites in the HIV-1 Vpu cytoplasmic domain is essential for Vpu-mediated ER degradation of CD4. Virology 232(1): 207-16

Peterson JR, A Ora, PN Van and A Helenius (1995) Transient, lectin-like association of calreticulin with folding intermediates of cellular and viral glycoproteins. Mol Biol Cell 6(9): 1173-84

Piguet V, YL Chen, A Mangasarian, M Foti, JL Carpentier and D Trono (1998) Mechanism of Nef-induced CD4 endocytosis: Nef connects CD4 with the mu chain of adaptor complexes. EMBO J 17(9): 2472-81

Pollard MG, KJ Travers and JS Weissman (1998) Ero1p: a novel and ubiquitous protein with an essential role in oxidative protein folding in the endoplasmic reticulum. Mol Cell 1(2): 171-82

Rehm A, P Stern, HL Ploegh and D Tortorella (2001) Signal peptide cleavage of a type I membrane protein, HCMV US11, is dependent on its membrane anchor. EMBO J 20(7): 1573-82

Rhee SS and JW Marsh (1994) Human immunodeficiency virus type 1 Nef-induced down-modulation of CD4 is due to rapid internalization and degradation of surface CD4. J Virol 68(8): 5156-63

Schrag JD, JJ Bergeron, Y Li, S Borisova, M Hahn, DY Thomas and M Cygler (2001) The structure of calnexin, an ER chaperone involved in quality control of protein folding. Mol Cell 8(3): 633-44

Schwartz O, V Marechal, S Le Gall, F Lemonnier and JM Heard (1996) Endocytosis of major histocompatibility complex class I molecules is induced by the HIV-1 Nef protein. Nat Med 2(3): 338-42

Senkevich TG, CL White, EV Koonin and B Moss (2002) Complete pathway for protein disulfide bond formation encoded by poxviruses. Proc Natl Acad Sci USA 99(10): 6667-72

Sevier CS, JW Cuozzo, A Vala, F Aslund and CA Kaiser (2001) A flavoprotein oxidase defines a new endoplasmic reticulum pathway for biosynthetic disulphide bond formation. Nat Cell Biol 3(10): 874–82

Sitia R, M Neuberger, C Alberini, P Bet, A Fra, C Valetti, G Williams and C Milstein (1990) Developmental regulation of IgM secretion: the role of the carboxy-terminal cysteine. Cell 60(5): 781–90

Su HL, CL Liao and YL Lin (2002) Japanese encephalitis virus infection initiates endoplasmic reticulum stress and an unfolded protein response. J Virol 76(9): 4162–71

Tardif KD, K Mori and A Siddiqui (2002) Hepatitis C virus subgenomic replicons induce endoplasmic reticulum stress activating an intracellular signaling pathway. J Virol 76(15): 7453–9

Tasab M, MR Batten and NJ Bulleid (2000) Hsp47: a molecular chaperone that interacts with and stabilizes correctly-folded procollagen. EMBO J 19(10): 2204–11

Tomazin R, J Boname, NR Hegde, DM Lewinsohn, Y Altschuler, TR Jones, P Cresswell, JA Nelson, SR Riddell and DC Johnson (1999) Cytomegalovirus US2 destroys two components of the MHC class II pathway, preventing recognition by CD4+ T cells. Nat Med 5(9): 1039–43

Tortorella D, BE Gewurz, MH Furman, DJ Schust and HL Ploegh (2000) Viral subversion of the immune system. Annu Rev Immunol 18:861–926

Travers KJ, CK Patil, L Wodicka, DJ Lockhart, JS Weissman and P Walter (2000) Functional and genomic analyses reveal an essential coordination between the unfolded protein response and ER-associated degradation. Cell 101(3): 249–58

Tsai B, Y Ye and TA Rapoport (2002) Retro-translocation of proteins from the endoplasmic reticulum into the cytosol. Nat Rev Mol Cell Biol 3(4): 246–55

Tu BP, SC Ho-Schleyer, KJ Travers and JS Weissman (2000) Biochemical basis of oxidative protein folding in the endoplasmic reticulum. Science 290(5496): 1571–4

Tu BP and JS Weissman (2002) The FAD- and O_2-dependent reaction cycle of Ero1-mediated oxidative protein folding in the endoplasmic reticulum. Mol Cell 10(5): 983–94

van der Wal FJ, M Kikkert and E Wiertz (2002) The HCMV gene products US2 and US11 target MHC class I molecules for degradation in the cytosol. Curr Top Microbiol Immunol 269:37–55

Vossen MT, EM Westerhout, C Soderberg-Naucler and EJ Wiertz (2002) Viral immune evasion: a masterpiece of evolution. Immunogenetics 54(8): 527–42

Wang L and B Dobberstein (1999) Oligomeric complexes involved in translocation of proteins across the membrane of the endoplasmic reticulum. FEBS Lett 457(3): 316–22

Ward CL and RR Kopito (1994) Intracellular turnover of cystic fibrosis transmembrane conductance regulator. Inefficient processing and rapid degradation of wild-type and mutant proteins. J Biol Chem 269(41): 25710–8

Wiertz EJ, TR Jones, L Sun, M Bogyo, HJ Geuze and HL Ploegh (1996) The human cytomegalovirus US11 gene product dislocates MHC class I heavy chains from the endoplasmic reticulum to the cytosol. Cell 84(5): 769–79

Williams M, JF Roeth, MR Kasper, RI Fleis, CG Przybycin and KL Collins (2002) Direct binding of human immunodeficiency virus type 1 Nef to the major histocompatibility complex class I (MHC-I) cytoplasmic tail disrupts MHC-I trafficking. J Virol 76(23): 12173-84

Wolin SL and P Walter (1988) Ribosome pausing and stacking during translation of a eukaryotic mRNA. EMBO J 7(11): 3559-69

York IA, C Roop, DW Andrews, SR Riddell, FL Graham and DC Johnson (1994) A cytosolic herpes simplex virus protein inhibits antigen presentation to CD8+ T lymphocytes. Cell 77(4): 525-35

Zhang Y, G Nijbroek, ML Sullivan, AA McCracken, SC Watkins, S Michaelis and JL Brodsky (2001) Hsp70 molecular chaperone facilitates endoplasmic reticulum-associated protein degradation of cystic fibrosis transmembrane conductance regulator in yeast. Mol Biol Cell 12(5): 1303-14

Ziegler H, R Thale, P Lucin, W Muranyi, T Flohr, H Hengel, H Farrell, W Rawlinson and UH Koszinowski (1997) A mouse cytomegalovirus glycoprotein retains MHC class I complexes in the ERGIC/cis-Golgi compartments. Immunity 6(1): 57-66

# Receptor Modulation in Viral replication: HIV, HSV, HHV-8 and HPV: Same Goal, Different Techniques to Interfere with MHC-I Antigen Presentation

V. Piguet (✉)

Department of Dermatology and Venerology, HUG, 24 Rue Micheli-du-Crest, 1211 Geneva, Switzerland
vincent.piguet@medecine.unige.ch

1	Introduction	200
2	Receptor Modulation During Viral Replication: Interference with MHC-I Antigen Presentation	201
2.1	Primate *Lentiviridae* Interference with MHC-I Presentation: Connecting the Receptor with the PACS-1-Dependent Pathways	201
2.2	HSV Downregulation of MHC-I Expression: Targeting the Transporter TAP	205
2.3	HHV-8 Downregulation of MHC-I Expression: Ubiquitination of the Tail	206
2.4	HPV Inhibition of MHC-I Expression: Blocking the Proton Pump	210
3	Perspectives	211
	References	212

Abstract Evasion of host immunity is a common objective of viruses that cause chronic infections. Viruses involved in sexually transmitted infections constitute no exception to this phenomenon. HIV, HPV, HSV, and HHV-8 subvert the class I major histocompatibility complex (MHC-I) antigen presentation pathway, thereby evading the cellular immune response. Although the goal of these viruses is the same and efficient MHC-I downregulation in infected cells is achieved, their techniques vary considerably. Whether viral inhibition occurs at the transcriptional level, during assembly of MHC-I complexes in the endoplasmic reticulum, during its journey to the cell surface, or after reaching the cell surface, each one of these viruses ingeniously achieves MHC-I downregulation and avoids the cellular immune response. Unraveling the mechanisms of interference with MHC-I antigen presentation employed by these viruses is not only crucial to understand their pathogenesis, but also reveals novel mechanisms of regulation of cellular receptors. When employed as modulators of cellular trafficking pathways, viruses become tools to dissect fundamental cell processes. In return, the precise dissection of these processes may offer new weapons against the ruses viruses employ to propagate and establish chronic infections.

1
Introduction

Cells require surface receptors for the capture of essential constituents and energy, for adaptation to their environment via specific signaling cascades, and for displaying information that is read by other cells or by extracellular factors. Surface receptors are at the core of the interactions that take place between a cell and the environment in which it resides. As a result, the expression of most cell surface receptors is tightly regulated through specialized processes (for recent review see Conner and Schmid 2003). Viruses often modulate the expression of cell surface receptors, in order to maximize their replication and survival in an infected host. A typical example is the modulation of class I major histocompatibility complex (MHC-I) surface levels by viruses that establish chronic infections, in order to avoid the cellular arm of the immune response. This review will focus on the mechanisms of regulation of MHC-I antigen presentation by viruses implicated in sexually transmitted infections, in other words: HIV, herpes simplex virus (HSV), human herpesvirus 8 (HHV-8), and HPV.

MHC-I consists of a highly polymorphic, membrane-anchored heavy chain noncovalently associated with β2-microglobulin (β2m). The assembly of the heavy chain with β2m takes place in the ER or in the cis-Golgi apparatus, where antigenic peptides are loaded (Bijlmakers and Ploegh 1993). A significant fraction of newly synthesized proteins, viral or cellular, undergoes proteasomal degradation generating peptides (Schubert et al. 2000). Antigenic peptides are supplied by the transporter associated with antigen processing, TAP. Without its peptide cargo, Class I complexes do not reach the cell surface, are unstable, and dissociate. The trimolecular complex (heavy chain with β2m and peptide) is subsequently addressed to the cell surface (Neefjes et al. 1990). MHC-I complexes are stably expressed at the cell surface, with only a minor fraction of the molecules being internalized spontaneously in T cells and in monocytes/macrophages (Neefjes et al. 1990; Reid and Watts 1990). In an attempt to paralyze the cellular immune response, viruses are able to target MHC-I at almost all steps of its trafficking: in the ER, on its way to the surface, and after it reaches the cell surface.

2
Receptor Modulation During Viral Replication: Interference with MHC-I Antigen Presentation

2.1
Primate *Lentiviridae* Interference with MHC-I Presentation: Connecting the Receptor with the PACS-1-Dependent Pathways

The human and simian immunodeficiency viruses (HIV and SIV) are lentiviruses, members of the retrovirus superfamily characterized by their genomic complexity. Besides the Gag, Pol, and Env genes found in all retroviruses, HIV and SIV each contain six additional reading frames. Two of these, Tat and Rev, encode crucial regulators of viral gene ex-pression. The four others, Nef, Vif, Vpr, and Vpu (in HIV-1) or Vpx (in HIV-2 and SIV), are dispensable for viral growth in most tissue culture systems. As a consequence, these genes are termed auxiliary or accessory. However, the so-called accessory proteins of primate lentiviruses represent critical virulence factors that manipulate several cellular pathways in order to maximize viral replication, including downregulation of MHC-I complexes from the cell surface. HIV can, at least to some extent, escape the attacks of cytotoxic T lymphocytes (CTLs) by down-regulating the expression of MHC-I on the surface of infected cells in vitro and in animal models (Collins et al. 1998; Munch et al. 2001; Schwartz et al. 1996). This manipulation of the cellular immune response might enable HIV to persist in the host for prolonged periods of time, which leads ultimately to a collapse of the immune system called acquired immunodeficiency syndrome (AIDS).

The main viral player that interferes with MHC-I antigen presentation is Nef (Schwartz et al. 1996; Fig. 1), even though two other viral proteins, Vpu and Tat, may occasionally play a minor role (Howcroft et al. 1993; Kerkau et al. 1997). Nef is a short cytoplasmic protein that associates with membranes through N-terminal myristoylation and exerts several additional effects on the infected cell. For instance, it is responsible for downregulating the cell surface expression of CD4, the main HIV receptor (Garcia and Miller 1991). Nef downmodulates MHC-I in a variety of cell types, including primary T lymphocytes (Kasper and Collins 2003; Mangasarian et al. 1999; Schwartz et al. 1996). Determinants necessary for Nef responsiveness are contained in the cytoplasmic domain of MHC-I and are centered around a critical tyrosine residue found in HLA-A and -B, but not in HLA-C (Le Gall et al. 1998). Correspondingly, HLA-C is unaffected by Nef (Cohen et al., 1999). This may be of physio-

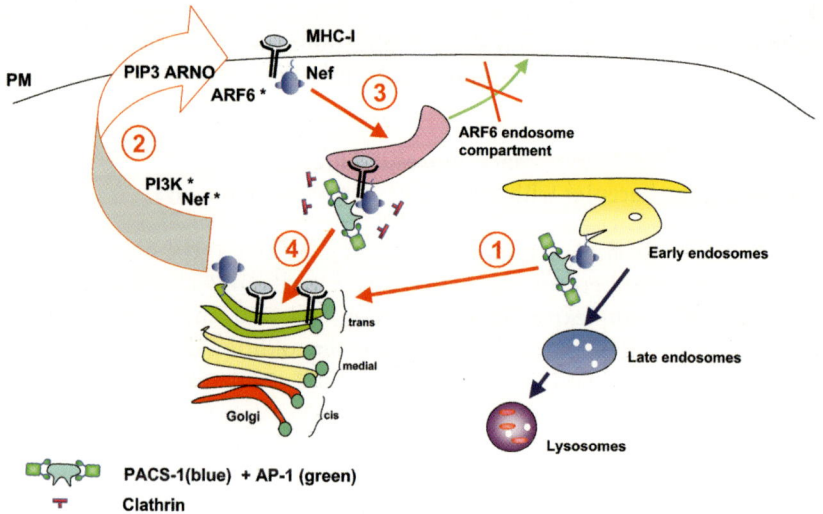

Fig. 1. HIV Nef connects MHC-I to the PACS-1 sorting pathway. HIV Nef interacts with MHC-I cytosolic domains and connects them to the PACS-1 sorting pathway. *1* Nef is first targeted to the *trans*-Golgi network (*TGN*) via its interaction with PACS-1 (phosphofurin acidic cluster sorting protein), where it activates phosphatidylinositol 3-kinase (*PI3K**) via its polyproline domain (*Nef**). *2* In turn, PI3K activates ADP ribosylation factor 6 (*ARF6**). *3* Nef interacts with MHC-I cytoplasmic domain and redirects MHC-I from the cell surface to the ARF-6 specific early endosomal compartment. *4* MHC-I is then redirected to the TGN by Nef via its acidic cluster domain and PACS-1 recruitment

logical importance, because HLA-C molecules are dominant inhibitory ligands that protect cells against lysis by natural killer (NK) lymphocytes, which normally destroy MHC-I-negative cells (Cohen et al. 1999). Recently, Nef has also been shown to interact with HLA-A and B both in vitro and in vivo, but not with HLA-C (Williams et al. 2002).

In the presence of Nef, MHC-I is normally synthesized and transported through the ER and *cis*-Golgi and can reach the cell surface. However, it is then diverted toward the endosomal pathway to accumulate in the *trans*-Golgi network (TGN), before undergoing degradation (Greenberg et al. 1998b; Le Gall et al. 1998; Schwartz et al. 1996). Some degree of interference also occurs during MHC-I transport from the TGN to the cell surface (Kasper and Collins 2003). Mutagenesis studies have highlighted the importance of specific determinants in the ability of HIV-1 Nef to regulate MHC-1. These include the Nef myristoylation signal, essential for its membrane association, an N-proximal α-helix, a centrally-located SH3-binding proline-based repeat, and a highly conserved cluster of acidic residues in the N-terminal third of the viral protein (EEEE[65]) (Greenberg et al. 1998b; Mangasarian et al. 1999). The acidic cluster (AC) of Nef is highly similar to the phosphofurin acidic cluster sorting protein-1 (PACS-1)-binding, TGN retrieval motif of furin and mannose-6-phosphate receptor (MPR). PACS-1 governs the endosome-to-Golgi trafficking of furin and MPR by connecting the AC-containing cytoplasmic domain of these molecules with the adaptor protein complex (AP-1) of endosomal clathrin-coated pits (CCP) (Wan et al. 1998). Likewise, the AC-mediated recruitment of PACS-1 is responsible for MHC-I downregulation and TGN targeting (Crump et al. 2001; Piguet et al. 2000). Nef and PACS-1 interact to hijack the ADP ribosylation factor 6 (ARF6)-dependent endocytic pathway, by a process involving phosphatidylinositol 3-kinase (PI3K), leading to a loss of cell surface MHC-I. This mechanism requires the sequential actions of three Nef motifs. The acidic cluster EEEE[65], the SH3 domain binding proline-based repeat, and a N-proximal α-helix are required one after the other in controlling PACS-1-dependent sorting to the TGN, ARF6 activation, and sequestration of internalized MHC-I to the TGN, respectively. As expected, inhibitors of PI3K block retrieval of MHC-I molecules to the TGN (Blagoveshchenskaya et al. 2002). The most likely route taken by MHC-I in Nef-expressing cells is the following: First, Nef is targeted to the TGN via its interaction with PACS-1, where it activates PI3K via Nef proline-based repeat. Second, PI3K activates ARF6. Third, Nef interacts with MHC-I cytoplasmic domain and redirects MHC-I from the cell surface to the ARF6 specific early endosomal compartment. Fourth, MHC-I is sent to the TGN by Nef via its AC domain and PACS-1 recruitment. Finally, MHC-I is retained in the TGN, via the N-proximal α-helix of Nef. Together, these results support a model in which Nef downregulates MHC-I by acting as a connector between the receptor cytoplasmic tail and the PACS-1 sorting pathway. However, a direct demonstration of complexes containing MHC-I, Nef, and PACS-1 in cells remains to be shown.

Nef acts also as a connector to downregulate CD4, the main HIV receptor. However, in this case both the downstream partners of Nef and the fate of its cellular target are different. In the Golgi and at the plasma membrane, Nef bridges the cytoplasmic tail of CD4 with the adaptor protein complex of CCP, thereby triggering the formation of CD4-specific endocytic vesicles. Then, in the early endosome, Nef links CD4 to a subset of COP-I coatomer proteins that then target the HIV receptor for degradation in lysosomes (Aiken et al. 1994; Greenberg et al. 1997; Le

Gall et al. 2000; Mangasarian et al. 1997; Piguet et al. 1998; Piguet et al. 1999; Rhee and Marsh 1994).

Interestingly, Nef can also recruit the catalytic subunit H of the vacuolar membrane ATPase (V-ATPase), which may facilitate CD4 internalization and the recruitment of components of CCP (Geyer et al. 2002b; Lu et al. 1998). A further effect of Nef-induced CD4 downregulation is to contribute to the fitness of viral replication, because it preserves the infectivity of HIV-1 virions by preventing interference between the receptor and particle release or envelope incorporation (Lama et al. 1999; Ross et al. 1999).

Interestingly, although Nef downregulates both CD4 and MHC-I, it can increase the surface levels of a dendritic cell-specific receptor called DC-SIGN. Nef inhibits the endocytosis of DC-SIGN (Sol-Foulon et al. 2002), thus increasing the surface levels of this receptor. DC-SIGN facilitates infection of T cells *in trans* and is also responsible for the interaction between dendritic cells and naïve T cells (Geijtenbeek et al. 2000a, 2000b). The mechanism of Nef inhibition of DC-SIGN endocytosis is partly elucidated. DC-SIGN contains a sorting signal characterized by a dileucine domain in its cytoplasmic domain (Engering et al. 2002). This domain is required for DC-SIGN sorting to the lysosomes in immature dendritic cells. Nef also requires its own dileucine domain to inhibit DC-SIGN endocytosis. The dileucine motif of Nef is required to interact with adaptor complexes at the plasma membrane (Bresnahan et al. 1998; Craig et al. 1998; Greenberg et al. 1998a). The simplest explanation of Nef-induced DC-SIGN upregulation is that Nef interferes with DC-SIGN sorting pathways because of a direct competition between Nef dileucine domain and DC-SIGN dileucine motif, both interacting with similar adaptor complexes at the plasma membrane. The consequence of Nef-mediated upregulation of DC-SIGN is an increase in viral replication in dendritic cell-T cell clusters (Sol-Foulon et al. 2002).

Manipulation of the surface levels of cellular receptors by HIV is complex but offers novel possibilities to understand the mechanisms that this virus employs to persist in the host. For instance, by understanding the precise role of Nef in enhancing viral replication and avoiding the cellular immune response, drugs could be developed that target domains in Nef involved in MHC-I downregulation. A potential benefit of Nef inhibitors would be an increase in viral recognition by the immune system and a decrease in viral replication.

2.2
HSV Downregulation of MHC-I Expression: Targeting the Transporter TAP

Herpes simplex virus (HSV) infections are the source of recurrent mucocutaneous infection, such as herpes labialis and herpes genitalis. Occasionally HSV infections are the cause of life-threatening or sight-impairing disease, especially in neonates and the immunocompromised patient population. After primary infection, the virus persists for life in a latent form in neurons of the host. Again, interference with the MHC-I antigen presentation pathway may play a central role in HSV persistence. Unlike HIV Nef, which targets MHC-I after its exit from the ER, HSV types 1 and 2 express an early gene product, ICP47, which retains MHC-I complexes in the ER (York et al. 1994; see also the chapter by Maggioni and Braakman, this volume). ICP47 is a cytosolic protein that inhibits the function of TAP by interacting stably with its cytosolic domain (Fruh et al. 1995; Fig. 2). This interaction blocks TAP-dependent peptide translocation into the ER (Ahn et al. 1996; Tomazin et al. 1996). Consequently, MHC-I complexes are (1) unstable, (2) retained in the ER, and (3) degraded. Indeed, when MHC-I molecules are unable to assemble properly, the misfolded MHC-1 heavy chain is removed from the ER to the cytosol, deglycosylated, and degraded by the proteasome (Hughes et al. 1997). ICP47 induces the accumulation of misfolded heavy chains in the ER, which leads to proteosomal degradation. The most likely mechanism by which ICP47 inhibits TAP is to act as a pseudo-substrate inhibitor. In favor of this hypothesis is the fact that the N-terminal domain of ICP47 (residues 2–35) is sufficient to interfere with TAP function (Neumann et al. 1997). However, direct competition with the peptide-binding domain of TAP might not show the whole picture. Indeed, ICP47 also inhibits the ATPase activity of TAP, which is required for peptide translocation (Gorbulev et al. 2001). Few viruses target TAP in order to inactivate MHC-I antigen presentation. The only other known example is the HCMV US6 protein. HCMV belongs to a subfamily of herpesviruses distinct from HSV. Unlike ICP47, US6 does not appear to block the binding of the peptides to TAP (from the cytosolic side) but inhibits the translocation of peptide into the ER (ER-luminal side) (Ahn et al. 1997; Hengel et al. 1997). As a consequence, US6 disturbs the assembly of MHC-I molecules (reviewed by Momburg and Hengel 2002). The benefit from these studies is to use viruses as tools to dissect the function of the peptide transporter TAP, as well as potentially identifying molecules that would block the viral protein, while leaving TAP function unaltered. Such compounds could facilitate viral clearance, thereby preventing persistence of herpesviruses in the host.

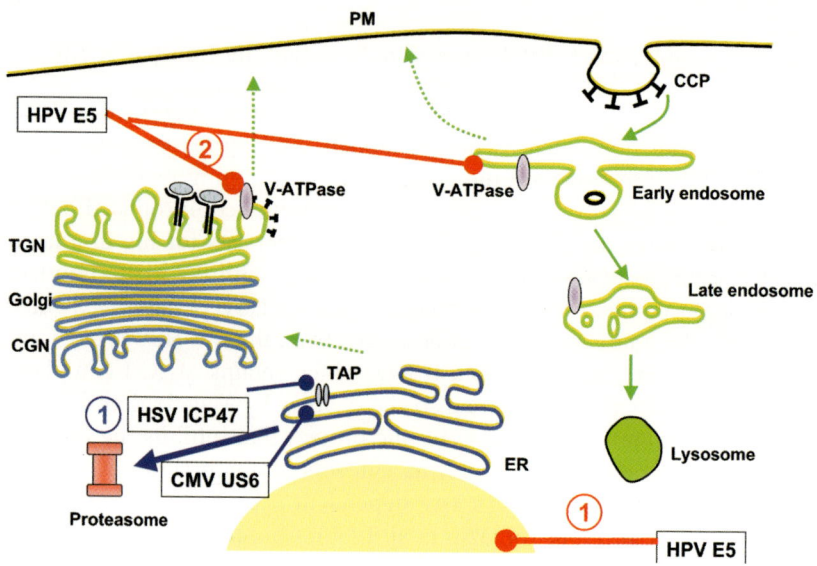

Fig. 2. Blocking transporters and pumps: TAP and the V-ATPase. *Blue*: HSV ICP47 blocks the peptide transporter TAP (transporter associated with antigen processing) on the cytosolic side of the ER. CMV U6 protein blocks TAP from the ER-luminal side. Both viral proteins induce an accumulation of unstable MHC-I molecules, which leads to their rapid degradation in the proteasome. *Red*: *1* HPV E5 protein decreases MHC-I expression at the transcriptional level. *2* HPV E5 interferes with the V-ATPase, a proton pump. This leads to a loss of acidification of the TGN and of the endosomes, leading to the accumulation of MHC-I in the TGN

2.3
HHV-8 Downregulation of MHC-I Expression: Ubiquitination of the Tail

Kaposi's sarcoma-associated herpesvirus (KHSV; HHV-8) is a lymphotropic γ2-herpesvirus that is strongly implicated in the pathogenesis of Kaposi's sarcoma and two AIDS-related lymphoproliferative syndromes: primary effusion lymphoma and multicentric Castleman's disease (reviewed in Hengge et al. 2002). DNA sequence analysis of the HHV-8 genome revealed, in addition to a number of homologs of cellular proteins [including a virus-encoded interleukin-6, MIP1-α and -β chemokines, Bcl-2, dihydrofolic reductase, and thymidylate synthetase (Nicholas et al. 1997)], several unique open reading frames, K3, K4.2, K5, and K7, that do not share homology with any known cellular genes (Nicholas et al. 1997; Russo et al. 1996). Interestingly, K3 and K5 [otherwise known as modulator of immune recognition (MIR)1 and MIR2] display 40%

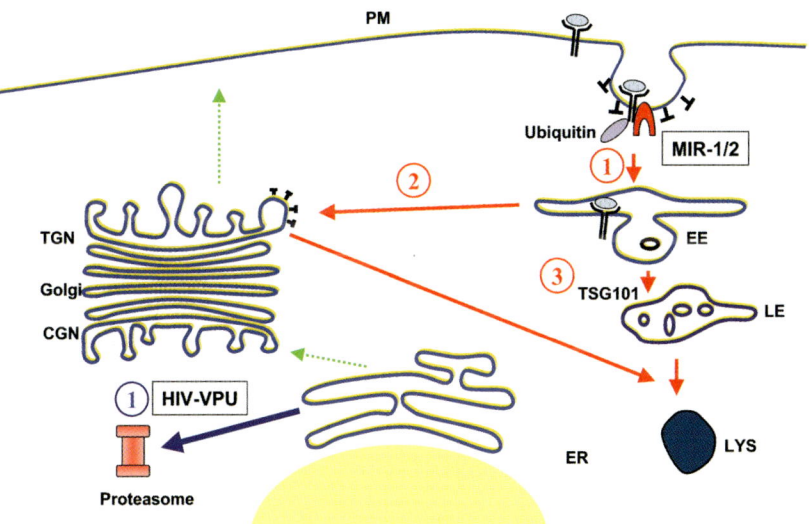

Fig. 3. Ubiquitination of receptor cytoplasmic tails. *Red*: *1* KHSV MIR1 and MIR2 induce the ubiquitination of MHC-I cytoplasmic domain, thereby inducing its rapid endocytosis via a clathrin dependent pathway. MIR1 and -2 function directly as E3 ubiquitin ligases. *2* MHC-I is then rerouted to the TGN and finally (*3*) degraded in the lysosomes. TSG101 is required for this last step (*3*). *Blue*: HIV VPU induces ubiquitination of CD4 cytoplasmic domain in the ER by acting as a connector between the receptor cytoplasmic tail and h-βTrCP, a protein that provides a link with the ubiquitin proteolysis machinery. Ubiquitinated CD4 is then is sent to the proteasome for destruction

amino acid homology with each other (Russo et al. 1996) and are expressed during the early lytic cycle of viral replication. MIR1 and -2 encode transmembrane proteins that downregulate cell surface MHC-I, as well as ICAM-1 and B7-2, ligands for NK cell-mediated cytotoxicity receptors. As a consequence, K5 expression drastically inhibits both MHC-I antigen presentation and NK cell-mediated cytotoxicity (Ishido et al. 2000a). HHV-8 interference of MHC-I antigen presentation is mediated by mechanisms that involve ubiquitination of the cytosolic domain of MHC-I (Coscoy et al. 2001; Fig. 3). After reaching the cell surface MHC-I complexes are rapidly internalized and targeted to lysosomes, where they are degraded (Ishido et al. 2000b). Endocytosis of MHC-I complexes may occur via clathrin-mediated endocytosis, because it can be inhibited by a dominant-negative form of dynamin (Coscoy and Ganem 2000). The motifs conferring sensitivity of MHC-I to MIR1 and MIR2 are lysines in the cytoplasmic tail of MHC-I (Coscoy et al. 2001). Furthermore, MHC I is ubiquitinated in MIR1- and MIR2-transfected cells and requires the presence of these lysines in the cytoplasmic tail of MHC-I. MIR1 and MIR2 both contain cytosolic zinc fingers of the PHD subfamily. PHD motifs share both sequence and structural homology with RING finger domains that are capable of E3 ubiquitin ligase activity (Joazeiro and Weissman 2000). E3 ubiquitin ligases give specificity to the ubiquitination process by recruiting the E2 ubiquitin-conjugating enzyme and directing ubiquitin onto the substrate. Indeed, proteins containing the MIR2 PHD domain can undergo auto-ubiquitination in the presence of E1, E2, ubiquitin, and ATP in vitro, which demonstrates that MIR2 functions as an E3 ubiquitin ligase that relies on its PHD domain (Coscoy et al. 2001). Mutations of the zinc-coordinating residues abrogate the PHD domain ubiquitin ligase activity as well as MHC-I downregulation.

Internalization of ubiquitinated MHC-I complexes by MIR1 is followed by the redirection of MHC-I to the TGN and finally to the lysosomes, where they are degraded (Means et al. 2002). Specific motifs in MIR1 are sequentially involved to remove MHC-I from the cell surface: An N-terminal zinc finger motif and a central sorting motif are involved in triggering internalization of MHC-I molecules and redirecting them to the TGN. Subsequently, the C-terminal diacidic cluster region of MIR1 is engaged in targeting MHC-I molecules to lysosomes (Means et al. 2002).

MIR1 targeting of MHC-I complexes to lysosomes requires another cellular partner, TSG101 (see also the chapter by Maggioni and Braakman, this volume). TSG101, the mammalian homolog of yeast Vps23 (Katzmann et al. 2001), is known to function in late endosome sorting in mammalian cells (Babst et al. 2000). Experiments using small interfering RNA (siRNA) have demonstrated that in the absence of TSG101 MIR1 is unable to target MHC-I complexes to lysosomal degradation, strongly suggesting that the viral protein can use the TSG101-dependent sorting pathway (Hewitt et al. 2002).

Ubiquitination of the cytosolic domain of a cellular receptor in order to target it to degradation is not unique to KHSV proteins MIR1 and -2. The lentiviral protein Vpu is also capable of mediating the ubiquitination of the cytosolic domain of CD4 (and possibly MHC-I) in order to target them to degradation. This downregulation occurs at the ER, before transport of newly synthesized proteins to the cell surface. This differentiates Vpu from the KHSV proteins MIR1 and -2, which act mainly after MHC-I has reached the cell surface. Vpu is a 16-kDa type I integral membrane phosphoprotein found exclusively in HIV-1 and SIV_{CPZ} but absent in HIV-2 and most SIV strains (Cohen et al. 1988; Strebel et al. 1988). Its 81-amino acid-long sequence encompasses an N-terminal hydrophobic membrane-spanning domain of 24 residues and a C-terminal cytoplasmic tail capable of forming oligomers. Vpu is expressed on intracellular membranes but not at the cell surface. Vpu induces the degradation of CD4 molecules retained in the ER as CD4-Env complexes (Willey et al. 1992). Like Nef, Vpu acts as a connector between CD4 and cellular degradative pathways. In this case, however, the proteasome rather than the lysosome is the entity responsible for CD4 destruction (Schubert et al. 1998). Vpu bridges the CD4 cytoplasmic tail, which it directly binds, with a protein known as h-βTrCP (Margottin et al. 1998). h-βTrCP contains an F-box motif and seven WD repeats. The WD repeat region of h-βTrCP interacts with Vpu in a phosphoserine-dependent manner while the F-box recruits Skp1p, a protein that provides a link with the ubiquitin proteolysis machinery. Whether Vpu action involves dislocation of CD4 from the ER into the cytoplasm, direct targeting of the cytosolic part of the glycoprotein by the proteasome, or another as yet undefined mechanism remains unclear. However, the model in which Vpu targets CD4 for proteosomal degradation is supported by the observation that Vpu action involves the ubiquitin pathway and is sensitive to inhibitors of cytosolic proteasome (Fujita et al. 1997). In some systems Vpu may interfere with MHC-I surface expression. This interference occurs at an early step in the biosynthesis of MHC-I molecules (Kerkau et al. 1997). It can be speculated that Vpu also induces the ubiquitination of MHC-I molecules, thereby targeting them to degradation before their arrival at the cell surface.

Ubiquitination of cellular receptors is not only a ploy used by viruses to remove receptors from the cell surface for their own benefit, but also a general mechanism to control surface levels of some receptors (Hicke 2001). For instance, a cellular homolog of MIR1 and MIR2 has been identified and called c-MIR (Goto et al. 2003). c-MIR induces a specific downregulation of B7-2 through ubiquitination, rapid endocytosis, and lysosomal degradation, processes very similar to MIR1 and -2 from KHSV. This is a clear example of how a process first identified as an anecdote linked with a viral infection may reveal more general mechanisms of regulation of endocytosis.

2.4
HPV Inhibition of MHC-I Expression: Blocking the Proton Pump

Human papillomaviruses are oncogenic viruses that cause benign hyperproliferative lesions in the skin and mucosae (warts and condylomas). These lesions tend to persist for prolonged periods of time and may occasionally progress toward tumors. Viral persistence is required for oncogenic transformation to occur (Tindle 2002). The papillomavirus genome is approximately 7.8 kb and is divided into early, late, and noncoding regions. At least two viral early proteins (E5 and E7) are responsible for removal of MHC-I from the cell surface. The early gene product E7 of the highly oncogenic strains HPV-16 and -18 downregulates the MHC-I heavy chain promoter and the TAP-1 and LMP-2 promoters (Georgopoulos et al. 2000), which leads to a decrease of MHC-I expression. Another HPV protein called E5 can also downregulate MHC-I expression at least at two levels. First, E5 decreases MHC-I expression at the transcriptional level. Second, expression of E5 causes retention of MHC-I in the Golgi apparatus, thus preventing its transport to the cell surface (Ashrafi et al. 2002; Marchetti et al. 2002). E5 therefore regulates MHC-I surface levels by different mechanisms, but prevention of MHC-I transport to the cell surface appears to be the dominant effect. The mechanisms of MHC-I retention in the Golgi are still unclear. E5 can interact with the subunit c of the vacuolar proton pump (V-ATPase; Goldstein et al. 1991). Furthermore, E5 expression leads to the inhibition of the acidification of endosomes and of the Golgi apparatus (Schapiro et al. 2000; Straight et al. 1995). E5 mutants that have lost their ability to bind the proton pump are unable to modify Golgi acidification (Schapiro et al. 2000) suggesting that E5 inhibits the V-ATPase directly. Whether this block of acidification leads to unstable MHC-I complexes that are retained in the Golgi apparatus remains to be established. E5 binding to the V-ATPase is to be compared to HIV-Nef interaction with the V-ATPase. In the case of E5, the viral protein seems to inactivate the proton pump in order to affect endosomal and TGN pH, which leads to perturbation in MHC-I trafficking. In the case of HIV-1 Nef, no functional perturbation of the V-ATPase has yet been identified and Nef might just use this interaction to enhance its association with the adaptor complexes while downregulating CD4 from the cell surface (Geyer et al. 2002a, 2002b; Lu et al. 1998). Nef expression in cells does not modify the pH of endosomes but redirects the trafficking of cellular receptors toward acidic compartments (Piguet et al., 1998). It is thus conceivable that HPV E5 and HIV-Nef proteins hijack the same cellular target, the V-ATPase, but then subvert two separate functions of the proton pump. Further experiments comparing the activities of the two viral proteins might provide a better understanding of the function of the V-ATPase that regulates the pH in the exo- and endocytic pathways.

3
Perspectives

A common necessity for viruses that persist in the host is interference with MHC-I antigen presentation. A common objective could imply similar mechanisms for achieving this goal. However, viruses have evolved complex strategies to paralyze MHC-I at all levels of its biogenesis, assembly, and trafficking. HPV E5 protein blocks MHC-I expression at the transcriptional level. HSV protein ICP47 inhibits TAP function, leading to retention of MHC-I complexes in the ER. HPV E5 blocks the acidification of the Golgi and of endosomes and leads to the accumulation of MHC-I in the Golgi. HIV Nef reroutes MHC-I complexes from the surface to the TGN, acting as a connector between the receptor cytoplasmic tail and the PACS-1 sorting pathway. Finally, the KHSV proteins MIR1 and MIR2 induce the ubiquitination of MHC-I complexes, leading to their rapid internalization from the cell surface, targeting to the TGN, and finally the lysosomes. The careful dissection of the ploys used by viruses to interfere with MHC-I presentation has led to the discovery of novel mechanisms of regulation of surface receptors, which are essential for a proper cell function, such as the PACS-1-dependent pathway or endocytosis regulated by ubiquitination. Furthermore, considerable progress has been achieved in the understanding of TAP and other cofactors necessary for the proper assembly of MHC-I complexes. Whether these studies will enable us to develop therapeutic strategies blocking the viral proteins that downregulate MHC-I and whether such drugs aimed at restoring normal surface levels of MHC-I on infected cells will lead to viral clearance remain to be demonstrated.

Acknowledgments This work was supported by the Geneva Cancer League and Swiss National Science Foundation grant No 3345-67200.01 to VP. VP is the recipient of a "Professor SNF" position (PP00A-68785).

References

Ahn, K., Gruhler, A., Galocha, B., Jones, T., Wiertz, E., Ploegh, H., Peterson, P., Yang, Y., and Fruh, K. (1997) The ER-luminal domain of the HCMV glycoprotein US6 inhibits peptide translocation by TAP. Immunity 6:613–621

Ahn, K., Meyer, T. H., Uebel, S., Sempe, P., Djaballah, H., Yang, Y., Peterson, P. A., Fruh, K., and Tampe, R. (1996) Molecular mechanism and species specificity of TAP inhibition by herpes simplex virus ICP47. EMBO J 15:3247–55

Aiken, C., Konner, J., Landau, N. R., Lenburg, M. E., and Trono, D. (1994) Nef induces CD4 endocytosis: requirement for a critical dileucine motif in the membrane-proximal CD4 cytoplasmic domain. Cell 76:853–64

Ashrafi, G. H., Tsirimonaki, E., Marchetti, B., O'Brien, P. M., Sibbet, G. J., Andrew, L., and Campo, M. S. (2002) Down-regulation of MHC class I by bovine papillomavirus E5 oncoproteins. Oncogene 21:248–59

Babst, M., Odorizzi, G., Estepa, E. J., and Emr, S. D. (2000) Mammalian tumor susceptibility gene 101 (TSG101) and the yeast homologue, Vps23p, both function in late endosomal trafficking. Traffic 1:248–58

Bijlmakers, M. J., and Ploegh, H. L. (1993) Putting together an MHC class I molecule. Curr Opin Immunol 5:21–6

Blagoveshchenskaya, A. D., Thomas, L., Feliciangeli, S. F., Hung, C. H., and Thomas, G. (2002) HIV-1 Nef downregulates MHC-I by a PACS-1- and PI3K-regulated ARF6 endocytic pathway. Cell 111:853–66

Bresnahan, P. A., Yonemoto, W., Ferrell, S., Williams-Herman, D., Geleziunas, R., and Greene, W. C. (1998) A dileucine motif in HIV-1 Nef acts as an internalization signal for CD4 downregulation and binds the AP-1 clathrin adaptor. Curr Biol 8:1235–1238

Cohen, E. A., Terwilliger, E. F., Sodroski, J. G., and Haseltine, W. A. (1988) Identification of a protein encoded by the vpu gene of HIV-1. Nature 334:532–534

Cohen, G. B., Gandhi, R. T., Davis, D. M., Mandelboim, O., Chen, B. K., Strominger, J. L., and Baltimore, D. (1999) The selective downregulation of class I major histocompatibility complex proteins by HIV-1 protects HIV-infected cells from NK cells. Immunity 10:661–71

Collins, K. L., Chen, B. K., Kalams, S. A., Walker, B. D., and Baltimore, D. (1998) HIV-1 Nef protein protects infected primary cells against killing by cytotoxic T lymphocytes. Nature 391:397–401

Conner, S. D., and Schmid, S. L. (2003) Regulated portals of entry into the cell. Nature 422:37–44

Coscoy, L., and Ganem, D. (2000) Kaposi's sarcoma-associated herpesvirus encodes two proteins that block cell surface display of MHC class I chains by enhancing their endocytosis. Proc Natl Acad Sci U S A 97:8051–6

Coscoy, L., Sanchez, D. J., and Ganem, D. (2001) A novel class of herpesvirus-encoded membrane-bound E3 ubiquitin ligases regulates endocytosis of proteins involved in immune recognition. J Cell Biol 155:1265–73

Craig, H. M., Pandori, M. W., and Guatelli, J. C. (1998) Interaction of HIV-1 nef with the cellular dileucine-based sorting pathway is required for CD4 down-regulation and optimal viral infectivity. Proc Natl Acad Sci U S A 95:11229–34

Crump, C. M., Xiang, Y., Thomas, L., Gu, F., Austin, C., Tooze, S. A., and Thomas, G. (2001) PACS-1 binding to adaptors is required for acidic cluster motif-mediated protein traffic. EMBO J 20:2191–201

Engering, A., Geijtenbeek, T. B., van Vliet, S. J., Wijers, M., van Liempt, E., Demaurex, N., Lanzavecchia, A., Fransen, J., Figdor, C. G., Piguet, V., and van Kooyk, Y. (2002) The dendritic cell-specific adhesion receptor DC-SIGN internalizes antigen for presentation to T cells. J Immunol 168:2118–26

Fruh, K., Ahn, K., Djaballah, H., Sempe, P., van Endert, P. M., Tampe, R., Peterson, P. A., and Yang, Y. (1995) A viral inhibitor of peptide transporters for antigen presentation. Nature 375:415–8

Fujita, K., Omura, S., and Silver, J. (1997) Rapid degradation of CD4 in cells expressing human immunodeficiency virus type 1 Env and Vpu is blocked by proteasome inhibitors. J Gen Virol 78:619–25

Garcia, J., and Miller, A. (1991) Serine phosphorylation independent downregulation of cell surface CD4 by Nef. Nature 350:508–511

Geijtenbeek, T. B., Kwon, D. S., Torensma, R., van Vliet, S. J., van Duijnhoven, G. C., Middel, J., Cornelissen, I. L., Nottet, H. S., KewalRamani, V. N., Littman, D. R., et al. (2000a) DC-SIGN, a dendritic cell-specific HIV-1-binding protein that enhances trans-infection of T cells. Cell 100:587–97

Geijtenbeek, T. B., Torensma, R., van Vliet, S. J., van Duijnhoven, G. C., Adema, G. J., van Kooyk, Y., and Figdor, C. G. (2000b) Identification of DC-SIGN, a novel dendritic cell-specific ICAM-3 receptor that supports primary immune responses. Cell 100:575–85

Georgopoulos, N. T., Proffitt, J. L., and Blair, G. E. (2000) Transcriptional regulation of the major histocompatibility complex (MHC) class I heavy chain, TAP1 and LMP2 genes by the human papillomavirus (HPV) type 6b, 16 and 18 E7 oncoproteins. Oncogene 19:4930–5

Geyer, M., Fackler, O. T., and Peterlin, B. M. (2002a) Subunit H of the V-ATPase involved in endocytosis shows homology to beta-adaptins. Mol Biol Cell 13:2045–56

Geyer, M., Yu, H., Mandic, R., Linnemann, T., Zheng, Y. H., Fackler, O. T., and Peterlin, B. M. (2002b) Subunit H of the V-ATPase binds to the medium chain of adaptor protein complex 2 and connects Nef to the endocytic machinery. J Biol Chem 277:28521–9

Goldstein, D. J., Finbow, M. E., Andresson, T., McLean, P., Smith, K., Bubb, V., and Schlegel, R. (1991) Bovine papillomavirus E5 oncoprotein binds to the 16 K component of vacuolar H^+-ATPases. Nature 352:347–9

Gorbulev, S., Abele, R., and Tampe, R. (2001) Allosteric crosstalk between peptide-binding, transport, and ATP hydrolysis of the ABC transporter TAP. Proc Natl Acad Sci U S A 98:3732–7

Goto, E., Ishido, S., Sato, Y., Ohgimoto, S., Ohgimoto, K., Nagano-Fujii, M., and Hotta, H. (2003) c-MIR, a human E3 ubiquitin ligase, is a functional homolog of herpesvirus proteins MIR1 and 2 and has similar activity. J Biol Chem 11:11

Greenberg, M., Bronson, S., Lock, M., Neumann, M., Pavlakis, G., and Skowronski, J. (1997) Co-localization of HIV-1 Nef with the AP-2 adaptor protein complex correlates with Nef-induced CD4 down-regulation. EMBO J 16:6964–6976

Greenberg, M., DeTulleo, L., Rapoport, I., Skowronski, J., and Kirchhausen, T. (1998a) A dileucine motif in HIV-1 Nef is essential for sorting into clathrin-coated pits and for downregulation of CD4. Curr Biol 8:1239–1242

Greenberg, M. E., Iafrate, A. J., and Skowronski, J. (1998b) The SH3 domain-binding surface and an acidic motif in HIV-1 Nef regulate trafficking of class I MHC complexes. EMBO J 17:2777–2789

Hengel, H., Koopmann, J. O., Flohr, T., Muranyi, W., Goulmy, E., Hammerling, G. J., Koszinowski, U. H., and Momburg, F. (1997) A viral ER-resident glycoprotein inactivates the MHC-encoded peptide transporter. Immunity 6:623–32

Hengge, U. R., Ruzicka, T., Tyring, S. K., Stuschke, M., Roggendorf, M., Schwartz, R. A., and Seeber, S. (2002) Update on Kaposi's sarcoma and other HHV8 associated diseases. Part 1: epidemiology, environmental predispositions, clinical manifestations, and therapy. Lancet Infect Dis 2:281–92

Hewitt, E. W., Duncan, L., Mufti, D., Baker, J., Stevenson, P. G., and Lehner, P. J. (2002) Ubiquitylation of MHC class I by the K3 viral protein signals internalization and TSG101-dependent degradation. EMBO J 21:2418–29

Hicke, L. (2001) A new ticket for entry into budding vesicles—ubiquitin. Cell 106:527–30

Howcroft, T. K., Strebel, K., Martin, M. A., and Singer, D. S. (1993) Repression of MHC class I gene promoter activity by two-exon Tat of HIV. Science 260:1320–2

Hughes, E. A., Hammond, C., and Cresswell, P. (1997) Misfolded major histocompatibility complex class I heavy chains are translocated into the cytoplasm and degraded by the proteasome. Proc Natl Acad Sci U S A 94:1896–901

Ishido, S., Choi, J. K., Lee, B. S., Wang, C., DeMaria, M., Johnson, R. P., Cohen, G. B., and Jung, J. U. (2000a) Inhibition of natural killer cell-mediated cytotoxicity by Kaposi's sarcoma-associated herpesvirus K5 protein. Immunity 13:365–74

Ishido, S., Wang, C., Lee, B. S., Cohen, G. B., and Jung, J. U. (2000b) Downregulation of major histocompatibility complex class I molecules by Kaposi's sarcoma-associated herpesvirus K3 and K5 proteins. J Virol 74:5300–9

Joazeiro, C. A., and Weissman, A. M. (2000) RING finger proteins: mediators of ubiquitin ligase activity. Cell 102:549–52

Kasper, M. R., and Collins, K. L. (2003) Nef-mediated disruption of HLA-A2 transport to the cell surface in T cells. J Virol 77:3041–9

Katzmann, D. J., Babst, M., and Emr, S. D. (2001) Ubiquitin-dependent sorting into the multivesicular body pathway requires the function of a conserved endosomal protein sorting complex, ESCRT-I. Cell 106:145–55

Kerkau, T., Bacik, I., Bennink, J., Yewdell, J., Hunig, T., Schimpl, A., and Schubert, U. (1997) The human immunodeficiency virus type 1 (HIV-1) Vpu protein interferes with an early step in the biosynthesis of major histocompatibility complex (MHC) class I molecules. J Exp Med 185:1295–1305

Lama, J., Mangasarian, A., and Trono, D. (1999) Cell-surface expression of CD4 reduces HIV-1 infectivity by blocking env incorporation in a nef- and vpu-inhibitable manner. Curr Biol 9:622–31

Le Gall, S., Buseyne, F., Trocha, A., Walker, B. D., Heard, J. M., and Schwartz, O. (2000) Distinct trafficking pathways mediate Nef-induced and clathrin-dependent major histocompatibility complex class I down-regulation. J Virol 74:9256–66

Le Gall, S., Erdtmann, L., Benichou, S., Berlioz-Torrent, C., Liu, L., Benarous, R., Heard, J.-M., and Schwarz, O. (1998) Nef interacts with the mu subunit of clathrin adaptor complexes and reveals a cryptic sorting signal in MHC-I molecules. Immunity 8:483–95

Lu, X., Yu, H., Liu, S., Brodsky, F. M., and Peterlin, B. M. (1998) Interactions between HIV1 Nef and vacuolar ATPase facilitate the internalization of CD4. Immunity 8:647–656

Mangasarian, A., Foti, M., Aiken, C., Chin, D., Carpentier, J. L., and Trono, D. (1997) The HIV-1 Nef protein acts as a connector with sorting pathways in the Golgi and at the plasma membrane. Immunity 6:67–77

Mangasarian, A., Piguet, V., Wang, J. K., Chen, Y., and Trono, D. (1999) Nef-induced CD4 and major histocompatibility complex class I (MHC-I) down-regulation are governed by distinct determinants: N-terminal alpha helix and proline repeat of Nef selectively regulate MHC-I trafficking. J Virol 73:1964–73

Marchetti, B., Ashrafi, G. H., Tsirimonaki, E., O'Brien, P. M., and Campo, M. S. (2002) The bovine papillomavirus oncoprotein E5 retains MHC class I molecules in the Golgi apparatus and prevents their transport to the cell surface. Oncogene 21:7808–16

Margottin, F., Bour, S. P., Durand, H., Selig, L., Benichou, S., Richard, V., Thomas, D., Strebel, K., and Benarous, R. (1998) A novel human WD protein, h-beta TrCp, that interacts with HIV-1 Vpu connects CD4 to the ER degradation pathway through an F-box motif. Mol Cell 1:565–74

Means, R. E., Ishido, S., Alvarez, X., and Jung, J. U. (2002) Multiple endocytic trafficking pathways of MHC class I molecules induced by a herpesvirus protein. EMBO J 21:1638–49

Momburg, F., and Hengel, H. (2002) Corking the bottleneck: the transporter associated with antigen processing as a target for immune subversion by viruses. Curr Top Microbiol Immunol 269:57–74

Munch, J., Stolte, N., Fuchs, D., Stahl-Hennig, C., and Kirchhoff, F. (2001) Efficient class I major histocompatibility complex down-regulation by simian immunodeficiency virus Nef is associated with a strong selective advantage in infected rhesus macaques. J Virol 75:10532–6

Neefjes, J. J., Stollorz, V., Peters, P. J., Geuze, H. J., and Ploegh, H. L. (1990) The biosynthetic pathway of MHC class II but not class I molecules intersects the endocytic route. Cell 61:171–83

Neumann, L., Kraas, W., Uebel, S., Jung, G., and Tampe, R. (1997) The active domain of the herpes simplex virus protein ICP47: a potent inhibitor of the transporter associated with antigen processing. J Mol Biol 272:484–92

Nicholas, J., Ruvolo, V., Zong, J., Ciufo, D., Guo, H. G., Reitz, M. S., and Hayward, G. S. (1997) A single 13-kilobase divergent locus in the Kaposi sarcoma-associated herpesvirus (human herpesvirus 8) genome contains nine open reading frames that are homologous to or related to cellular proteins. J Virol 71:1963–74

Piguet, V., Chen, Y.-L., Mangasarian, A., Foti, M., Carpentier, J., and Trono, D. (1998) Mechanism of Nef induced CD4 endocytosis: Nef connects CD4 with the mu chain of adaptor complexes. EMBO J 17:2472–2481

Piguet, V., Gu, F., Foti, M., Demaurex, N., Gruenberg, J., Carpentier, J. L., and Trono, D. (1999) Nef-induced CD4 degradation: a diacidic-based motif in Nef functions as a lysosomal targeting signal through the binding of beta-COP in endosomes. Cell 97:63–73

Piguet, V., Wan, L., Borel, C., Mangasarian, A., Demaurex, N., Thomas, G., and Trono, D. (2000) HIV-1 Nef protein binds to the cellular protein PACS-1 to downregulate class I major histocompatibility complexes. Nat Cell Biol 2:163–167

Reid, P. A., and Watts, C. (1990) Cycling of cell-surface MHC glycoproteins through primaquine-sensitive intracellular compartments. Nature 346:655–7

Rhee, S. S., and Marsh, J. W. (1994) Human immunodeficiency virus type 1 Nef-induced down-modulation of CD4 is due to rapid internalization and degradation of surface CD4. J Virol 68:5156–63

Ross, T. M., Oran, A. E., and Cullen, B. R. (1999) Inhibition of HIV-1 progeny virion release by cell-surface CD4 is relieved by expression of the viral nef protein. Curr Biol 9:613–21

Russo, J. J., Bohenzky, R. A., Chien, M. C., Chen, J., Yan, M., Maddalena, D., Parry, J. P., Peruzzi, D., Edelman, I. S., Chang, Y., and Moore, P. S. (1996) Nucleotide sequence of the Kaposi sarcoma-associated herpesvirus (HHV8) Proc Natl Acad Sci U S A 93:14862–7

Schapiro, F., Sparkowski, J., Adduci, A., Suprynowicz, F., Schlegel, R., and Grinstein, S. (2000) Golgi alkalinization by the papillomavirus E5 oncoprotein. J Cell Biol 148:305–15

Schubert, U., Anton, L. C., Bacik, I., Cox, J. H., Bour, S., Bennink, J. R., Orlowski, M., Strebel, K., and Yewdell, J. W. (1998) CD4 glycoprotein degradation induced by human immunodeficiency virus type 1 Vpu protein requires the function of proteasomes and the ubiquitin-conjugating pathway. J Virol 72:2280–8

Schubert, U., Anton, L. C., Gibbs, J., Norbury, C. C., Yewdell, J. W., and Bennink, J. R. (2000) Rapid degradation of a large fraction of newly synthesized proteins by proteasomes. Nature 404:770–4

Schwartz, O., Maréchal, V., Le Gall, S., Lemonnier, F., and Heard, J. M. (1996) Endocytosis of major histocompatibility complex class I molecules is induced by the HIV-1 Nef protein. Nat Med 2:338–42

Sol-Foulon, N., Moris, A., Nobile, C., Boccaccio, C., Engering, A., Abastado, J. P., Heard, J. M., van Kooyk, Y., and Schwartz, O. (2002) HIV-1 Nef-induced upregulation of DC-SIGN in dendritic cells promotes lymphocyte clustering and viral spread. Immunity 16:145–55

Straight, S. W., Herman, B., and McCance, D. J. (1995) The E5 oncoprotein of human papillomavirus type 16 inhibits the acidification of endosomes in human keratinocytes. J Virol 69:3185–92

Strebel, K., Klimkait, T., and Martin, M. A. (1988) A novel gene of HIV-1:vpu, and its 16-kilodalton product. Science 241:1221–1223

Tindle, R. W. (2002) Immune evasion in human papillomavirus-associated cervical cancer. Nat Rev Cancer 2:59–65

Tomazin, R., Hill, A. B., Jugovic, P., York, I., van Endert, P., Ploegh, H. L., Andrews, D. W., and Johnson, D. C. (1996) Stable binding of the herpes simplex virus ICP47 protein to the peptide binding site of TAP. EMBO J 15:3256–66

Wan, L., Molloy, S. S., Thomas, L., Liu, G., Xiang, Y., Rybak, S. L., and Thomas, G. (1998) PACS-1 defines a novel gene family of cytosolic sorting proteins required for trans-Golgi network localization. Cell 94:205–16

Willey, R. L., Maldarelli, F., Martin, M. A., and Strebel, K. (1992) Human immunodeficiency virus type 1 Vpu protein induces rapid degradation of CD4. J Virol 66:7193–200

Williams, M., Roeth, J. F., Kasper, M. R., Fleis, R. I., Przybycin, C. G., and Collins, K. L. (2002) Direct binding of human immunodeficiency virus type 1 Nef to the major histocompatibility complex class I (MHC-I) cytoplasmic tail disrupts MHC-I trafficking. J Virol 76:12173–84

York, I. A., Roop, C., Andrews, D. W., Riddell, S. R., Graham, F. L., and Johnson, D. C. (1994) A cytosolic herpes simplex virus protein inhibits antigen presentation to CD8+ T lymphocytes. Cell 77:525–35

Trafficking of Viral Membrane Proteins

R. Byland · M. Marsh (✉)

Cell Biology Unit, MRC-LMCB and Department of Biochemistry
and Molecular Biology, University College London, Gower Street,
London, WC1E 6BT, UK
m.marsh@ucl.ac.uk

1	Introduction	219
2	The Cellular Protein Sorting and Trafficking Machinery	221
3	Traffic of Viral Envelope Proteins	227
3.1	Retroviruses	227
3.2	Herpesviruses	233
3.3	Orthomyxoviruses	237
3.4	Rhabdoviruses	238
3.5	Poxviruses	239
4	Virus Assembly	240
4.1	ESCRTing Virus Release	240
4.2	Other Late Domain Sequences	242
5	Conclusions	243
References		244

Abstract Many viruses express membrane proteins. For enveloped viruses in particular, membrane proteins are frequently structural components of the virus that mediate the essential tasks of receptor recognition and membrane fusion. The functional activities of these proteins require that they are sorted correctly in infected cells. These sorting events often depend on the ability of the virus to mimic cellular protein trafficking signals and to interact with the cellular trafficking machinery. Importantly, loss or modification of these signals can influence virus infectivity and pathogenesis.

1
Introduction

The final steps in the assembly of enveloped viruses occur in the context of a cellular membrane when the nascent particle undergoes a budding reaction that simultaneously generates the viral envelope and releases the free virion. The cellular membrane can be the plasma membrane, leading to virus release directly to the extracellular space, or an intracellular membrane (e.g. the ER, Golgi apparatus or endosomal system), in which case the virions are delivered into intracellular vacuoles from which they are released to the extracellular space by a secretory-type mechanism. During the budding process membrane proteins are incorporated into the viral envelope. These can include integral membrane proteins and peripheral proteins, and both types can be virally encoded or of cellular origin. The number of envelope proteins varies greatly between different virus families. Genetically relatively simple viruses, such as rhabdoviruses, encode a single integral membrane glycoprotein that mediates the key entry functions of receptor recognition and membrane fusion. More complex viruses, such as herpesviruses, encode up to 60 putative membrane proteins, many of which can be found in the viral envelope (Britt and Mach 1996; Spaete et al. 1994).

To ensure that fully infectious virions are produced, the components of a mature virus must be brought together in infected cells in a temporally and spatially co-ordinated manner. The proteins must be synthesized at an appropriate time and transported to the membrane system where they will be incorporated into domains that will become viral envelopes. As discussed in the chapter by Maggioni and Braakman, this volume, integral membrane proteins are synthesized on the ER and use cellular mechanisms to ensure correct folding, quality control and export to the Golgi apparatus. Subsequently these proteins must be transported, directly or indirectly, to sites in the cell where budding occurs. The transport events may require transit through specific cellular compartments where, for example, glycosylation is completed or proteolytic cleavage occurs. For this transport, viruses exploit cellular trafficking machineries and use signals that frequently mimic those found in cellular proteins.

Although most studies have focussed on integral membrane proteins, similar types of processes must occur for viral peripheral membrane proteins, such as retroviral Gag polyproteins (the precursor for viral matrix and capsid proteins) that associate with the cytoplasmic side of cellular membranes and are subsequently located on the interior of assembled virions. These proteins are synthesized on free polysomes and targeted to membranes by a N-terminal myristic acid moiety plus, in many cases, a second motif such as an adjacent stretch of basic amino acids (reviewed in Bijlmakers and Marsh 2003). For many peripheral membrane proteins, association with a specific membrane system may be direct, but for others targeting may involve vesicular transport. Currently little is known of the mechanisms involved in the trafficking of these peripheral proteins.

In addition to structural proteins of the virion, a number of viruses encode membrane proteins with other roles. These can include proteins that modify the cell surface expression of MHC antigens or clear chemokines or antibodies from the environment around infected cells, thereby modifying the efficacy of the host's immune response to infection. Again, these proteins use host cell trafficking machineries and signals to effect their functions.

Here we discuss some of the molecular signals used by viruses to distribute their membrane proteins within infected cells and how these signals contribute to virus assembly and/or pathogenesis. For the most part the discussion will focus on integral membrane proteins and reference to peripheral proteins will be limited. In addition, we also review the role of the cellular ESCRT machinery in the assembly of certain enveloped viruses.

2
The Cellular Protein Sorting and Trafficking Machinery

A considerable amount of information has now emerged on the trafficking pathways in eukaryotic cells and the cellular and molecular mechanisms through which these pathways operate. The vacuolar apparatus can be considered as a series of functionally overlapping membrane-bound compartments:

- The ER/Golgi systems, in which glycoprotein synthesis, folding, oligomerization, glycosylation, acylation and quality control occur
- The TGN/endosomal systems, in which many of the sorting reactions that control constitutive and regulated secretion, polarity and endocytosis occur
- The plasma membrane
- The late endosomal/lysosomal systems, in which many receptors and their ligands, plus other membrane proteins and material internalized by endocytosis, are degraded

A network of trafficking pathways (see Fig. 1) mediates transport between the different compartments in a highly regulated manner (Bonifacino and Glick 2004; Bonifacino et al. 1996; Hirst and Robinson 1998; Pelkmans and Helenius 2002; Sorkin 2000). After synthesis, the default

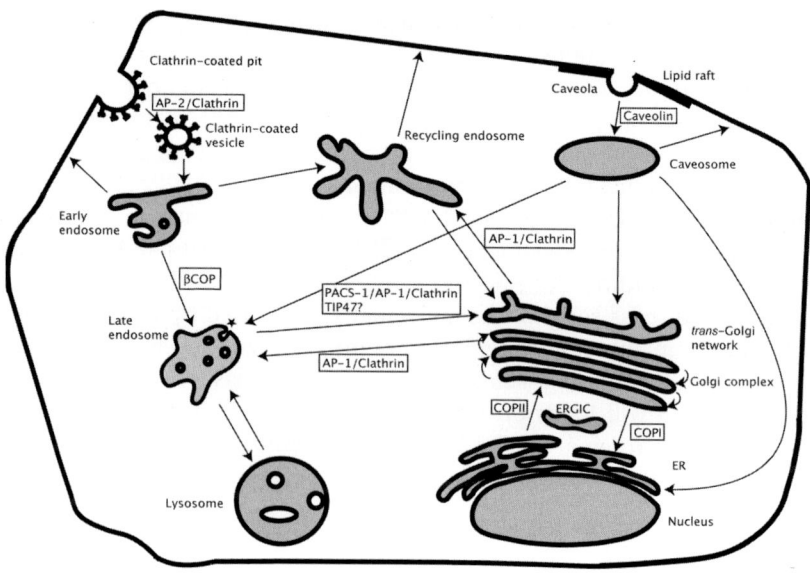

Fig. 1. Cellular endocytic and exocytic trafficking pathways. *Arrows* depict known transport pathways between different organelles. Where known, relevant coat and adaptor proteins are indicated in *boxes*. Multivesicular body (*MVB*) formation as an example of an outwardly directed vesiculation process is indicated (*)

pathway for secretory proteins leads them from the ER to the Golgi apparatus and then via the TGN to the plasma membrane. Many plasma membrane proteins follow a similar route. Some proteins, for example the cation-independent mannose-6-phosphate receptor (CI-M6PR), that carries lysosomal enzymes, are directed from the TGN to late endosomes via early endosomes (Kornfeld and Mellman 1989; Tikkanen et al. 2000; Trowbridge et al. 1993). Proteins delivered to the plasma membrane can be reinternalized by clathrin-dependent or clathrin-independent/lipid raft-dependent pathways to endosomal organelles from where they may be recycled, targeted to lysosomes or sorted to other cellular compartments including the Golgi apparatus, ER or alternative plasma membrane domains. In polarized cells, membrane and secretory proteins can be targeted to either basolateral or apical domains and sorting to these domains can occur at the TGN or in endosomes. Some of the signals involved in these latter sorting steps are related to endocytosis signals and appear to be interpreted by similar machineries. These pathways may also exist in non-polarized cells (Scheiffele et al. 1997), although epithelial-specific sorting reactions do occur (Sugimoto et al. 2002).

The trafficking of membrane proteins in the vacuolar apparatus involves sorting signals and a set of cellular machineries to interpret the information in these signals. Very different molecular features can influence protein sorting, and the distribution of a protein is dependent on a cell's ability to decipher multiple signals operating at different stations in the vacuolar system. Several excellent reviews cover much of this information (Bonifacino et al. 1996; Bonifacino and Traub 2003; Hirst and Robinson 1998; Schmid 1997; Sorkin 2000; Trowbridge et al. 1993). Here we give only superficial coverage of a few examples that are particularly relevant to viral protein trafficking.

Aside from the sequences that specify ER translocation, protein folding and ER quality control (see the chapter by Maggioni and Braakman, this volume), specific signals regulate protein export from the ER and transport through the stations of the vacuolar pathway. These signals act as markers either for transport or for retention. Signals for transport allow proteins to be incorporated into transport vesicles and move efficiently from one compartment to another. These signals often operate by binding the proteins in which they are located to coat proteins that form the vesicle, or by ensuring proteins are located in the membrane domains that are incorporated into vesicles. Retention signals operate in the opposite way. They prevent incorporation of proteins into transport vesicles and cause them to be retained in a specific membrane system. These signals may operate by linking membrane proteins to cytoskeletal elements, or keeping proteins in membrane domains that do not form vesicles. Note that the well-characterized C-terminal KKxx (where x = any amino acid) retrieval signal, that prevents export of ER proteins, is in fact a transport signal that specifies incorporation of proteins into retrograde transport vesicles operating between the Golgi and the ER (Letourneur et al. 1994). Many proteins contain multiple signals, several of which may allow efficient transport between several compartments while others restrict transport once the protein has been delivered to a specific target destination. Proteins that lack signals are transported inefficiently between compartments through non-specific inclusion in transport vesicles.

The molecular features of sorting signals are varied and can include:

1. *Short linear amino acid sequences.* Best exemplified by the tyrosine-based signals that mediate endocytosis. Two types of Y-based motif have been identified. YxxØ type signals (where Ø = a large hydrophobic amino acid) are found in the cytoplasmic domains (at least 7 residues from the transmembrane domain) of many cell surface receptors and other

proteins that undergo endocytosis in clathrin-coated vesicles (CCVs). These motifs interact with the $\mu 2$ subunit of the CCV AP2 adaptor complex (see below). A second Y-based motif, FxNPxY, is found in the low-density lipoprotein receptor and some other proteins that also undergo endocytosis in CCVs. However, other adaptors such as ARH or Numb may be involved in recruiting these proteins into CCVs (Aridor and Traub 2002; He et al. 2002). In addition to mediating sorting at the plasma membrane, Y-based motifs can also operate at the TGN and may influence sorting in polarized cells, delivery to endosomes and other specific transport steps. Motifs such as di-leucine sequences can also mediate endocytosis and TGN sorting. Motifs such as DxE appear to be required for efficient export of proteins from the ER (for detailed list of ER export signals, see Bonifacino and Glick 2004). Others, such as acidic clusters with a pair of adjacent serines that can be phosphorylated by casein kinase II (CKII) (Molloy et al. 1994; Schafer et al. 1995; Takahashi et al. 1995), can specify other sorting steps (Aridor and Traub 2002). Examples of motifs involved in sorting are listed in Table 1.

2. *Transmembrane domains.* The number of amino acids in the transmembrane domain of a single-pass integral membrane protein can influence its association with membrane sub-domains of different lipid composition. Because of the ability of cholesterol to render the acyl chains of sphingolipids more rigid, cholesterol-rich membrane domains are a few Å thicker. Proteins with transmembrane domains of approximately 22 aa show a propensity to associate with these cholesterol-rich domains, whereas proteins with a transmembrane domain of around 18 aa tend to be excluded (Bretscher and Munro 1993). Conserved residues in the transmembrane domains of some membrane proteins have also been linked to association with detergent-insoluble membrane domains (DIMs) and apical sorting (Lin et al. 1998).

3. *Lipid modifications.* A number of cell surface proteins are linked to the plasma membrane by glycophosphatidyl inositol (GPI) moieties that are added to certain transmembrane proteins during their synthesis in the ER. The GPI moiety replaces the transmembrane domain. Other proteins are post-translationally modified by the addition of palmitic acid to cysteine residues in the cytoplasmic domain or the transmembrane domain/cytoplasmic domain junction. Myristoylated proteins can also be palmitoylated on cysteines close to the N-terminus. These lipid modified proteins can all show some propensity to interact with cholesterol-rich microdomains, which may influence their distribution in the plasma membrane and other membrane systems.

Table 1. Motifs involved in sorting

Signal	Cellular protein	Viral protein	Trafficking function	Molecules binding to cellular proteins	References
YxxØ	Transferrin receptor MPR	HIV gp41 VZV gE/gB/gH HSV gE/gB SIV gp41 HTLV-1 gp Vaccinia B5R Moloney MuLV	Endocytosis Sorting to TGN Basolateral sorting	AP2 AP1	Marks et al. 1996 Collawn et al. 1990 Canfield et al. 1991 Jadot et al. 1992
FxNPxY	LDL-receptor		Endocytosis Basolateral sorting	ARH Numb	Davis et al. 1987 Chen et al. 1990) He et al. 2002
Phosphoserine	Activated GPCRs	Viral GPCRs	Endocytosis	AP2 Clathrin β-arrestin	
LL	CD4 MPR	HIV gp41 SIV gp41 Vaccinia B5R	Endocytosis TGN sorting	AP1 AP2	Marks et al. 1996 Johnson and Kornfeld 1992 Pitcher et al. 1999
Acidic cluster +Ser/Thr	Furin	VZV gE HSV gE HCMV gB	Endocytosis Phosphorylation by CKII	PACS-1	Jones et al. 1995 Voorhees et al. 1995
Large hydrophobic	Signalling proteins in rafts	Influenza HA	Association with DIMs Apical sorting	Lipids	Lin et al. 1998
Di-aromatic	MPR	HCMV gB HIV gp41	Endosomes to TGN	TIP47	Diaz and Pfeffer 1998 Nishimura and Balch 1997
DxE	LAP	VSV-G	ER-export (non-cycling proteins)	COPII	Nishimura et al., 1999

4. *Glycosylation.* Carbohydrates have been proposed to act as apical targeting signals for some secretory and membrane proteins (see Scheiffele et al. 1995; Yeaman et al. 1997).
5. *Oligomerization and cross-linking.* Cross-linking or oligomerization of membrane proteins can lead to a change of trafficking routes (see, for example, Ukkonen et al. 1986).

Trafficking between membrane compartments is mediated by transport vesicles, many of which are transiently coated with protein complexes. Three distinct coats have been characterized to date. COP II coats are associated with vesicles that bud from the ER en route to the Golgi apparatus, COP I coats are associated with retrograde transport vesicles from the Golgi to the ER (Robinson 1987), and CCVs are associated with endocytosis, transport from the TGN and some pathways from endosomes. Additional coats have been implicated in other transport events but remain to be characterized in detail (Seaman and Williams 2002).

The sorting information contained within membrane and other proteins is interpreted by the cell's transport and sorting machineries. The best understood of these machineries are the clathrin-associated adaptor complexes, and in particular AP2, for which detailed structural information now exists (Collins et al. 2002). AP adaptors, of which four have been identified (AP1–4), are heterotetrameric complexes composed of two large (α,β,γ or δ), one medium μ (~50 kDa) and one small σ (~20 kDa) adaptin (Hirst and Robinson 1998). AP2 is associated with endocytic CCVs (cf. AP1 and AP3, which are associated with TGN and/or endosomally-derived CCVs). The large adaptins contain a core domain and a so-called C-terminal ear domain connected to the core by a flexible linker. The linker of the β-subunits contains a 'clathrin box' sequence that binds the adaptor to the N-terminal β propeller domain of clathrin (ter Haar et al. 1998). The μ2 subunit is responsible for binding YxxØ type signals. The hydrophobic side chains of the Tyr and Y+3 residues bind into hydrophobic pockets located in the μ2 β sheet C-terminal domain. However, the binding site is only accessible after phosphorylation of threonine 156 by adaptor-associated kinase 1 (AAK1), indicating that access is regulated (Conner and Schmid 2002; Ricotta et al. 2002). AAK1 activity is governed by clathrin (Conner et al. 2003; Jackson et al. 2003) linking recruitment of YxxØ-containing cargo to clathrin assembly at the plasma membrane. Di-leucine motifs with an upstream acidic residue (e.g. ExxxLL), appear to bind the γ- and σ1-subunits of AP1 rather than the μ-subunit (Janvier et al. 2003).

Similar modes of recognition, and perhaps regulation, may occur for other clathrin adaptor complexes (Ghosh and Kornfeld 2003a). Whether the same applies for other adaptors, including GGA proteins (for review, see Bonifacino 2004), and coats that are now being identified remains to be established (Aridor and Traub 2002; Ghosh and Kornfeld 2003b). How other types of sorting signals that do not rely on short peptide sequences are interpreted also awaits additional study.

3
Traffic of Viral Envelope Proteins

Viruses, in particular enveloped viruses, have learned to exploit the cellular trafficking pathways to facilitate their replication. In some viruses the trafficking itineraries of the envelope proteins are relatively simple and may lead to the proteins being exported from the ER and delivered to the cell surface. By contrast, for other viruses more complex itineraries involving multiple signals have been identified. Viral proteins have adopted many of the sorting signals found in cellular proteins, and they exploit the cellular sorting machineries. In some cases viruses encode adaptors, for example HIV Nef, to couple proteins to trafficking machineries with which they do not normally interact, or interact in a different manner, thus sorting cellular proteins to different sites in infected cells. Here we discuss several different families of viruses and the trafficking properties of some of their membrane proteins. Relevant targeting motifs are outlined in Tables 1 and 2.

3.1
Retroviruses

Retroviruses encode a single envelope precursor protein (Env), which is synthesized on the ER, undergoes oligomerization (usually to trimers) and is then transported via the Golgi apparatus to the cell surface. En route, each Env protein is proteolytically cleaved by furin or a furin-like protease to form a heterodimer consisting of a so-called surface unit (SU) and a transmembrane unit (TM). In the case of the human immunodeficiency virus (HIV), and the related simian immunodeficiency virus (SIV), Env is synthesized as a 160-kDa precursor that is cleaved to produce gp120 (SU) and gp41 (TM) (see the chapter by Maggioni and Braakman, this volume). Env incorporation into virions has largely been considered to be a function of the cytoplasmic domain of TM and its in-

Table 2. Targeting motifs

Retroviruses	
HIV-1 Subtype B	RVRQGYSPLSFQTHLPTPRGPDRPEGIEEEGGERDRDRSIRLVNGSLALIWDDLRSLCLFSY HRLRDLLLIVTRIVELLGRRGWEALKYWWNLLQYWSQELKNSAVSLLNATAIAVAEGTDR VIEVVQGACRAIRHIPRRIRQGLERILL
SIV mac 239	KLRQGYRPVFSSPPSYFQQTHIQQDPALPTREGKERDGGEGGGNSSWPWQIEYIHFLIRQLI RLLTWLFSNCRTLLSRVYQILQPILQRLSATLQRIREVLRTELTYLQYGWSYFHEAVQAVWR SATETLAGAWGDLWETLRRGGRWILAIPRRIRQGLELTLL
HTLV-1	RHLPSRVRYPHYSLIKPESSL
HTLV-2	QALPQRLQNRHNQYSLINPETML
FIV	DCIRNCIHKILGYTVIAMPEVEGEEIQPQMELRRNGRQCGMSEKEEE
Herpesviruses	
HSV-1 gB	RYVMRLQSNPMKALYPLTTKELKNPTNPDASGEGEEGGDFDEAKLAEAREMIRYMALVSA MERTEHKAKKKGTSALLSAKVTDMVMRKRRNTNYTQVPNKDGDADEDDL
HSV-2 gB	RYVLQLQRNPMKALYPLTTKELKTSDPGGVGGEGEEGAEGGGFDEAKLAEAREMIRYMAL VSAMERTEHKARKKGTSALLSSKVTNMVLRKRNKARYSPLHNEDEAGDEDEL
VZV gB	YRYVLKLKTSPMKALYPLTTKGLKQLPEGMDPFAEKPNATDTPIEEIGDSQNTEPSVNSGFD PDKFREAQEMIKYMTLVSAAERQESKARKKNKTSALLTSRLTGLALRNRRGYSRVRTENVT GV
HCMV gB	YTRQRRLCTQPLQNLFPYLVSADGTTVTSGSTKDTSLQAPPSYEESVYNSGRKGPGPPSS DASTAAPPYTNEQAYQMLLALARLDAEQRAQQNGTDSLDG QTGTQDKGQKPNLLDRLR HRKNGYRHLKDSDEEENV
HSV-1 gE	ACMTCWRRRAWRAVKSRASGKGPTYIRVADSELYADWSSDSEGERDQVPWLAPPERPDS PSTNGSGFEILSPTAPSVYPRSDGHQSRRQLTTFGSGRPDRRYSQASDSSVFW
VZV gE	KRMRVKAYRVDKSPYNQSMYYAGLPVDDFEDSESTDTEEEFGNAIGGSHGGSSYTVYIDK TR
VZV gH	(WMLCGN)SRLREYNKIPLT
HSV-1 gH	KVLRTSVPFFWRRE
Orthomyxoviruses	
INFLUENZA HA	...MGVYQILAIYATVAGSLSLAIMMAGISFWMCS NGSLQCRICI
Rhabdoviruses	
VSV-G	RVGIYLCIKLKHTKKRQIYTDIEMNRLGK
Poxviruses	
Vaccinia B5R	CSCDKNNDQYKFHKLLP

teraction with Gag (Cosson 1996; Vincent et al. 1999), but the fact that most retroviruses and retroviral Env proteins will form pseudotypes and also incorporate cellular membrane proteins argues against high-fidelity sorting during retroviral assembly. For the most part, retroviral Envs have short cytoplasmic domains. HIV and SIV are unusual in that their

TM cytoplasmic domain varies from 150 to 200 amino acids for different strains of virus.

HIV and SIV Env Trafficking. HIV Env trafficking has been studied with respect to the formation of virions. Thus most models propose that Env must be transported to the cell surface, where in the main HIV assembly occurs. In many cell types HIV does bud from the plasma membrane, but the degrees to which Env is incorporated into these particles is unclear. Early studies suggested that much of the newly synthesized HIV Env exported from the ER is transported to lysosomes (Willey et al. 1988). Recent work has suggested a complex itinerary for Env and that this trafficking activity is crucial for viral pathogenesis (Blot et al. 2003; Fultz et al. 2001).

All HIV and SIV Envs contain a conserved YxxØ type signal close to the junction of the cytoplasmic and transmembrane domains (see Table 2). This sequence functions as an endocytosis signal in HIV and SIV Env and in Env cytoplasmic domain reporter constructs (Bowers et al. 2000; LaBranche et al. 1995; Rowell et al. 1995; Wyss et al. 2001), and as a basolateral sorting motif in polarized epithelial cells (LaBranche et al. 1995; Lodge et al. 1997b; Owens et al. 1991; Rowell et al. 1995; Sauter et al. 1996). SIV Envs, with short cytoplasmic domains, can spontaneously gain a Y/C mutation in the YxxØ sequence when the virus is maintained in tissue culture (LaBranche et al. 1995). This mutation destroys the sorting information in the motif and the cell surface expression of Env is dramatically upregulated, arguing that endocytosis normally keeps cell surface Env levels low (LaBranche et al. 1995). When SIV, containing an *Env* gene in which the same Tyr codon is deleted or replaced by that of another amino acid, is used to infect non-human primates the animals become infected but do not develop AIDS unless the mutation reverts (Fultz et al. 2001). Thus, though not required for growth in tissue culture, the trafficking activity of the YxxØ signal is required for pathogenesis in vivo. In addition, the bulk of the SIV cytoplasmic domain is not required for growth in tissue culture as the *Env* gene frequently gains a mutation that places a premature stop codon about 20 amino acids into the cytoplasmic domain of TM. However, these truncations revert when the viruses are introduced into primate hosts. The functional activities of the full-length cytoplasmic domain are unclear as, for SIV at least, truncated Envs can still be incorporated into virions. Thus the requirements for the full-length tail are intimately tied into the biology of these viruses in vivo and may involve key roles for trafficking.

Full-length HIV and SIV Env, or Env cytoplasmic domain containing reporter constructs, in which the membrane proximal YxxØ signal is inactivated by mutation of the Tyr retain some capacity for endocytosis and AP2 binding and show only modest increases in cell surface levels (Bowers et al. 2000; Wyss et al. 2001) but no longer show basolateral sorting (Lodge et al. 1994, 1997b). This indicates (1) that additional endocytosis information exists downstream of the membrane proximal signal and (2) that the membrane proximal signal is the only component of Env responsible for polarized sorting. The nature of the additional endocytic activity is obscure. In HIV Env, a C-terminal di-leucine has been implicated in AP1 binding and mutation of this motif in conjunction with mutation of the membrane proximal Tyr has additive effects in increasing Env cell surface expression (Wyss et al. 2001). Mutation of analogous sequences in SIV Env had no effect on cell surface expression or endocytosis (Bowers et al. 2000). However, truncation of SIVmac Env at residue 767 increased Env incorporation into virions even in the presence of an intact YxxØ motif (Yuste and Desrosiers 2003). How this mutation affects the trafficking of Env remains to be established.

In addition to the endocytosis and basolateral sorting information, other sorting signals must exist in Env. A significant amount of HIV Env, or an Env cytoplasmic domain reporter protein, is located in the Golgi apparatus and /or the TGN (Berlioz-Torrent et al. 1999; Blot et al. 2003). Antibody feeding experiments show that at least some of this material is internalized from the cell surface and that Env cycles between the plasma membrane and the Golgi/TGN (Blot et al. 2003). EM observations of HIV-infected macrophages also show prominent labelling for Env on viruses in late endosomes, where the majority of infectious virus assembles in these cells (Pelchen-Matthews et al. 2003; Raposo et al. 2002). Thus HIV-Env trafficking may involve transit from the cell surface to early and late endosomes, from where it is directed to the Golgi apparatus or the TGN. From this site it is either returned to the cell surface or perhaps cycles to endosomes. Little is known of how HIV Env traffics to late endosomes, although the fact that the membrane proximal Tyr is preceded by a Gly (GYxxØ) as in a number of late endosomal/lysosomal proteins may be relevant (Bonifacino and Traub 2003).

The cellular machinery responsible for the trafficking of Env has been identified to some extent. The membrane proximal GYxxØ in HIV and SIV Env can bind $\mu 2$ and the AP2 complex (Bowers et al. 2000; Wyss et al. 2001). This signal may also bind AP1 and AP3 (Bonifacino and Traub 2003; Ohno et al. 1997). Binding to AP1 appears to be weaker than to AP2, and its functional relevance (and that of AP3 binding) is unclear, although a role in sorting Env from the Golgi to endosomes and/or recycling internalized Env to the TGN is possible. The membrane proximal Tyr is crucial for both AP2 and AP1 binding. In addition, for AP2 binding, the G at Y-1, P at Y+2 and Ø at Y+3 all influence endocytosis (Boge et al. 1998; Bowers et al. 2000; Sauter et al. 1996). In this respect, the HIV/SIV membrane proximal signal conforms to a consensus μ chain binding motif (Bonifacino and Dell'Angelica 1999). Other possible Y-based motifs in HIV and SIV Envs have not been found to have a major role in sorting or endocytosis, although they may bind adaptors or adaptor components in vitro. TIP47 (tail interacting protein of 47 kDa), a protein that plays a role in recycling from late endosomes to the TGN and binds a di-aromatic motif in M6PRs (Diaz and Pfeffer 1998), has also been implicated in sorting HIV and SIV Envs, where a similar motif ($Y_{802}\,W_{803}$ in HIV-1_{IIIB}) is conserved. This interaction has been suggested to be crucial for Env incorporation into virions (Blot et al. 2003).

The key component that drives retrovirus assembly is Gag. Virus-like particles (VLPs) can be formed in many cell types when Gag is expressed with or without other viral components. Gag has been reported to be targeted directly to the plasma membrane and in some cases to associate with detergent-insoluble microdomains (Lindwasser and Resh 2001; Ono and Freed 2001; Suomalainen 2002). Although in many cell systems HIV particles and VLPs bud from the plasma membrane, in some cells, for example macrophages, the majority of particles assemble on endocytic membranes (Pelchen-Matthews et al. 2003; Raposo et al. 2002). How exactly Gag is targeted to these membranes is unclear. One possibility is that Gag may interact with AP2 and be carried from the plasma membrane on endocytic vesicles (Batonick et al. 2003). Alternatively Gag may be targeted directly to endosomal membranes. Perturbation of these membranes by overexpressing a dominant-negative form of Vps4, a AAA ATPase involved in ESCRT recycling (see below), can increase Gag association with endosomal compartments (von Schwedler et al. 2003) as can modulation of phosphatidyl inositol phosphate distribution (Ono and Freed 2003). The traffic of Env from the plasma membrane through endosomal compartments and back to the plasma membrane may then enable Env to be captured by Gag particles budding at different cellular locations.

Other Retroviruses. The presence of endocytosis signals in the Env cytoplasmic domain is not uncommon in retroviruses. The human T cell leukaemia virus (HTLV-I), a δ-retrovirus, has been studied in some detail. HTLV-I Env is required for infectivity and cell-cell fusion but is only expressed at low levels on the surfaces of infected cells and when Env is expressed in the absence of other viral components (Delamarre et al. 1997; Derse et al. 2001; Jassal et al. 2001; Nagy et al. 1983). The cytoplasmic domain of Env appears to play important roles in virus transmission as its truncation can increase cell-cell fusion activity (syncytium formation) and decrease cell-cell transmission of the virus (Kim et al. 2003; Pique et al. 1993). In common with many other retroviruses, the cytoplasmic domain of HTLV-I Env is short (28 aa), but it contains Tyr residues at positions 476 and 479, both of which are found in a YxxØ context. Mutation of either, or both, Tyr residues leads to loss of basolateral targeting when this Env is expressed in polarized cells (Lodge et al. 1997a). In the absence of other viral proteins, HTLV-I Env is located on intracellular membranes close to the nucleus as well as at the plasma membrane. Mutation of Y479, but not Y476, increases Env cell surface expression and the fusogenic activity of transfected cells. The HTLV-I Env cytoplasmic domain binds both AP1 and AP2 in pull-down experiments and interacts with the μ1- and μ2-subunits in yeast two-hybrid assays. Significantly, the interaction with μ1 can be abolished by mutation of Y479 only, whereas binding of μ2 is affected by mutations in either or both tyrosines (Berlioz-Torrent et al. 1999).

Thus, as with HIV and SIV, a putative endocytosis signals appears to limit HTLV-I Env cell surface expression, but in other respects the trafficking information encoded in this protein is less complex than that of HIV/SIV Envs. The transfer of HTLV-I from infected to uninfected cells has suggested a role for a so-called 'virological synapse' (Bangham 2003). In infected cells it appears that Gag is sorted to these regions of cell-cell interaction. It remains to be seen whether this Gag is associated with preassembled particles in membrane-bound compartments, or whether viral particles bud into the 'synapse' or are transferred from one cell to another by another mechanism, perhaps involving cell-cell fusion (Igakura et al. 2003).

In addition to HIV/SIV and HTLV-I, YxxØ sequences have been found in the Env proteins of, for example, the avian α-retrovirus, Rous sarcoma virus (RSV) and Mason-Pfizer monkey virus (MPMV). In RSV Env the signal is functionally silent but is activated in Env proteins that are deficient for palmitoylation or have modified transmembrane domain sequences (Ochsenbauer et al. 2000). When the signal is active, Env exhibits rapid endocytosis through CCVs. Whether this signal can be modulated in infected cells at different stages of the virus life cycle is unclear. Beta-retroviruses such as MPMV differ from other retroviruses in that their capsid assembly occurs in the cytoplasm of infected cells before Gag membrane association. A motif has been identified in the MPMV Gag that targets the newly synthesized protein to the area of the cell around the microtubule organizing centre (MTOC), where core assembly appears to occur (Sfakianos and Hunter 2003). Subsequently, assembled cores traffic to the plasma membrane for envelopment, budding and release. A YxxØ type motif in MPMV Env is required for these events and suggests that trafficking of Env through the recycling endosome compartment is required to bring the core particles to the plasma membrane (Sfakianos et al. 2003). Whether the core particles interact with Env in recycling endosomes and traffic to the cell surface on recycling vesicles is unclear.

3.2
Herpesviruses

Herpesviruses express multiple envelope glycoproteins. Alpha-herpesviruses, such as herpes simplex virus (HSV), carry ~12 envelope glycoproteins, whereas β-herpesviruses, such as human cytomegalovirus (HCMV), encode ~60 putative glycoproteins, many of which may be carried in the viral envelope. Some glycoproteins, for example gB, gH and gL, are essential for viral entry and are thought to be conserved in all herpesviruses. Others appear to be included to facilitate the infection and replication of specific viruses. These may encode additional receptor binding molecules, to broaden the range of susceptible cells, or proteins that have been implicated in the signalling events occurring early after fusion that prime cells for replicating the incoming virus (Kledal et al. 1998).

Glycoprotein B (gB) is a type I integral membrane protein that is essential for the attachment and fusion of herpesviruses with their host cells (Spear and Longnecker 2003). The proteins form homodimers or trimers in some herpesviruses and, in HCMV, are proteolytically cleaved to generate two associated polypeptides. Immunofluorescence analysis of the subcellular distribution of HSV gB shows low levels of surface expression and intracellular vesicular staining that co-labels for the early endosome antigen, EEA1. This intracellular localization is linked to the presence of a YxxL motif that is conserved in the C-terminal cytoplasmic domain of all sequenced gBs (see Table 2) (with the exception of Epstein-Barr virus gB) (Heineman and Hall 2001; Nixdorf et al. 2000; Radsak et al. 1996; Tirabassi and Enquist 1998; Tugizov et al. 1999). Removal of the YxxL motif in HSV gB increases the expression of the protein at the cell surface (Fan et al. 2002). Other potential trafficking motifs can also be identified, for example HSV gB has a second YxxØ motif closer to the transmembrane domain, as well as a di-leucine signal between the two Y-containing motifs. However, mutation of either one or both of these motifs has no obvious effect on HSV gB distribution. In another α-herpesvirus, varicella zoster virus (VZV), gB is primarily seen in the TGN and to a lesser extent at the plasma membrane (Heineman and Hall 2001; Heineman et al. 2000; Wang et al. 1998). As for HSV gB, the cytoplasmic tail of VZV gB contains a di-leucine and two YxxØ motifs (Heineman and Hall 2001). Mutational analysis indicated that the membrane proximal YxxØ motif is required for Golgi localization and the distal YxxØ motif for endocytosis. Mutations in the di-leucine sequence do not affect protein distribution (Heineman and Hall 2001). Whether these subtle differences in gB trafficking contribute to the fact that VZV-infected cells form syncytia is unclear.

HCMV gB undergoes endocytosis and recycling in fibroblasts and epithelial cells. Internalization is believed to be clathrin dependent (Tugizov et al. 1999), and the internalized protein is seen to co-localize with β-COP, Rab4, Rab5, Rab11, AP1, VIP-21 and TGN46, suggesting it may traffic through early and recycling endosomes and the TGN (Jarvis et al. 2002; Tugizov et al. 1998). The cytosolic domain of gB contains several potential signals for sorting to endocytic vesicles, including tyrosine- and di-leucine-based motifs as well as a cluster of acidic residues (Tugizov et al. 1999). Mutations in the acidic cluster impair internalization and abolish recycling (Tugizov et al. 1999). In contrast to the α-herpesviruses, HCMV gB is transported to apical membranes in epithelial cells independently of other viral glycoproteins, demonstrating that the protein also contains autonomous information for vectorial sorting (Tugizov et al. 1998). Apical transport is abolished by partial deletion of the 20-aa transmembrane domain of gB, and stretches of large hydrophobic amino acids in this domain have been implicated in raft-dependent gB sorting (Tugizov et al. 1998). In addition, a deletion of the acidic cluster in the cytoplasmic domain leads to mis-sorting of the protein to basolateral membranes in polarized cells (Tugizov et al. 1998). After the deletion, the potential of the cytoplasmic domain YxxØ and di-leucine motifs as basolateral sorting determinants may become apparent. However, the cytoplasmic domain of gB might be involved in other functions in certain cell types. In U373 cells, gB induces syncytium formation regulated by its cytoplasmic domain. It appears to be dependent on rapid concentration of gB into endocytic vesicles at the plasma membrane (Tugizov et al. 1994, 1995). Whether this indicates a need to generate regions of concentrated gB for fusion complex formation is unclear. Endocytic trafficking of gB may also play an important role in virus assembly. The gB incorporated into virions appeared to have trafficked over the plasma membrane (Radsak et al. 1996), and envelopment of the virus has been proposed to happen in endocytic organelles (Fraile-Ramos et al. 2002; Jarvis et al. 2002; Radsak et al. 1996). Thus endocytosis of gB may be crucial for targeting the protein to sites of virus assembly.

Glycoprotein E (gE) is found in a number of herpesviruses. Together with gI, gE forms a high-affinity receptor for immunoglobulin Fc domains (Johnson and Feenstra 1987). In HSV-infected epithelial cells, the gE-gI complex is found at adherens junctions on the lateral surface of cells and is able to bind junction components to promote HSV spread (Dingwell and Johnson 1998; Wisner et al. 2000). HSV lacking gE/gI, or expressing a gE-gI complex lacking the cytoplasmic domains, is unable to spread in epithelial cell cultures (Johnson et al. 2001; McMillan and Johnson 2001; Wisner et al. 2000).

HSV gE cycles between the TGN and the cell surface. In the early stages of infection, the bulk of the protein is in the TGN and only small amounts are found at the plasma membrane (Alconada et al. 1999). Later in infection, the Golgi complex fragments and increased levels of gE/gI are seen at cell junctions together with the TGN marker TGN46 (Alconada et al. 1998, 1999; McMillan and Johnson 2001). The cytoplasmic domains of gEs in herpesviruses have four regions of sequence similarity. A YxxØ motif and an acidic cluster are highly conserved. The acidic cluster contains one or more potential CKII phosphorylation sites (Alconada et al. 1996,, 1999; Zhu et al. 1996). A further Tyr-containing tetrapeptide, and a conserved aromatic residue surrounded by hydrophobic amino acids, may also contribute to trafficking. Truncation of the acidic cluster in HSV gE leads to its partial mislocalization to the plasma membrane, but a role for phosphorylation has not yet been found. Mutation of the Tyr in the YxxØ motif also leads to partial missorting, and truncation of all four putative signals leads to exclusive surface localization (Alconada et al. 1999; Jones et al. 1995; Takahashi et al. 1995). The association of gE with gI also influences trafficking. In epithelial cells infected with gI-negative HSV mutants, neither gE nor TGN46 is found on lateral surfaces, suggesting that although gE possesses information necessary for internalization and recycling, transport to cell junctions is dependent on information in gI (McMillan and Johnson 2001). Thus, although the intracellular accumulation of gE/gI may be associated with virus assembly, transport of gE/gI to junctional domains may facilitate cell-cell spread of the virus (McMillan and Johnson 2001).

In VZV, gE and gI form a complex shortly after their synthesis through interactions that require two conserved Cys-rich regions in the gE ectodomain (Alconada et al. 1998; Yao et al. 1993). As with the HSV homologues, VZV gE/gI accumulates in the TGN but also cycles over the cell surface (Alconada et al. 1996, 1999; Zhu et al. 1995, 1996). Three potential sorting signals similar to those found in HSV gE have been identified in the VZV gE cytoplasmic domain; two Tyr-containing tetrapeptides and an acidic cluster containing putative CKII phosphorylation sites (Alconada et al. 1996; Zhu et al. 1996). The concerted action of all three of these motifs is believed to be required for the steady-state localization of gE to the TGN and for cycling between the TGN, the cell surface and endosomes. Interaction with AP1 at least is involved in this trafficking (Olson and Grose 1997; Zhu et al. 1995, 1996). When VZV gI is expressed without gE it is exclusively localized to the plasma membrane, while gE alone is internalized. Expression of both proteins leads to co-localization in the TGN, indicating that gI does not contain sorting signals and that its distribution is determined by sorting information in gE (Alconada et al. 1998). Examples of the distribution of one protein being dependent on signals in an associated protein are not restricted to herpesviruses. The bunyavirus envelope protein G2 is transported to the plasma membrane when expressed alone but is retained in the Golgi when expressed with G1 which contains a Golgi retention signal (Chen et al. 1991; Matsuoka et al. 1991; Melin et al. 1995; Ronnholm 1992). Similarly in rubella virus, E1 is targeted to the Golgi through its interaction with E2 (Hobman et al. 1995).

Glycoprotein H (gH) is the third most abundant envelope protein in VZV after gE and gB. It is required for membrane fusion and forms heterodimeric complexes with gL. Formation of this complex is essential for maturation and cell surface expression of gH (Forghani et al. 1994; Hutchinson et al. 1992; Montalvo and Grose 1986; Rodriguez et al. 1993). gH is also efficiently internalized from the surface of VZV-infected cells and localizes to the TGN. Co-expression of recombinant gH and its chaperone gL showed that internalization is independent of other viral glycoproteins, suggesting that gH contains its own sorting information (Pasieka et al. 2003). The cytoplasmic domain of VZV gH is only 12–14 aa long. Nevertheless, gH internalization is clathrin mediated and involves a YxxØ motif (Pasieka et al. 2003). All gH proteins of α-herpesviruses may be internalized in this way, except for HSV gH, where an endocytosis signal appears to be missing. In HSV, gH can form complexes with gB, which may provide the missing endocytosis information. The cytoplasmic domains of β- and γ-herpesvirus gH proteins are only

6–10 amino acids long and may be too short to contain functional trafficking information (Gompels et al. 1988; Pasieka et al. 2003). Little is known about the trafficking of gH's chaperone gL.

Other Glycoproteins. In addition to the key proteins required for virus entry, recent studies have suggested that other viral and cellular proteins are incorporated into herpesviral membranes. The cytomegaloviruses are particularly intriguing as these viruses encode a number of heptahelical putative heterotrimeric G protein-coupled receptors (GPCRs). HCMV encodes three, of which at least two (UL33 and US27) are incorporated into the viral envelope (Fraile-Ramos et al. 2002). The third protein (US28) has been demonstrated to bind a range of chemokines including the unusual membrane-linked chemokine fractalkine (CX_3CL1). US28 is constitutively active (Kledal et al. 1998). It is also located primarily in late endosome-associated vesicles in infected cells, together with UL33 and US27, and may well be incorporated into virions budding into these vesicles (Fraile-Ramos et al. 2001, 2002). US28 traffics over the cell surface. Its internalization is clathrin dependent and appears to require constitutive phosphorylation (Mokros et al. 2002), but, in contrast to many other GPCRs, it is not dependent on β-arrestins (Fraile-Ramos et al. 2003). These trafficking activities may be co-ordinated not only to permit inclusion of US28 into viral membranes (and subsequent signalling functions in the membrane of the target cells after fusion) but also to allow US28 to scavenge chemokines from around infected cells, thereby modulating the efficacy of immune responses to viral infection (Beisser et al. 2002).

How exactly trafficking information is co-ordinated in herpesviruses to allow the components of these complex viral membranes to be brought together at the sites of viral budding remains to be established. Proteins that may shepherd key viral membrane proteins to assembly sites have been proposed, for example U6 in HSV (see Newcomb et al. 2003), but analogous proteins have yet to be identified in viruses such as HCMV.

3.3
Orthomyxoviruses

Influenza A viruses are the prototype viruses of the orthomyxovirus sub-family. These viruses contain two main surface glycoproteins, haemagglutinin (HA) and neuraminidase (NA). HA has been widely used as a model for studies of protein folding, protein quality control and membrane fusion (see chapters by Earp et al. and Salonen et al., this volume). It is a non-covalently linked homotrimer. Each monomer has a short 10-amino acid C-terminal cytoplasmic domain containing three cysteines, which can be palmitoylated. By contrast, NA is a homotetramer of a type II membrane protein.

A primary site for influenza infection is the epithelium of the respiratory tract. In infected polarized epithelial cells, both HA and NA are expressed on the apical surface, where they are associated with lipid rafts (Scheiffele et al. 1997). The transmembrane domain of HA and NA can specify apical targeting through association with DIMs. Detailed mutational analysis of the HA transmembrane domain indicates that although partitioning into DIMs is required for apical sorting it is not sufficient and that certain conserved residues within the transmembrane domain may be required for interaction with the DIM-associated apical sorting machinery (Lin et al. 1998). No internalization or recycling activity has been detected for HA.

3.4
Rhabdoviruses

Vesicular stomatitis virus (VSV), the prototype rhabdovirus, encodes a single type I integral envelope glycoprotein G. G has been used extensively to study protein export from the ER and transit through the Golgi. Truncation of the cytoplasmic domain leads to slower export without affecting folding or trimerization of the protein, suggesting the presence of export signals (Doms et al. 1988). Mutation of a DxE motif (see Table 1) does not affect transport, but mutation of a six-residue sequence (YTDIEM, position 19–24 in the cytoplasmic domain) inhibits efficient export (Nishimura and Balch 1997).

In epithelial cells, G is targeted to basolateral surfaces, from where the virus buds (Boulan and Pendergast 1980). G also localizes to coated pits in these domains through which it is internalized (Matlin et al. 1983). A YxxØ motif in the G cytoplasmic domain has been implicated in basolateral targeting as mutation of the single tyrosine (Y501) and/or I504 (positions 19 and 22 in the cytoplasmic domain, respectively) cause the protein to appear on both the apical and basolateral surfaces (Thomas et al. 1993). In common with some other basolateral sorting motifs, Y501 could not be replaced by any other amino acid but I504 could be replaced by other hydrophobic amino acids (Thomas and Roth 1994). G also undergoes endocytosis, although the rate is slower than that observed for other proteins using Y-based signals. The VSV Tyr motif does not play a role in internalization and the nature of the endocytosis signal remains unclear (Thomas et al. 1993).

3.5
Poxviruses

Vaccinia virus (VV) is the prototype member of the poxvirus family. VV particles can exist in several different forms. The intracellular mature form (IMV) found in the cytoplasm is wrapped by two tightly apposed membranes derived from the ERGIC (ER to Golgi Intermediate Compartment). IMVs can acquire two additional membranes from the TGN. These particles can be transported to the plasma membrane, where they are released by fusion of the outer membrane with the plasma membrane to generate extracellular enveloped virions (EEV) (reviewed in Smith et al. 2002). The membranes acquired from the TGN contain five glycosylated and one non-glycosylated virally encoded membrane proteins. Four of these are not essential for EEV formation, but two, F13L (Blasco and Moss 1991) and B5R (Engelstad and Smith 1993; Wolffe et al. 1993), are required for envelopment and high-level production of infectious EEV (Blasco and Moss 1991). B5R, a 42-kDa glycoprotein, has also been implicated in cell entry, because antibodies against its extracellular domain can neutralize the virus (Galmiche et al. 1999). In infected cells both F13L and B5R concentrate in the membranes of the TGN (Schmelz et al. 1994); F13L is a cytosolic peripheral protein that requires acylation for membrane association when expressed alone. B5R is a type I membrane protein (Isaacs et al. 1992) that is believed to contain its own sorting information. These sorting signals reside in the transmembrane and/or cytoplasmic domains of the protein, as they are sufficient to direct chimeric proteins to the Golgi and ensure their integration into EEVs (Katz et al. 1997). Studies with chimeric proteins also indicate that the B5R cytoplasmic domain prevents accumulation of the protein in the plasma membrane, by either Golgi retention or endocytic retrieval (Ward and Moss 2000). This domain displays two motifs that might regulate surface expression, a tyrosine at position 310 and a di-leucine signal at positions 315/316 (see Table 2). Mutation of either or both of these motifs increases the levels of B5R on the plasma membrane. Antibody uptake experiments demonstrate that B5R is cycling from the Golgi apparatus to the plasma membrane and back with the stage of transit through the TGN being the slowest part of the cycle. Mutations of Y310 or the di-leucine motif impaired retrograde transport, whereas a double mutation abrogated it (Ward and Moss 2000). However, the exact contribution of each signal and the mechanism of transport remain to be fully established.

4
Virus Assembly

Assembly of new viruses requires the temporally and spatially co-ordinated co-localization of all the structural components required to form an infectious virus. For most enveloped viruses, assembly is often seen as a budding process in which the assembling virion progressively deforms the cellular membrane into a bud that eventually pinches off as a free virus particle. For viruses with rapid replication cycles these events often occur at the cell surface (Rowell et al. 1995), but for viruses that establish long-term sustained infections assembly must be balanced to ensure the survival of infected cells. In this case, a high cell surface expression of viral proteins is likely to attract the attention of the immune system and alternative routes for assembly may be taken by the virus.

In many cases the localization of envelope proteins appears to determine the site of virus assembly. For other viruses the process is more complex. For example, it has long been believed that HIV and other retroviruses bud exclusively from the plasma membrane. However, HIV Env is internalized efficiently from the surface, either requiring that Env must be relocated during assembly or suggesting that not all assembly occurs at the cell surface. It is possible that Gag might mask the internalization motifs on Env, and allow Envs to accumulate at the cell surface, but in some cells at least infectious HIV is assembled intracellularly (Pelchen-Matthews et al. 2003).

It is becoming increasingly clear that cellular components are required to facilitate the assembly and release of many enveloped viruses. One of the most extensively characterized of these machineries is the eESCRT (Endosomal Sorting Complex Required for Transport) (Katzmann et al. 2002), a set of protein complexes involved in sorting membrane proteins to lysosomes.

4.1
ESCRTing Virus Release

In contrast to the inwardly directed formation of membrane transport vesicles, such as CCVs, which, regardless of the compartment on which they are formed, are released into the cytoplasm of the cell, viruses are released into the extracellular space or into the lumen of an organelle (which is topologically outside the cell). The majority of studies of vesiculation have involved inwardly driven processes, i.e. vesiculation into the cytoplasm. Recently it has become apparent that cells have a highly conserved machinery for outward vesiculation, i.e. vesiculation towards the outside of the cell or lumen of intracellular compartments (see Fig. 1) (Katzmann et al. 2002). This machinery, initially characterized through the analysis of yeast vacuolar protein sorting (Vps) mutants is responsible for the sorting of enzymes and substrates destined for lysosomes into membrane domains which form small vesicles in the lumina of endosomes generating organelles termed multivesicular bodies (MVBs). Once delivered to lysosomes, these internal membranes are preferential targets for hydrolytic degradation (Katzmann et al. 2002). In addition to sorting to lysosomes, other outward vesiculation events have now been recognized and may involve the same machinery. One particular example is the generation of exosomes, small 50- to 90-nm-diameter membrane vesicles formed in antigen-presenting cells (APC) and loaded with a range of molecules including MHC class II-antigen complexes (Stoorvogel et al. 2002). These vesicles can be released from APC and are thought to play an as yet undefined role in immune responses. The formation of MVBs requires ESCRT complexes (ESCRT 0–3) that are recruited sequentially to endosomal membranes. These proteins select cargo (frequently mono-ubiquitinated membrane proteins) and sort this cargo into membrane domains that bud into the endosomal lumen. The ESCRT machinery is recycled through the actions of a AAA-type ATPase, Vps4 (Katzmann et al. 2002).

The topology of these events is identical to virion budding, and it has emerged that the ESCRT machinery is also essential for the release of HIV and a number of other retroviruses, as well as rhabdoviruses, filoviruses (e.g. Ebola) and possibly also influenza viruses (see Marsh and Thali 2003; reviewed in Pornillos et al. 2002). The molecular mechanism through which the ESCRT machinery facilitates assembly is best understood for retroviruses, including HIV. The Gag proteins of these viruses contain sequences that are essential for late steps in the assembly/budding process. These so-called late (L) domains are autonomous signals that can be moved around within a Gag or matrix protein, can function *in trans* or be swapped between viruses. Mutations in these L-domains cause inhibition of virus release after the onset of assembly but before scission of particles from the donor membrane. With HIV L-domain mutants, budding profiles can be seen aligned along the plasma membrane like a set of lollipops (Pornillos et al. 2002). In HIV the L domain sequences are located in the C-terminal p6 domain of Gag. A sequence PTAP has been identified as key for L-domain function (Freed 2002; Pornillos et al. 2002).

Searches for the interacting partners revealed that the PTAP motifs bind the protein Tsg 101 (tumour susceptibility gene 101), a component of ESCRT-1. Further analyses have shown that depletion of Tsg101 and ESCRT-1 from cells by RNA interference (RNAi) inhibits HIV release and that virus release is blocked at the same stage of assembly as L-domain mutants. Thus the notion has developed that, through Gag, HIV recruits the cellular machinery required for outward vesiculation. Interestingly the PTAP sequence appears to mimic the sequence PSAP found in the cellular protein Hrs. Hrs is a component of ESCRT-0 that usually recruits ESCRT-1 to the membrane (Bache et al. 2003; Pornillos et al. 2003).

4.2
Other Late Domain Sequences

In addition to the PTAP motif, other p6 sequences may contribute to the activities of the HIV L-domain. Recent studies have implicated a sequence, which lies downstream of the PTAP motif as having L-domain activity. This sequence (YxxL) can recruit a protein called AIP1/ALIX, which interacts with both Tsg101 and the CHMP4 component of the ESCRT-3 complex. Thus HIV appears to have adopted a belt and braces ability to recruit ESCRT-3 either through Tsg101 or through AIP1/ALIX. However, the significance of the later interaction is unclear, as mutations in the PTAP motif, or Tsg101 knock down, alone inhibit release of infectious virus.

The non-primate lentivirus, equine infectious anaemia virus (EIAV), appears to use a YxxL as its primary L-domain motif. As with HIV, the motif interacts with AIP1/ALIX to recruit ESCRT-3. This sequence can be exchanged with the HIV PTAP motif, indicating that although the two motifs recruit different effectors they both operate through the same pathway.

Late domain sequences have been found in other enveloped viruses. In the murine leukaemia virus (MLV) the sequence PPPY is required for budding. The PPPY motif does not bind Tsg101, but instead appears to interact with a Nedd4-like ubiquitin E3 ligase BUL I (Yasuda et al. 2002). As Tsg101 also binds ubiquitinated proteins it has been speculated that this might be an alternative means to recruit the ESCRT machinery, perhaps by-passing ESCRT-1. Indeed, Tsg101 knock down by RNAi does not affect MLV release. Although a number of different viruses use different L-domain motifs and different means to recruit ESCRT components, assembly of all of them is inhibited by dominant-negative Vps4. Vps4 appears to be required to recycle membrane-bound ESCRT complexes. In the presence of dominant-negative Vps4 the ESCRT machinery becomes irreversibly associated with endosomal membranes and is no longer available to facilitate virus release.

5
Conclusions

The trafficking of viral proteins has long been of interest to virologists and cell biologists. Indeed, viral membrane proteins have been crucial models used for dissecting key aspects of membrane trafficking machineries. More recent studies have indicated that many of these proteins contain multiple trafficking signals that allow them to adopt complex trafficking itineraries in the endocytic and exocytic pathways. The importance of these signals in the biology of different viruses is being increasingly recognized. One of the key observations to emerge is that not only can specific signals affect the distribution of a viral protein in a cell in tissue culture but they are also crucial for the pathogenesis of the virus in vivo. Indeed, detailed studies of the Tyr-based signals in SIV and HIV Envs are shedding new light on the biology of these viruses. Much more remains to be learned: It is likely that novel signals remain to be discovered and that information on the trafficking of viral integral and peripheral proteins will continue to emerge. Some of this information may prove to be useful in deriving novel strategies to combat viral infection. Whether or not this is the case, it is nevertheless important to understand the trafficking itinerary of the proteins in order to develop a more complete understanding of the cellular mechanisms that underlie virus infection.

Acknowledgements R.B. is supported by grants from the Roche Research Foundation and the NIH (AI-49784) and M.M. by the UK Medical Research Council. We thank Annegret Pelchen-Matthews for critical reading and colleagues who have contributed ideas and comments on the manuscript.

References

Alconada, A., Bauer, U., Baudoux, L., Piette, J., and Hoflack, B. (1998) Intracellular transport of the glycoproteins gE and gI of the varicella-zoster virus. gE accelerates the maturation of gI and determines its accumulation in the trans-Golgi network. J Biol Chem 273:13430–13436

Alconada, A., Bauer, U., and Hoflack, B. (1996) A tyrosine-based motif and a casein kinase II phosphorylation site regulate the intracellular trafficking of the varicella-zoster virus glycoprotein I, a protein localized in the trans-Golgi network. EMBO J 15:6096–6110

Alconada, A., Bauer, U., Sodeik, B., and Hoflack, B. (1999) Intracellular traffic of herpes simplex virus glycoprotein gE: characterization of the sorting signals required for its trans-Golgi network localization. J Virol 73:377–387

Aridor, M., and Traub, L. M. (2002) Cargo selection in vesicular transport: the making and breaking of a coat. Traffic 3:537–546

Bache, K. G., Brech, A., Mehlum, A., and Stenmark, H. (2003) Hrs regulates multivesicular body formation via ESCRT recruitment to endosomes. J Cell Biol 162:435–442

Bangham, C. R. (2003) The immune control and cell-to-cell spread of human T-lymphotropic virus type 1. J Gen Virol 84:3177–89

Batonick, M., Zampieri, C., Honing, S., Spearman, P., and Thali, M. (2003) HIV-1 particle release is increased in cells expressing non-functional clathrin-associated adaptor AP-2. Paper presented at: Retroviruses (Cold Spring Harbor Laboratory, Cold Spring Harbor, NY), p33

Beisser, P. S., Goh, C. S., Cohen, F. E., and Michelson, S. (2002) Viral chemokine receptors and chemokines in human cytomegalovirus trafficking and interaction with the immune system. CMV chemokine receptors. Curr Top Microbiol Immunol 269:203–234

Berlioz-Torrent, C., Shacklett, B. L., Erdtmann, L., Delamarre, L., Bouchaert, I., Sonigo, P., Dokhelar, M. C., and Benarous, R. (1999) Interactions of the cytoplasmic domains of human and simian retroviral transmembrane proteins with components of the clathrin adaptor complexes modulate intracellular and cell surface expression of envelope glycoproteins. J Virol 73:1350–1361

Bijlmakers, M. J., and Marsh, M. (2003) The on-off story of protein palmitoylation. Trends Cell Biol 13:32–42

Blasco, R., and Moss, B. (1991) Extracellular vaccinia virus formation and cell-to-cell virus transmission are prevented by deletion of the gene encoding the 37,000-Dalton outer envelope protein. J Virol 65:5910–5920

Blot, G., Janvier, K., Le Panse, S., Benarous, R., and Berlioz-Torrent, C. (2003) Targeting of the human immunodeficiency virus type 1 envelope to the trans-Golgi network through binding to TIP47 is required for env incorporation into virions and infectivity. J Virol 77:6931–6945

Boge, M., Wyss, S., Bonifacino, J. S., and Thali, M. (1998) A membrane-proximal tyrosine-based signal mediates internalization of the HIV-1 envelope glycoprotein via interaction with the AP-2 clathrin adaptor. J Biol Chem 273:15773–15778

Bonifacino, J. S. (2004) The GGA proteins: adaptors on the move. Nat Rev Mol Cell Biol 5:23–32

Bonifacino, J. S., and Dell'Angelica, E. C. (1999) Molecular bases for the recognition of tyrosine-based sorting signals. J Cell Biol 145:923–926

Bonifacino, J. S., and Glick, B. S. (2004) The mechanisms of vesicle budding and fusion. Cell 116:153–166

Bonifacino, J. S., Marks, M. S., Ohno, H., and Kirchhausen, T. (1996) Mechanisms of signal-mediated protein sorting in the endocytic and secretory pathways. Proc Assoc Am Physicians 108:285–295

Bonifacino, J. S., and Traub, L. M. (2003) Signals for sorting of transmembrane proteins to endosomes and lysosomes. Annu Rev Biochem 72:395–447

Boulan, E. R., and Pendergast, M. (1980) Polarized distribution of viral envelope proteins in the plasma membrane of infected epithelial cells. Cell 20:45–54

Bowers, K., Pelchen-Matthews, A., Honing, S., Vance, P. J., Creary, L., Haggarty, B. S., Romano, J., Ballensiefen, W., Hoxie, J. A., and Marsh, M. (2000) The simian immunodeficiency virus envelope glycoprotein contains multiple signals that regulate its cell surface expression and endocytosis. Traffic 1:661–674

Bretscher, M. S., and Munro, S. (1993) Cholesterol and the Golgi apparatus. Science 261:1280–1281

Britt, W. J., and Mach, M. (1996) Human cytomegalovirus glycoproteins. Intervirology 39:401–412

Canfield, W. M., Johnson, K. F., Ye, R. D., Gregory, W., and Kornfeld, S. (1991) Localization of the signal for rapid internalization of the bovine cation-independent mannose 6-phosphate/insulin-like growth factor-II receptor to amino acids 24–29 of the cytoplasmic tail. J Biol Chem 266:5682–5688

Chen, S. Y., Matsuoka, Y., and Compans, R. W. (1991) Golgi complex localization of the Punta Toro virus G2 protein requires its association with the G1 protein. Virology 183:351–365

Chen, W. J., Goldstein, J. L., and Brown, M. S. (1990) NPXY, a sequence often found in cytoplasmic tails, is required for coated pit-mediated internalization of the low density lipoprotein receptor. J Biol Chem 265:3116–3123

Collawn, J. F., Stangel, M., Kuhn, L. A., Esekogwu, V., Jing, S. Q., Trowbridge, I. S., and Tainer, J. A. (1990) Transferrin receptor internalization sequence YXRF implicates a tight turn as the structural recognition motif for endocytosis. Cell 63:1061–1072

Collins, B. M., McCoy, A. J., Kent, H. M., Evans, P. R., and Owen, D. J. (2002) Molecular architecture and functional model of the endocytic AP2 complex. Cell 109:523–535

Conner, S. D., and Schmid, S. L. (2002) Identification of an adaptor-associated kinase, AAK1, as a regulator of clathrin-mediated endocytosis. J Cell Biol 156:921–929

Conner, S. D., Schroter, T., and Schmid, S. L. (2003) AAK1-Mediated micro2 phosphorylation is stimulated by assembled clathrin. Traffic 4:885–890

Cosson, P. (1996) Direct interaction between the envelope and matrix proteins of HIV-1. EMBO J 15:5783–5788

Davis, C. G., van Driel, I. R., Russell, D. W., Brown, M. S., and Goldstein, J. L. (1987) The low density lipoprotein receptor. Identification of amino acids in cytoplasmic domain required for rapid endocytosis. J Biol Chem 262:4075–4082

Delamarre, L., Rosenberg, A. R., Pique, C., Pham, D., and Dokhelar, M. C. (1997) A novel human T-leukemia virus type 1 cell-to-cell transmission assay permits definition of SU glycoprotein amino acids important for infectivity. J Virol 71:259–266

Derse, D., Hill, S. A., Lloyd, P. A., Chung, H., and Morse, B. A. (2001) Examining human T-lymphotropic virus type 1 infection and replication by cell-free infection with recombinant virus vectors. J Virol 75:8461–8468

Diaz, E., and Pfeffer, S. R. (1998) TIP47: a cargo selection device for mannose 6-phosphate receptor trafficking. Cell 93:433–443

Dingwell, K. S., and Johnson, D. C. (1998) The herpes simplex virus gE-gI complex facilitates cell-to-cell spread and binds to components of cell junctions. J Virol 72:8933–8942

Doms, R. W., Ruusala, A., Machamer, C., Helenius, J., Helenius, A., and Rose, J. K. (1988) Differential effects of mutations in three domains on folding, quaternary structure, and intracellular transport of vesicular stomatitis virus G protein. J Cell Biol 107:89–99

Engelstad, M., and Smith, G. L. (1993) The vaccinia virus 42-kDa envelope protein is required for the envelopment and egress of extracellular virus and for virus virulence. Virology 194:627–637

Fan, Z., Grantham, M. L., Smith, M. S., Anderson, E. S., Cardelli, J. A., and Muggeridge, M. I. (2002) Truncation of herpes simplex virus type 2 glycoprotein B increases its cell surface expression and activity in cell-cell fusion, but these properties are unrelated. J Virol 76:9271–9283

Forghani, B., Ni, L., and Grose, C. (1994) Neutralization epitope of the varicella-zoster virus gH:gL glycoprotein complex. Virology 199:458–462

Fraile-Ramos, A., Kledal, T. N., Pelchen-Matthews, A., Bowers, K., Schwartz, T. W., and Marsh, M. (2001) The human cytomegalovirus US28 protein is located in endocytic vesicles and undergoes constitutive endocytosis and recycling. Mol Biol Cell 12:1737–1749

Fraile-Ramos, A., Kohout, T. A., Waldhoer, M., and Marsh, M. (2003) Endocytosis of the viral chemokine receptor US28 does not require beta-arrestins but is dependent on the clathrin-mediated pathway. Traffic 4:243–253

Fraile-Ramos, A., Pelchen-Matthews, A., Kledal, T. N., Browne, H., Schwartz, T. W., and Marsh, M. (2002) Localization of HCMV UL33 and US27 in endocytic compartments and viral membranes. Traffic 3:218–232

Freed, E. O. (2002) Viral late domains. J Virol 76:4679–4687

Fultz, P. N., Vance, P. J., Endres, M. J., Tao, B., Dvorin, J. D., Davis, I. C., Lifson, J. D., Montefiori, D. C., Marsh, M., Malim, M. H., and Hoxie, J. A. (2001) In vivo attenuation of simian immunodeficiency virus by disruption of a tyrosine-dependent sorting signal in the envelope glycoprotein cytoplasmic tail. J Virol 75:278–291

Galmiche, M. C., Goenaga, J., Wittek, R., and Rindisbacher, L. (1999) Neutralizing and protective antibodies directed against vaccinia virus envelope antigens. Virology 254:71–80

Ghosh, P., and Kornfeld, S. (2003a) AP-1 binding to sorting signals and release from clathrin-coated vesicles is regulated by phosphorylation. J Cell Biol 160:699–708

Ghosh, P., and Kornfeld, S. (2003b) Phosphorylation-induced conformational changes regulate GGAs 1 and 3 function at the trans-Golgi network. J Biol Chem 278:14543–14549

Gompels, U. A., Craxton, M. A., and Honess, R. W. (1988) Conservation of glycoprotein H (gH) in herpesviruses: nucleotide sequence of the gH gene from herpesvirus saimiri. J Gen Virol 69: 2819–2829

He, G., Gupta, S., Yi, M., Michaely, P., Hobbs, H. H., and Cohen, J. C. (2002) ARH is a modular adaptor protein that interacts with the LDL receptor, clathrin, and AP-2. J Biol Chem 277:44044–44049

Heineman, T. C., and Hall, S. L. (2001) VZV gB endocytosis and Golgi localization are mediated by YXXphi motifs in its cytoplasmic domain. Virology 285:42–49

Heineman, T. C., Krudwig, N., and Hall, S. L. (2000) Cytoplasmic domain signal sequences that mediate transport of varicella-zoster virus gB from the endoplasmic reticulum to the Golgi. J Virol 74:9421–9430

Hirst, J., and Robinson, M. S. (1998) Clathrin and adaptors. Biochim Biophys Acta 1404:173–193

Hobman, T. C., Woodward, L., and Farquhar, M. G. (1995) Targeting of a heterodimeric membrane protein complex to the Golgi: rubella virus E2 glycoprotein contains a transmembrane Golgi retention signal. Mol Biol Cell 6:7-20

Hutchinson, L., Browne, H., Wargent, V., Davis-Poynter, N., Primorac, S., Goldsmith, K., Minson, A. C., and Johnson, D. C. (1992) A novel herpes simplex virus glycoprotein, gL, forms a complex with glycoprotein H (gH) and affects normal folding and surface expression of gH. J Virol 66:2240–2250

Igakura, T., Stinchcombe, J. C., Goon, P. K., Taylor, G. P., Weber, J. N., Griffiths, G. M., Tanaka, Y., Osame, M., and Bangham, C. R. (2003) Spread of HTLV-I between lymphocytes by virus-induced polarization of the cytoskeleton. Science 299:1713–1716

Isaacs, S. N., Wolffe, E. J., Payne, L. G., and Moss, B. (1992) Characterization of a vaccinia virus-encoded 42-kilodalton class I membrane glycoprotein component of the extracellular virus envelope. J Virol 66:7217–7224

Jackson, A. P., Flett, A., Smythe, C., Hufton, L., Wettey, F. R., and Smythe, E. (2003) Clathrin promotes incorporation of cargo into coated pits by activation of the AP2 adaptor micro2 kinase. J Cell Biol 163:231–236

Jadot, M., Canfield, W. M., Gregory, W., and Kornfeld, S. (1992) Characterization of the signal for rapid internalization of the bovine mannose 6-phosphate/insulin-like growth factor-II receptor. J Biol Chem 267:11069–11077

Janvier, K., Kato, Y., Boehm, M., Rose, J. R., Martina, J. A., Kim, B. Y., Venkatesan, S., and Bonifacino, J. S. (2003) Recognition of dileucine-based sorting signals from HIV-1 Nef and LIMP-II by the AP-1 γ-σ1 and AP-3 δ-σ3 hemicomplexes. J Cell Biol 163:1281–1290

Jarvis, M. A., Fish, K. N., Soderberg-Naucler, C., Streblow, D. N., Meyers, H. L., Thomas, G., and Nelson, J. A. (2002) Retrieval of human cytomegalovirus glycoprotein B from cell surface is not required for virus envelopment in astrocytoma cells. J Virol 76:5147–5155

Jassal, S. R., Lairmore, M. D., Leigh-Brown, A. J., and Brighty, D. W. (2001) Soluble recombinant HTLV-1 surface glycoprotein competitively inhibits syncytia formation and viral infection of cells. Virus Res 78:17-34

Johnson, D. C., and Feenstra, V. (1987) Identification of a novel herpes simplex virus type 1-induced glycoprotein which complexes with gE and binds immunoglobulin. J Virol 61:2208-2216

Johnson, D. C., Webb, M., Wisner, T. W., and Brunetti, C. (2001) Herpes simplex virus gE/gI sorts nascent virions to epithelial cell junctions, promoting virus spread. J Virol 75:821-833

Johnson, K. F., and Kornfeld, S. (1992) The cytoplasmic tail of the mannose 6-phosphate/insulin-like growth factor-II receptor has two signals for lysosomal enzyme sorting in the Golgi. J Cell Biol 119:249-257

Jones, B. G., Thomas, L., Molloy, S. S., Thulin, C. D., Fry, M. D., Walsh, K. A., and Thomas, G. (1995) Intracellular trafficking of furin is modulated by the phosphorylation state of a casein kinase II site in its cytoplasmic tail. EMBO J 14:5869-5883

Katz, E., Wolffe, E. J., and Moss, B. (1997) The cytoplasmic and transmembrane domains of the vaccinia virus B5R protein target a chimeric human immunodeficiency virus type 1 glycoprotein to the outer envelope of nascent vaccinia virions. J Virol 71:3178-3187

Katzmann, D. J., Odorizzi, G., and Emr, S. D. (2002) Receptor downregulation and multivesicular-body sorting. Nat Rev Mol Cell Biol 3:893-905

Kim, F. J., Manel, N., Boublik, Y., Battini, J. L., and Sitbon, M. (2003) Human T-cell leukemia virus type 1 envelope-mediated syncytium formation can be activated in resistant mammalian cell lines by a carboxy-terminal truncation of the envelope cytoplasmic domain. J Virol 77:963-969

Kledal, T. N., Rosenkilde, M. M., and Schwartz, T. W. (1998) Selective recognition of the membrane-bound CX3C chemokine, fractalkine, by the human cytomegalovirus-encoded broad-spectrum receptor US28. FEBS Lett 441:209-214

Kornfeld, S., and Mellman, I. (1989) The biogenesis of lysosomes. Annu Rev Cell Biol 5:483-525

LaBranche, C. C., Sauter, M. M., Haggarty, B. S., Vance, P. J., Romano, J., Hart, T. K., Bugelski, P. J., Marsh, M., and Hoxie, J. A. (1995) A single amino acid change in the cytoplasmic domain of the simian immunodeficiency virus transmembrane molecule increases envelope glycoprotein expression on infected cells. J Virol 69:5217-5227

Letourneur, F., Gaynor, E. C., Hennecke, S., Demolliere, C., Duden, R., Emr, S. D., Riezman, H., and Cosson, P. (1994) Coatomer is essential for retrieval of dilysine-tagged proteins to the endoplasmic reticulum. Cell 79:1199-1207

Li, F., Goila-Gaur, R., Salzwedel, K., Kilgore, N. R., Reddick, M., Matallana, C., Castillo, A., Zoumplis, D., Martin, D. E., Orenstein, J. M., Allaway, G. P., Freed, E. O., and Wild, C. T. (2003) PA-457: a potent HIV inhibitor that disrupts core condensation by targeting a late step in Gag processing. Proc Natl Acad Sci USA 100:13555-13560

Lin, S., Naim, H. Y., Rodriguez, A. C., and Roth, M. G. (1998) Mutations in the middle of the transmembrane domain reverse the polarity of transport of the influenza virus hemagglutinin in MDCK epithelial cells. J Cell Biol 142:51-57

Lindwasser, O. W., and Resh, M. D. (2001) Multimerization of human immunodeficiency virus type 1 Gag promotes its localization to barges, raft-like membrane microdomains. J Virol 75:7913–7924

Lodge, R., Delamarre, L., Lalonde, J. P., Alvarado, J., Sanders, D. A., Dokhelar, M. C., Cohen, E. A., and Lemay, G. (1997a) Two distinct oncornaviruses harbor an intracytoplasmic tyrosine-based basolateral targeting signal in their viral envelope glycoprotein. J Virol 71:5696–5702

Lodge, R., Gottlinger, H., Gabuzda, D., Cohen, E. A., and Lemay, G. (1994) The intracytoplasmic domain of gp41 mediates polarized budding of human immunodeficiency virus type 1 in MDCK cells. J Virol 68:4857–4861

Lodge, R., Lalonde, J. P., Lemay, G., and Cohen, E. A. (1997b) The membrane-proximal intracytoplasmic tyrosine residue of HIV-1 envelope glycoprotein is critical for basolateral targeting of viral budding in MDCK cells. EMBO J 16:695–705

Marks, M. S., Woodruff, L., Ohno, H., and Bonifacino, J. S. (1996) Protein targeting by tyrosine- and di-leucine-based signals: evidence for distinct saturable components. J Cell Biol 135:341–354

Marsh, M., and Thali, M. (2003) HIV's great escape. Nat Med 9:1262–1263

Matlin, K., Bainton, D. F., Pesonen, M., Louvard, D., Genty, N., and Simons, K. (1983) Transepithelial transport of a viral membrane glycoprotein implanted into the apical plasma membrane of Madin-Darby canine kidney cells. I. Morphological evidence. J Cell Biol 97:627–637

Matsuoka, Y., Chen, S. Y., and Compans, R. W. (1991) Bunyavirus protein transport and assembly. Curr Top Microbiol Immunol 169:161–179

McMillan, T. N., and Johnson, D. C. (2001) Cytoplasmic domain of herpes simplex virus gE causes accumulation in the trans-Golgi network, a site of virus envelopment and sorting of virions to cell junctions. J Virol 75:1928–1940

Melin, L., Persson, R., Andersson, A., Bergstrom, A., Ronnholm, R., and Pettersson, R. F. (1995) The membrane glycoprotein G1 of Uukuniemi virus contains a signal for localization to the Golgi complex. Virus Res 36:49–66

Mokros, T., Rehm, A., Droese, J., Oppermann, M., Lipp, M., and Hopken, U. E. (2002) Surface expression and endocytosis of the human cytomegalovirus-encoded chemokine receptor US28 is regulated by agonist-independent phosphorylation. J Biol Chem 277:45122–45128

Molloy, S. S., Thomas, L., VanSlyke, J. K., Stenberg, P. E., and Thomas, G. (1994) Intracellular trafficking and activation of the furin proprotein convertase: localization to the TGN and recycling from the cell surface. EMBO J 13:18–33

Montalvo, E. A., and Grose, C. (1986) Neutralization epitope of varicella zoster virus on native viral glycoprotein gp118 (VZV glycoprotein gpIII) Virology 149:230–241

Nagy, K., Clapham, P., Cheingsong-Popov, R., and Weiss, R. A. (1983) Human T-cell leukemia virus type I: induction of syncytia and inhibition by patients' sera. Int J Cancer 32:321–328

Newcomb, W. W., Thomsen, D. R., Homa, F. L., and Brown, J. C. (2003) Assembly of the herpes simplex virus capsid: identification of soluble scaffold-portal complexes and their role in formation of portal-containing capsids. J Virol 77:9862–9871

Nishimura, N., and Balch, W. E. (1997) A di-acidic signal required for selective export from the endoplasmic reticulum. Science 277:556–558

Nishimura, N., Bannykh, S., Slabough, S., Matteson, J., Altschuler, Y., Hahn, K., and Balch, W. E. (1999) A di-acidic (DXE) code directs concentration of cargo during export from the endoplasmic reticulum. J Biol Chem 274:15937–15946

Nixdorf, R., Klupp, B. G., Karger, A., and Mettenleiter, T. C. (2000) Effects of truncation of the carboxy terminus of pseudorabies virus glycoprotein B on infectivity. J Virol 74:7137–7145

Ochsenbauer, C., Dubay, S. R., and Hunter, E. (2000) The Rous sarcoma virus Env glycoprotein contains a highly conserved motif homologous to tyrosine-based endocytosis signals and displays an unusual internalization phenotype. Mol Cell Biol 20:249–260

Ohno, H., Aguilar, R. C., Fournier, M. C., Hennecke, S., Cosson, P., and Bonifacino, J. S. (1997) Interaction of endocytic signals from the HIV-1 envelope glycoprotein complex with members of the adaptor medium chain family. Virology 238:305–315

Olson, J. K., and Grose, C. (1997) Endocytosis and recycling of varicella-zoster virus Fc receptor glycoprotein gE: internalization mediated by a YXXL motif in the cytoplasmic tail. J Virol 71:4042–4054

Ono, A., and Freed, E. O. (2001) Plasma membrane rafts play a critical role in HIV-1 assembly and release. Proc Natl Acad Sci U S A 98:13925–13930

Ono, A. and Freed, E. O. (2003) Evidence for a role for PI(4,5)P2 in the targeting of HIG-1 GAG to the plasma membrane. Paper presented at: Retroviruses (Cold Spring Harbor Laboratory, Cold Spring Harbor, NY), p101

Owens, R. J., Dubay, J. W., Hunter, E., and Compans, R. W. (1991) Human immunodeficiency virus envelope protein determines the site of virus release in polarized epithelial cells. Proc Natl Acad Sci U S A 88:3987–3991

Pasieka, T. J., Maresova, L., and Grose, C. (2003) A functional YNKI motif in the short cytoplasmic tail of varicella-zoster virus glycoprotein gH mediates clathrin-dependent and antibody-independent endocytosis. J Virol 77:4191–4204

Pelchen-Matthews, A., Kramer, B., and Marsh, M. (2003) Infectious HIV-1 assembles in late endosomes in primary macrophages. J Cell Biol 162:443–455

Pelkmans, L., and Helenius, A. (2002) Endocytosis via caveolae. Traffic 3:311–320

Pique, C., Pham, D., Tursz, T., and Dokhelar, M. C. (1993) The cytoplasmic domain of the human T-cell leukemia virus type I envelope can modulate envelope functions in a cell type-dependent manner. J Virol 67:557–561

Pitcher, C., Honing, S., Fingerhut, A., Bowers, K., and Marsh, M. (1999) Cluster of differentiation antigen 4 (CD4) endocytosis and adaptor complex binding require activation of the CD4 endocytosis signal by serine phosphorylation. Mol Biol Cell 10:677–691

Pornillos, O., Garrus, J. E., and Sundquist, W. I. (2002) Mechanisms of enveloped RNA virus budding. Trends Cell Biol 12:569–579

Pornillos, O., Higginson, D. S., Stray, K. M., Fisher, R. D., Garrus, J. E., Payne, M., He, G. P., Wang, H. E., Morham, S. G., and Sundquist, W. I. (2003) HIV Gag mimics the Tsg101-recruiting activity of the human Hrs protein. J Cell Biol 162:425–434

Radsak, K., Eickmann, M., Mockenhaupt, T., Bogner, E., Kern, H., Eis-Hubinger, A., and Reschke, M. (1996) Retrieval of human cytomegalovirus glycoprotein B from the infected cell surface for virus envelopment. Arch Virol 141:557–572

Raposo, G., Moore, M., Innes, D., Leijendekker, R., Leigh-Brown, A., Benaroch, P., and Geuze, H. (2002) Human macrophages accumulate HIV-1 particles in MHC II compartments. Traffic 3:718–729

Ricotta, D., Conner, S. D., Schmid, S. L., von Figura, K., and Honing, S. (2002) Phosphorylation of the AP2 mu subunit by AAK1 mediates high affinity binding to membrane protein sorting signals. J Cell Biol 156:791–795

Robinson, M. S. (1987) Coated vesicles and protein sorting. J Cell Sci 87 : 203–204

Rodriguez, J. E., Moninger, T., and Grose, C. (1993) Entry and egress of varicella virus blocked by same anti-gH monoclonal antibody. Virology 196:840–844

Ronnholm, R. (1992) Localization to the Golgi complex of Uukuniemi virus glycoproteins G1 and G2 expressed from cloned cDNAs. J Virol 66:4525–4531

Rowell, J. F., Stanhope, P. E., and Siliciano, R. F. (1995) Endocytosis of endogenously synthesized HIV-1 envelope protein. Mechanism and role in processing for association with class II MHC. J Immunol 155:473–488

Sauter, M. M., Pelchen-Matthews, A., Bron, R., Marsh, M., LaBranche, C. C., Vance, P. J., Romano, J., Haggarty, B. S., Hart, T. K., Lee, W. M., and Hoxie, J. A. (1996) An internalization signal in the simian immunodeficiency virus transmembrane protein cytoplasmic domain modulates expression of envelope glycoproteins on the cell surface. J Cell Biol 132:795–811

Schafer, W., Stroh, A., Berghofer, S., Seiler, J., Vey, M., Kruse, M. L., Kern, H. F., Klenk, H. D., and Garten, W. (1995) Two independent targeting signals in the cytoplasmic domain determine trans-Golgi network localization and endosomal trafficking of the proprotein convertase furin. EMBO J 14:2424–2435

Scheiffele, P., Peranen, J., and Simons, K. (1995) N-glycans as apical sorting signals in epithelial cells. Nature 378:96–98

Scheiffele, P., Roth, M. G., and Simons, K. (1997) Interaction of influenza virus haemagglutinin with sphingolipid-cholesterol membrane domains via its transmembrane domain. EMBO J 16:5501–5508

Schmelz, M., Sodeik, B., Ericsson, M., Wolffe, E. J., Shida, H., Hiller, G., and Griffiths, G. (1994) Assembly of vaccinia virus: the second wrapping cisterna is derived from the trans Golgi network. J Virol 68:130–147

Schmid, S. L. (1997) Clathrin-coated vesicle formation and protein sorting: an integrated process. Annu Rev Biochem 66:511–548

Seaman, M. N., and Williams, H. P. (2002) Identification of the functional domains of yeast sorting nexins Vps5p and Vps17p. Mol Biol Cell 13:2826–2840

Sfakianos, J. N., and Hunter, E. (2003) M-PMV capsid transport is mediated by Env/Gag interactions at the pericentriolar recycling endosome. Traffic 4:671–680

Sfakianos, J. N., LaCasse, R. A., and Hunter, E. (2003) The M-PMV cytoplasmic targeting-retention signal directs nascent Gag polypeptides to a pericentriolar region of the cell. Traffic 4:660–670

Smith, G. L., Vanderplasschen, A., and Law, M. (2002) The formation and function of extracellular enveloped vaccinia virus. J Gen Virol 83:2915–2931

Sorkin, A. (2000) The endocytosis machinery. J Cell Sci 113:4375–4376

Spaete, R. R., Gehrz, R. C., and Landini, M. P. (1994) Human cytomegalovirus structural proteins. J Gen Virol 75: 3287-3308

Spear, P. G., and Longnecker, R. (2003) Herpesvirus entry: an update. J Virol 77:10179-10185

Stoorvogel, W., Kleijmeer, M. J., Geuze, H. J., and Raposo, G. (2002) The biogenesis and functions of exosomes. Traffic 3:321-330

Sugimoto, H., Sugahara, M., Folsch, H., Koide, Y., Nakatsu, F., Tanaka, N., Nishimura, T., Furukawa, M., Mullins, C., Nakamura, N., Mellman, I., and Ohno, H. (2002) Differential recognition of tyrosine-based basolateral signals by AP-1B subunit mu1B in polarized epithelial cells. Mol Biol Cell 13:2374-2382

Suomalainen, M. (2002) Lipid rafts and assembly of enveloped viruses. Traffic 3:705-709

Takahashi, S., Nakagawa, T., Banno, T., Watanabe, T., Murakami, K., and Nakayama, K. (1995) Localization of furin to the trans-Golgi network and recycling from the cell surface involves Ser and Tyr residues within the cytoplasmic domain. J Biol Chem 270:28397-28401

ter Haar, E., Musacchio, A., Harrison, S. C., and Kirchhausen, T. (1998) Atomic structure of clathrin: a beta propeller terminal domain joins an alpha zigzag linker. Cell 95:563-573

Thomas, D. C., Brewer, C. B., and Roth, M. G. (1993) Vesicular stomatitis virus glycoprotein contains a dominant cytoplasmic basolateral sorting signal critically dependent upon a tyrosine. J Biol Chem 268:3313-3320

Thomas, D. C., and Roth, M. G. (1994) The basolateral targeting signal in the cytoplasmic domain of glycoprotein G from vesicular stomatitis virus resembles a variety of intracellular targeting motifs related by primary sequence but having diverse targeting activities. J Biol Chem 269:15732-15739

Tikkanen, R., Obermuller, S., Denzer, K., Pungitore, R., Geuze, H. J., von Figura, K., and Honing, S. (2000) The dileucine motif within the tail of MPR46 is required for sorting of the receptor in endosomes. Traffic 1:631-640

Tirabassi, R. S., and Enquist, L. W. (1998) Role of envelope protein gE endocytosis in the pseudorabies virus life cycle. J Virol 72:4571-4579

Trowbridge, I. S., Collawn, J. F., and Hopkins, C. R. (1993) Signal-dependent membrane protein trafficking in the endocytic pathway. Annu Rev Cell Biol 9:129-161

Tugizov, S., Maidji, E., Xiao, J., and Pereira, L. (1999) An acidic cluster in the cytosolic domain of human cytomegalovirus glycoprotein B is a signal for endocytosis from the plasma membrane. J Virol 73:8677-8688

Tugizov, S., Maidji, E., Xiao, J., Zheng, Z., and Pereira, L. (1998) Human cytomegalovirus glycoprotein B contains autonomous determinants for vectorial targeting to apical membranes of polarized epithelial cells. J Virol 72:7374-7386

Tugizov, S., Navarro, D., Paz, P., Wang, Y., Qadri, I., and Pereira, L. (1994) Function of human cytomegalovirus glycoprotein B: syncytium formation in cells constitutively expressing gB is blocked by virus-neutralizing antibodies. Virology 201:263-276

Tugizov, S., Wang, Y., Qadri, I., Navarro, D., Maidji, E., and Pereira, L. (1995) Mutated forms of human cytomegalovirus glycoprotein B are impaired in inducing syncytium formation. Virology 209:580-591

Ukkonen, P., Lewis, V., Marsh, M., Helenius, A., and Mellman, I. (1986) Transport of macrophage Fc receptors and Fc receptor-bound ligands to lysosomes. J Exp Med 163:952–971

Vincent, M. J., Melsen, L. R., Martin, A. S., and Compans, R. W. (1999) Intracellular interaction of simian immunodeficiency virus Gag and Env proteins. J Virol 73:8138–8144

von Schwedler, U. K., Stuchell, M., Muller, B., Ward, D. M., Chung, H. Y., Morita, E., Wang, H. E., Davis, T., He, G. P., Cimbora, D. M., Scott, A., Krausslich, H. G., Kaplan, J., Morham, S. G., and Sundquist, W. I. (2003) The protein network of HIV budding. Cell 114:701–713

Voorhees, P., Deignan, E., van Donselaar, E., Humphrey, J., Marks, M. S., Peters, P. J., and Bonifacino, J. S. (1995) An acidic sequence within the cytoplasmic domain of furin functions as a determinant of trans-Golgi network localization and internalization from the cell surface. EMBO J 14:4961–4975

Wang, Z., Gershon, M. D., Lungu, O., Panagiotidis, C. A., Zhu, Z., Hao, Y., and Gershon, A. A. (1998) Intracellular transport of varicella-zoster glycoproteins. J Infect Dis 178 Suppl 1:S7–S12

Ward, B. M., and Moss, B. (2000) Golgi network targeting and plasma membrane internalization signals in vaccinia virus B5R envelope protein. J Virol 74:3771–3780

Willey, R. L., Bonifacino, J. S., Potts, B. J., Martin, M. A., and Klausner, R. D. (1988) Biosynthesis, cleavage, and degradation of the human immunodeficiency virus 1 envelope glycoprotein gp160. Proc Natl Acad Sci U S A 85:9580–9584

Wisner, T., Brunetti, C., Dingwell, K., and Johnson, D. C. (2000) The extracellular domain of herpes simplex virus gE is sufficient for accumulation at cell junctions but not for cell-to-cell spread. J Virol 74:2278–2287

Wolffe, E. J., Isaacs, S. N., and Moss, B. (1993) Deletion of the vaccinia virus B5R gene encoding a 42-kilodalton membrane glycoprotein inhibits extracellular virus envelope formation and dissemination. J Virol 67:4732–4741

Wyss, S., Berlioz-Torrent, C., Boge, M., Blot, G., Honing, S., Benarous, R., and Thali, M. (2001) The highly conserved C-terminal dileucine motif in the cytosolic domain of the human immunodeficiency virus type 1 envelope glycoprotein is critical for its association with the AP-1 clathrin adaptor. J Virol 75:2982–2992

Yao, Z., Jackson, W., Forghani, B., and Grose, C. (1993) Varicella-zoster virus glycoprotein gpI/gpIV receptor: expression, complex formation, and antigenicity within the vaccinia virus-T7 RNA polymerase transfection system. J Virol 67:305–314

Yasuda, J., Hunter, E., Nakao, M., and Shida, H. (2002) Functional involvement of a novel Nedd4-like ubiquitin ligase on retrovirus budding. EMBO Rep 3:636–640

Yeaman, C., Le Gall, A. H., Baldwin, A. N., Monlauzeur, L., Le Bivic, A., and Rodriguez-Boulan, E. (1997) The O-glycosylated stalk domain is required for apical sorting of neurotrophin receptors in polarized MDCK cells. J Cell Biol 139:929–940

Yuste, E. and Desrosiers, R. C. (2003) Mutations in the cytoplasmic domain of SIV gp41 can dramatically increase envelope content in virions, infectivity and neutralization. Paper presented at: Retroviruses (Cold Spring Harbor Laboratory, Cold Spring Harbor, NY), p276

Zhu, Z., Gershon, M. D., Hao, Y., Ambron, R. T., Gabel, C. A., and Gershon, A. A. (1995) Envelopment of varicella-zoster virus: targeting of viral glycoproteins to the trans-Golgi network. J Virol 69:7951–7959

Zhu, Z., Hao, Y., Gershon, M. D., Ambron, R. T., and Gershon, A. A. (1996) Targeting of glycoprotein I (gE) of varicella-zoster virus to the trans-Golgi network by an AYRV sequence and an acidic amino acid-rich patch in the cytosolic domain of the molecule. J Virol 70:6563–6575

Subject Index

AAA ATPase 231
actin 74,89
– cortex 72, 81
– cytoskeleton 94
– filament 71, 74, 94
– polymerisation 78
adaptor-associated kinase 1 (AAK1) 226
adeno-associated virus (AAV) 118
adenovirus 123
– capsid 91
– cell entry 90
ADP ribosylation factor 6 203
aggresome 71
alphavirus 141
amantadine 16
amphipatic α-helix 140
antigenic
– drift 188
– shift 188
– variation 188
antiviral drugs 16
AP1/2 225
arterivirus 162

baculovirus 94, 95
brome mosaic virus (BMV) 150
bunyavirus 128
butanedione monoxime 79

canine parvovirus (CPV) 91
caveolae 11
cell surface receptor 4
cellular trafficking 220, 222
cerulenin 158
chaperones 179
chemokine 237
chloroplast envelope 150

CHMP4 242
cholesterol 48, 224
clathrin 9, 222, 234
clathrin coated vesicles (CCV) 225
CNX 182
comovirus 158
conformational change 36, 39, 51
COP
– I 226
– II 155, 226
– β-COP 234
coreceptor 6, 17
coronavirus 162
cowpea mosaic comovirus 158
coxsackie-adenovirus receptor (CAR) 6
cross-linking 226
CRT 182
C-terminal helix 35
cytochalasin 78, 88
cytopathic vacuoles type I 141
cytoplasm 122
cytoplasmic
– tail 47
– transport 82, 84, 117
cytoskeletal
– filament 81
– proteins 87
cytoskeleton 69, 70, 81, 95
– cooperation 80
cytosol 123
cytosolic transport 83, 85
cytotoxic T lymphocyte 201

DC-SIGN 204
detergent insoluble membrane domain (DIM) 224
dextran 118

disulfide bonds   184
DMV   162
DNA retrovirus   119
Drosophila   77, 158
DTT   185
dynactin   76, 82
dynamitin   87, 89, 93
dynein   76, 82, 87, 89

ectodomain   32, 46, 47
EDEM   187
endocytic entry   2
endocytosis   7, 8, 13, 80, 90, 117, 207
endosomal marker   143
endosome   222, 224, 230
Env gene   229
envelope protein   227
Epstein-Barr virus (EBV)   11, 233
equine
– arteritis virus (EAV)   162
– infectious anaemia virus (EIAV)   242
ER   157, 221
ERAD   186
Ero1p   184
ERp57   183
Erv2   184
Escherichia coli   120
ESCRT   221, 231, 240
extracellular
– enveloped virion (EEV)   239
– matrix   110

F protein   37
FHV   163
Flaviviridae   159
flock house virus   158
fluorescent tag   80
folding   178
fusion
– peptide   42, 45
– protein
– – class I   30, 33, 34, 52
– – class II   31, 32, 38, 52

G protein-coupled receptor (GPCR)   237
Gag   220, 228

GGA protein   227
glycophosphatidyl inositol (GPI)   224
glycoprotein   26, 86, 237
– B   233
– E   235
– G   238
– H   236
– oligomer   27
glycosylation   182, 220, 221, 226
Golgi apparatus   157, 210, 222, 226, 239
GTP hydrolysis   114
GTPase Ran   113
GTP-tubulin   73
guanidine   157
guanylyltransferase   142

helix bundle formation   50
hepadnavirus   119
heparan sulfate   4
hepatitis
– B virus   119
– C virus   140, 159
herpes simplex virus (HSV)   82, 205, 233
– type 1   120
herpesvirus   5, 86, 87, 124, 205, 233
– β   233
heterodimer   33
hexon   124
HPV protein   210
human
– cytomegalovirus   233
– immunodeficiency virus (HIV)   121, 9, 88, 200, 201
– – C-terminal helix   50
– – entry   49
– – Env trafficking   229, 230
– – envelope glycoprotein (Env)   26, 34
– – fusion   35
– papillomavirus   210
– T cell leukemia virus (HTLV-I)   231

importin β   117
influenza
– hemagglutinin (HA)   26, 30, 33, 43, 52
– virus   5, 10, 72, 127, 237

integral membrane protein   139
intermediate filament protein   71
internal fusion peptide   44
Italian ringspot virus   161

jasplakinolide   78

Kaposi's sarcoma   206
kinesin   75, 87
– N-type   75
Kunjin virus   159

L domain   241
latrunculin   78, 88
lentivirus   121, 201
lipid
– modification   224
– raft   48, 222, 238
– stalk   40
low pH activation   28
lysosomal marker   143
lysosome   157

macropinocytosis   12
MAP kinase ERK   116
Mason-Pfizer monkey virus (MPMV)   232
measles virus   6
membrane
– dynamics   40
– fusion   11
–proteins   27
– penetration   14, 16
methyltransferase   142
MHC
– class I   190, 207
– class II   190, 241
microfilament   71
microtubule   73, 90, 91
– transport   84
mitochondrial membrane   158
molecular crowding   179
MTOC   93
multivesicular body (MVB)   222
murine leukemia virus (MLV)   242
– Env glycoprotein   46
myosin   74, 89

– II   75
myristic acid   220

Nedd4-like ubiquitin ligase   242
Nef   202, 203, 227
negative-sense RNA virus   127
nepovirus   158
neurotropic virus   82
neutral pH   4, 29
nidovirus   162
nocodazole   88
nodavirus   158
non-clathrin mediated endocytosis   10
nonstructural protein   139, 142
N-proximal α helix   203
N-terminal
– fusion peptide   43
– myristoylation   201
– peptide   50
– zinc finger   208
NTPase   142
nuclear
– envelope   111
– export signal (NES)   113
– import   121, 126
– localization signal (NLS)   114
– pore complex (NPC)   111, 114
nucleocapsid morphogenesis   95
nucleo-cytoplasmic transport   113
nucleoporin   112, 114, 115
nucleus   113, 122

oligomerization   226
orthomyxovirus   127, 237

palmitoylation   144
papilloma virus   125
papovavirus   125
paramyxovirus F protein   36, 37
parvovirus   15, 118
– entry   91
PDI   184
peptide translocation   205
PHD   208
phosphorylation   13
Picornaviridae   10, 14
PKC   14

plant virus   158
plasma membrane   8, 68, 72, 92, 93, 124
polarized
– cell   222, 232
– epithelia
– – infection   7
– – receptor   7
poliovirus   140, 154, 164
polyoma virus   125
Polyomaviridae   11
poxvirus   92
protein sorting   221
proton pump   210
PTAP motif   242

Rab protein   13, 234
rabies virus   77
Ran-independent transport   116
receptor   2, 6, 17
– activation   29
– cytoplasmic tail   207
redundancy   181
retention   186
retrotransposon   120
retrovirus   120, 227, 231
– alpha-retrovirus   29, 232
– budding   89
– δ-retrovirus   231
reverse transcription   123
rhabdovirus   238
RNA
– capping   142
– helicase   142
– polymerase subunit   142
– replication   139
– triphosphatase   143
– virus   139
Rous sarcoma virus (RSV)   232
rubella virus   150

Semliki Forest virus (SFV) 140, 164
– fusion protein   38
signal peptidase   177
simian immunodeficiency virus (SIV)   201
– Env trafficking   229

Sindbis virus   141

snurportin   117
sphingolipid   48
stalk model   40

TAP   205
targeting   176
TBE E   44
– homodimer   38
– protein   39
tegument protein   125
TGN46   234
TIP47   225, 231
tobacco
– etch potyvirus   158
– mosaic virus (TMV)   150
Togaviridae   150
Tombusvirus   161
tonoplast   150
trafficking   221
– signal   243
trans-Golgi network (TGN) 155, 221
translocation   114, 176
transmembrane domain   41, 45, 224
transposition   121
tubulin   74
tumour susceptibility gene 101 (TSG 101)   242
turnip yellow mosaic virus   150

ubiquitination   208, 209
UL33   237
unfolded protein response (UPR)   187
US27   237
US28   237

vaccinia virus (VV)   92, 239
vacuolar membrane ATPase   204
varicella zoster virus (VZV)   234
vesicular stomatitis virus (VSV)   238
VIP-21   234
viral
– capsid transport   86
– envelope   175
– genome, nuclear import   117
– protein trafficking   223

Subject Index

– subversion   189
virion   70, 88, 126
virus
– assembly   240
– entry   3, 16, 83
– receptor   4

– recombination   189
virus-like particle (VLP)   231
Vps4   231
Vpu   209

Wiskott-Aldrich syndrome protein   94

Current Topics in Microbiology and Immunology

Volumes published since 1989 (and still available)

Vol. 244: **Daëron, Marc; Vivier, Eric (Eds.)**: Immunoreceptor Tyrosine-Based Inhibition Motifs. 1999. 20 figs. VIII, 179 pp. ISBN 3-540-65789-4

Vol. 245/I: **Justement, Louis B.; Siminovitch, Katherine A. (Eds.)**: Signal Transduction and the Coordination of B Lymphocyte Development and Function I. 2000. 22 figs. XVI, 274 pp. ISBN 3-540-66002-X

Vol. 245/II: **Justement, Louis B.; Siminovitch, Katherine A. (Eds.)**: Signal Transduction on the Coordination of B Lymphocyte Development and Function II. 2000. 13 figs. XV, 172 pp. ISBN 3-540-66003-8

Vol. 246: **Melchers, Fritz; Potter, Michael (Eds.)**: Mechanisms of B Cell Neoplasia 1998. 1999. 111 figs. XXIX, 415 pp. ISBN 3-540-65759-2

Vol. 247: **Wagner, Hermann (Ed.)**: Immunobiology of Bacterial CpG-DNA. 2000. 34 figs. IX, 246 pp. ISBN 3-540-66400-9

Vol. 248: **du Pasquier, Louis; Litman, Gary W. (Eds.)**: Origin and Evolution of the Vertebrate Immune System. 2000. 81 figs. IX, 324 pp. ISBN 3-540-66414-9

Vol. 249: **Jones, Peter A.; Vogt, Peter K. (Eds.)**: DNA Methylation and Cancer. 2000. 16 figs. IX, 169 pp. ISBN 3-540-66608-7

Vol. 250: **Aktories, Klaus; Wilkins, Tracy, D. (Eds.)**: Clostridium difficile. 2000. 20 figs. IX, 143 pp. ISBN 3-540-67291-5

Vol. 251: **Melchers, Fritz (Ed.)**: Lymphoid Organogenesis. 2000. 62 figs. XII, 215 pp. ISBN 3-540-67569-8

Vol. 252: **Potter, Michael; Melchers, Fritz (Eds.)**: B1 Lymphocytes in B Cell Neoplasia. 2000. XIII, 326 pp. ISBN 3-540-67567-1

Vol. 253: **Gosztonyi, Georg (Ed.)**: The Mechanisms of Neuronal Damage in Virus Infections of the Nervous System. 2001. approx. XVI, 270 pp. ISBN 3-540-67617-1

Vol. 254: **Privalsky, Martin L. (Ed.)**: Transcriptional Corepressors. 2001. 25 figs. XIV, 190 pp. ISBN 3-540-67569-8

Vol. 255: **Hirai, Kanji (Ed.)**: Marek's Disease. 2001. 22 figs. XII, 294 pp. ISBN 3-540-67798-4

Vol. 256: **Schmaljohn, Connie S.; Nichol, Stuart T. (Eds.)**: Hantaviruses. 2001, 24 figs. XI, 196 pp. ISBN 3-540-41045-7

Vol. 257: **van der Goot, Gisou (Ed.)**: Pore-Forming Toxins, 2001. 19 figs. IX, 166 pp. ISBN 3-540-41386-3

Vol. 258: **Takada, Kenzo (Ed.)**: Epstein-Barr Virus and Human Cancer. 2001. 38 figs. IX, 233 pp. ISBN 3-540-41506-8

Vol. 259: **Hauber, Joachim, Vogt, Peter K. (Eds.)**: Nuclear Export of Viral RNAs. 2001. 19 figs. IX, 131 pp. ISBN 3-540-41278-6

Vol. 260: **Burton, Didier R. (Ed.)**: Antibodies in Viral Infection. 2001. 51 figs. IX, 309 pp. ISBN 3-540-41611-0

Vol. 261: **Trono, Didier (Ed.)**: Lentiviral Vectors. 2002. 32 figs. X, 258 pp. ISBN 3-540-42190-4

Vol. 262: **Oldstone, Michael B.A. (Ed.)**: Arenaviruses I. 2002, 30 figs. XVIII, 197 pp. ISBN 3-540-42244-7

Vol. 263: **Oldstone, Michael B. A. (Ed.)**: Arenaviruses II. 2002, 49 figs. XVIII, 268 pp. ISBN 3-540-42705-8

Vol. 264/I: **Hacker, Jörg; Kaper, James B. (Eds.)**: Pathogenicity Islands and the Evolution of Microbes. 2002. 34 figs. XVIII, 232 pp. ISBN 3-540-42681-7

Vol. 264/II: **Hacker, Jörg; Kaper, James B. (Eds.)**: Pathogenicity Islands and the Evolution of Microbes. 2002. 24 figs. XVIII, 228 pp. ISBN 3-540-42682-5

Vol. 265: **Dietzschold, Bernhard; Richt, Jürgen A. (Eds.)**: Protective and Pathological Immune Responses in the CNS. 2002. 21 figs. X, 278 pp. ISBN 3-540-42668-X

Vol. 266: **Cooper, Koproski (Eds.)**: The Interface Between Innate and Acquired Immunity, 2002, 15 figs. XIV, 116 pp. ISBN 3-540-42894-1

Vol. 267: **Mackenzie, John S.; Barrett, Alan D. T.; Deubel, Vincent (Eds.)**: Japanese Encephalitis and West Nile Viruses. 2002. 66 figs. X, 418 pp. ISBN 3-540-42783-X

Vol. 268: **Zwickl, Peter; Baumeister, Wolfgang (Eds.)**: The Proteasome-Ubiquitin Protein Degradation Pathway. 2002, 17 figs. X, 213 pp. ISBN 3-540-43096-2

Vol. 269: **Koszinowski, Ulrich H.; Hengel, Hartmut (Eds.)**: Viral Proteins Counteracting Host Defenses. 2002, 47 figs. XII, 325 pp. ISBN 3-540-43261-2

Vol. 270: **Beutler, Bruce; Wagner, Hermann (Eds.)**: Toll-Like Receptor Family Members and Their Ligands. 2002, 31 figs. X, 192 pp. ISBN 3-540-43560-3

Vol. 271: **Koehler, Theresa M. (Ed.)**: Anthrax. 2002, 14 figs. X, 169 pp. ISBN 3-540-43497-6

Vol. 272: **Doerfler, Walter; Böhm, Petra (Eds.)**: Adenoviruses: Model and Vectors in Virus-Host Interactions. Virion and Structure, Viral Replication, Host Cell Interactions. 2003, 63 figs., approx. 280 pp. ISBN 3-540-00154-9

Vol. 273: **Doerfler, Walter; Böhm, Petra (Eds.)**: Adenoviruses: Model and Vectors in Virus-Host Interactions. Immune System, Oncogenesis, Gene Therapy. 2004, 35 figs., approx. 280 pp. ISBN 3-540-06851-1

Vol. 274: **Workman, Jerry L. (Ed.)**: Protein Complexes that Modify Chromatin. 2003, 38 figs., XII, 296 pp. ISBN 3-540-44208-1

Vol. 275: **Fan, Hung (Ed.)**: Jaagsiekte Sheep Retrovirus and Lung Cancer. 2003, 63 figs., XII, 252 pp. ISBN 3-540-44096-3

Vol. 276: **Steinkasserer, Alexander (Ed.)**: Dendritic Cells and Virus Infection. 2003, 24 figs., X, 296 pp. ISBN 3-540-44290-1

Vol. 277: **Rethwilm, Axel (Ed.)**: Foamy Viruses. 2003, 40 figs., X, 214 pp. ISBN 3-540-44388-6

Vol. 278: **Salomon, Daniel R.; Wilson, Carolyn (Eds.)**: Xenotransplantation. 2003, 22 figs., IX, 254 pp. ISBN 3-540-00210-3

Vol. 279: **Thomas, George; Sabatini, David; Hall, Michael N. (Eds.)**: TOR. 2004, 49 figs., X, 364 pp. ISBN 3-540-00534-X

Vol. 280: **Heber-Katz, Ellen (Ed.)**: Regeneration: Stem Cells and Beyond. 2004, 42 figs., XII, 194 pp. ISBN 3-540-02238-4

Vol. 281: **Young, John A. T. (Ed.)**: Cellular Factors Involved in Early Steps of Retroviral Replication. 2003, 21 figs., IX, 240 pp. ISBN 3-540-00844-6

Vol. 282: **Stenmark, Harald (Ed.)**: Phosphoinositides in Subcellular Targeting and Enzyme Activation. 2003, 20 figs., X, 210 pp. ISBN 3-540-00950-7

Vol. 283: **Kawaoka, Yoshihiro (Ed.)**: Biology of Negative Strand RNA Viruses: The Power of Reverse Genetics. 2004, 24 figs., IX, 350 pp. ISBN 3-540-40661-1

Vol. 284: **Harris, David (Ed.)**: Mad Cow Disease and Related Spongiform Encephalopathies. 2004, 34 figs., IX, 219 pp. ISBN 3-540-20107-6

Vol. 285: **Marsh, Mark (Ed.)**: Membrane Trafficking in Viral Replication. 2004, 19 figs., IX, 259 pp. ISBN 3-540-21430-5

Vol. 286: **Madshus, Inger H. (Ed.)**: Signalling from Internalized Growth Factor Receptors. 2004, 19 figs., IX, 187 pp. ISBN 3-540-21038-5

Vol. 287: **Enjuanes, Luis (Ed.)**: Coronavirus Replication and Reverse Genetics. 2005, 49 figs., XI, 257 pp. ISBN 3-540-21494-1

Printing: Saladruck, Berlin
Binding: Stein+Lehmann, Berlin